AF849312

Use of Toxicological and Chemical Models to Prioritize DBP Research

Prepared by:
Richard J. Bull, MoBull Consulting

David A. Reckhow, University of Massachusetts Amherst

Vincent Rotello, University of Massachusetts Amherst

Owen M. Bull, YAMS LLC

and

Junsung Kim, University of Massachusetts Amherst

Jointly sponsored by:
Awwa Research Foundation
6666 West Quincy Avenue, Denver, CO 80235-3098
and

U.S. Environmental Protection Agency
Washington D.C.

Published by:

Awwa Research Foundation

American Water Works Association

IWA Publishing

DISCLAIMER

This study was jointly funded by the Awwa Research Foundation (AwwaRF) and the U.S. Environmental Protection Agency (USEPA) under Cooperative Agreement No. R-82940901. AwwaRF and USEPA assume no responsibility for the content of the research study reported in this publication or for the opinions or statements of fact expressed in the report. The mention of trade names for commercial products does not represent or imply the approval or endorsement of AwwaRF or USEPA. This report is presented solely for informational purposes.

Use of Toxicological and Chemical Models to Prioritize DBP Research

Awwa Research Foundation

About the Awwa Research Foundation

The Awwa Research Foundation (AwwaRF) is a member-supported, international, nonprofit organization that sponsors research to enable water utilities, public health agencies, and other professionals to provide safe and affordable drinking water to consumers.

The Foundation's mission is to advance the science of water to improve the quality of life. To achieve this mission, the Foundation sponsors studies on all aspects of drinking water, including supply and resources, treatment, monitoring and analysis, distribution, management, and health effects. Funding for research is provided primarily by subscription payments from approximately 1,000 utilities, consulting firms, and manufacturers in North America and abroad. Additional funding comes from collaborative partnerships with other national and international organizations, allowing for resources to be leveraged, expertise to be shared, and broad-based knowledge to be developed and disseminated. Government funding serves as a third source of research dollars.

From its headquarters in Denver, Colorado, the Foundation's staff directs and supports the efforts of more than 800 volunteers who serve on the board of trustees and various committees. These volunteers represent many facets of the water industry, and contribute their expertise to select and monitor research studies that benefit the entire drinking water community.

The results of research are disseminated through a number of channels, including reports, the Web site, conferences, and periodicals.

For subscribers, the Foundation serves as a cooperative program in which water suppliers unite to pool their resources. By applying Foundation research findings, these water suppliers can save substantial costs and stay on the leading edge of drinking water science and technology. Since its inception, AwwaRF has supplied the water community with more than $300 million in applied research.

More information about the Foundation and how to become a subscriber is available on the Web at **www.awwarf.org.**

CONTENTS

TABLES

FIGURES

FOREWORD

The Awwa Research Foundation is a nonprofit corporation that is dedicated to the implementation of a research effort to help utilities respond to regulatory requirements and traditional high-priority concerns of the industry. The research agenda is developed through a process of consultation with subscribers and drinking water professionals. Under the umbrella of a Strategic Research Plan, the Research Advisory Council prioritizes the suggested projects based upon current and future needs, applicability, and past work; the recommendations are forwarded to the Board of Trustees for final selection. The foundation also sponsors research projects through the unsolicited proposal process; the Collaborative Research, Research Applications, and Tailored Collaboration programs; and various joint research efforts with organizations such as the U.S. Environmental Protection Agency, the U.S. Bureau of Reclamation, and the Association of California Water Agencies.

This publication is a result of one of these sponsored studies, and it is hoped that its findings will be applied in communities throughout the world. The following report serves not only as a means of communicating the results of the water industry's centralized research program but also as a tool to enlist the further support of the nonmember utilities and individuals. Projects are managed closely from their inception to the final report by the foundation's staff and large cadre of volunteers who willingly contribute their time and expertise. The foundation serves a planning and management function and awards contracts to other institutions such as water utilities, universities, and engineering firms. The funding for this research effort comes primarily from the Subscription Program, through which water utilities subscribe to the research program and make an annual payment proportionate to the volume of water they deliver and consultants and manufacturers subscribe based on their annual billings. The program offers a cost-effective and fair method for funding research in the public interest.

A broad spectrum of water supply issues is addressed by the foundation's research agenda: resources, treatment and operations, distribution and storage, water quality and analysis, toxicology, economics, and management. The ultimate purpose of the coordinated effort is to assist water suppliers to provide the highest possible quality of water economically and reliably. The true benefits are realized when the results are implemented at the utility level. The foundation's trustees are pleased to offer this publication as a contribution toward that end.

Walter J. Bishop
Chair, Board of Trustees
Awwa Research Foundation

Robert C. Renner, P.E.
Executive Director
Awwa Research Foundation

ACKNOWLEDGMENTS

The authors of this report are indebted to the following water utilities and individuals for their cooperation and participation in this project.

East Bay Municipal Water District - Ron Hunsinger, Jerry Ongerth, and Laurie Work
Municipal Water District of Southern California - Stuart Krasner

Ron Hunsinger, Jerry Ongerth and Laurie Work provided critical comment on the organization of the report and the presentation of particular facts. They are in large part responsible for making a potentially dry report more readable and for providing some excellent context to the bladder cancer risks. Stuart Krasner gave valuable assistance by providing occurrence information on a variety of disinfection by-products which are rarely measured in routine analysis of drinking water. These data substantially affected the rankings of classes of DBPs for further research. Russell Christman, of the University of North Carolina, reviewed and commented on the predictive chemistry efforts.

The authors particularly appreciated the efforts of the Project Advisory Committee. John Lipscomb of the National Center for Environmental Assessment of the USEPA provided detailed peer review comments related to the toxicology sections of the report. These comments were particularly useful in improving the interpretation of the QSTR analysis. Bruce Macler of the Region IX Office of the USEPA provided a great deal of encouragement for moving the project forward. We appreciated his review of the draft report; his suggestions helped to guide the development of the final product. Gary Amy of the University of Colorado provided some very valuable input into the initial phases of the project. Gary moved to the UNESCO-IHE Institute for Water Education in the Netherlands during the course of the project, which made communication a little more difficult. We thank him for his support.

Roberta D. Bull provided reviews of written materials throughout the duration of the project and invaluable editorial input into the structure of the final report. We very much thank her for her efforts.

The original concept of the project was to attempt the placement of the results of the analysis into a more quantitative framework. Some outstanding suggestions were obtained from George Backus of the Policy Assessment Corporation for putting together a framework for analysis of risks from disinfection by-products. The data obtained from QSTR analyses did not lend themselves to incorporation into such a framework, but it remains a suggestion that should be pursued as more hard data become available.

EXECUTIVE SUMMARY

INTRODUCTION

Epidemiological data have, more or less consistently, indicated that an increased risk of bladder cancer is associated with the consumption of chlorinated water. Less consistent associations have been observed for colorectal cancer, pancreatic cancer, and brain cancer. Toxicological studies have found that selected disinfection by-products (DBPs) are capable of producing cancer in experimental animals under controlled circumstances. However, the risks identified in epidemiological studies differ significantly from those found in animal studies. There are important differences in target organ. More importantly, the magnitude of risks from summing the carcinogenic activity found in animal studies is approximately 1/20 to 1/1000 less than is calculated from epidemiological studies. The differences in the size of the projected risks continue to grow for several reasons. First, the trihalomethanes (THMs) that have been studied (chloroform and bromodichloromethane) were found to induce tumors when administered by corn oil gavage; subsequent studies, however, showed that tumors do not occur when similar doses were administered in drinking water. Second, the mechanisms by which members of the THM and haloacetic acid (HAA) classes produce cancer do not appear to involve mutagenic effects. Rather they are either secondary to toxicity or provide selective growth advantage to cells that are tumor precursors. These are mechanisms that are generally thought to behave in a non-linear fashion as extrapolations are made to low doses. Similar difficulties arise when bringing together epidemiological and toxicological studies of reproductive and/or developmental toxicities.

One of several possible explanations for the differences in projected risks is that the critical DBPs have not been studied toxicologically. Epidemiological studies necessarily deal with the entire mixture of DBPs that are produced with chlorination. Toxicological studies, on the other hand, focus upon a very small number of the DBPs, primarily those that are produced at the highest concentrations. The narrow focus of the toxicological studies conducted to date has made it difficult to confirm or deny epidemiological associations of risk with specific cancers. It would be possible to confirm the results of epidemiological studies if a means were developed to identify, among several hundred DBPs, those chemicals with sufficient potency as carcinogens to account for the cancers observed at the concentrations that are found in drinking water. It is impractical to think that a substantial fraction of the DBPs produced by chlorination could be subjected to sufficiently sophisticated toxicological testing to help establish a causal relationship between chlorinated drinking water and increased cancer risk. Therefore, a process for identifying those DBPs that are most likely to have the appropriate toxicological properties and to be present at sufficient concentrations in drinking water to make a contribution to the risk is needed in order to establish research priorities.

The problem is exacerbated by the fact that only a fraction of the chemicals that are produced in disinfection have been identified. Specifically, non-halogenated by-products have been poorly characterized chemically. On the other hand, some of the by-products are of little interest because they are chemicals commonly viewed as nutrients (e.g. a variety of fatty acids). Because of this complexity, an important part of an effort to establish causality is a means for identifying novel by-products that may contribute to the risk.

APPROACH

The Carcinogen Potency Database (CPDB) was used to identify those chemicals that have been shown to be carcinogenic and to generate distributions in the potency of chemicals that fall into different chemical classes. In addition, a subset of this information was used to set up a secondary database of chemicals that have been established as a cause of bladder cancer in one or more species, including humans. Among those chemicals reported to cause bladder cancer that appear most closely related to DBPs that might be formed from natural organic matter (NOM), were aromatic and heterocyclic amines, nitrosamines, and quinones. An overwhelming number of bladder carcinogens are amines. This fact resulted in an initial focus on nitrogen-containing precursor substructures and attempts to identify analogous compounds as potential DBPs. This was termed the feed-back process.

A more conventional feed-forward process was used for the bulk of the quantitative structure toxicity relationship (QSTR) investigations. The formation of novel products was predicted considering reactions of substructures in NOM with commonly used disinfectants using principles of synthetic organic chemistry. These chemicals, along with those that have been identified as DBPs in surveys of drinking water and a number of products that have been identified as products of HOCl reactions in the biomedical literature, were examined using QSTR to identify the chemicals that are of most interest from a health effects point of view. As several QSTR tools are available, it was necessary to select one for this project. TOPKAT® (Accelrys) was chosen because 1) it has a suite of QSTR models that predict chronic lowest adverse effect levels (LOAELs), 2) it taps several databases for predicting carcinogenic activity, and 3) it includes a developmental toxicity potential model and an Ames' mutagenicity model.

A set of decision criteria was developed to rectify diverse predictions among the eight individual carcinogenicity models included within TOPKAT®. These criteria were generally biased to exclude chemicals for which negative or marginal predictions of carcinogenic activity are made. However, a strong positive prediction of carcinogenicity from Food and Drug Administration (FDA) or National Toxicology Program (NTP) databases generally trumped predictions of non-carcinogenicity in the same species or strain in the other database because the negative prediction was generally found to result from a lack of appropriately related structures in the training set for the specific model queried. Additional criteria were based upon the predicted or actual LOAELs, a positive prediction in at least one sex of two species, and a similarity of the query structure to chemicals in each model's training set. Finally, predicted Ames' mutagenic activity was used for prioritizing within classes of DBPs. This resulted in a relatively small group of chemicals that should be targeted for research. The next step was to experimentally confirm whether or not these chemicals have the appropriate toxicological properties to contribute to diseases that have been associated with chlorinated drinking water.

Despite the fact that this project demonstrates the feasibility of identifying candidates among DBPs that may account for adverse health effects related to disinfection processes, it is very important to realize that its success is not heavily vested in the actual QSTR programs used. In the case of TOPKAT®, it is clear that a problem as complex as identifying the toxic properties of DBPs cannot simply rely upon established QSTR programs. There are classes of DBPs that were not dealt with effectively by TOPKAT®, either because the chemicals in the training sets were not closely related to query structures or because certain molecular fragments were not represented in any of the databases. The prediction of carcinogenic activity was rarely consistent among the eight models as the models were developed using different databases. Other QSTR programs may have performed better in this project, but all will have varying degrees of the same problem. In part, this is because all are dependent upon the limited amount of empirical data that are available at the time the program was created and marketed. For this reason, it was necessary to critically evaluate outputs of the models using expert judgment. It is essential that some of the original studies of compounds within the databases on which predictions were made are consulted to be certain that the predictions are relevant to drinking water exposures. Finally, some of the more substantive findings of the study came from examining the basic biomedical literature and synthesizing arguments for why a particular by-product or class of by-products might be important independent of the QSTR program. Projects of this type will require the expert judgment of chemists and toxicologists for the foreseeable future.

FINDINGS

Aside from methodological issues discussed above, there were five general conclusions drawn from the present work. Embedded in these general findings are a number of specific findings. These are provided here in summary form and are discussed in more detail in the subsequent text.

- Quantitative structure-toxicity relationship (QSTR) analyses of chemicals identified as DBPs in the literature failed to identify candidates with sufficient predicted potency to be credible as potential causes of the epidemiological association between chlorinated drinking water and carcinogenic or reproductive/developmental toxicities.
- Several reaction pathways of chlorine were described with reasonably anticipated precursor substructures in NOM that have the potential for yielding by-products of sufficient predicted potency (and perhaps specificity) to contribute to a bladder cancer risk of the magnitude associated with chlorinated drinking water.
  - Reaction products of disinfectants with substructures in NOM were found to have predicted chronic LOAELs in the low μg/kg day^{-1} range. Representatives within these groups do have sufficient potency, if not the selectivity, to account for epidemiological associations of chlorinated drinking water with adverse health effects, but only if these chemicals occur in drinking water. Thus, several classes of putative DBPs were identified for future research. These include:
    - Haloquinones
    - Halocyclopentenoic acids
    - Organic N-haloamines
    - Nitrosamines and nitrosamides
    - Halonitriles and haloamides
  - Among groups identified as potential by-products of reactions of chlorine or chloramine, the haloquinone group appears the most capable of inducing bladder cancer, as well as other cancers.
    - 2,6-Dichloro-3-methyl-1,4-benzoquinone was identified as the putative chlorination by-product with the greatest toxicological potential be a bladder carcinogen. However, this compound has not been shown to occur in drinking water. It is more likely to be stable in chloraminated water than in the presence of free chlorine.

- Halocyclopentenoic acids are structurally related to halofuranones, such as 3-chloro-4-(dichloromethyl)-5-hydroxy-2[5H]furanone (MX). By itself, MX, does not appear sufficiently potent to be the sole cause of bladder cancer at the concentrations found in drinking water. However, a group of compounds with similar modes of action, present at concentrations comparable to those found for MX in more recent surveys, may be important collectively.
    - 3,5-Dichloro-1-hydroxylcyclopent-2-enoic acid is a representative of this group of putative DBPs.
- Members of the N-haloamine class may also be important. There are insufficient toxicological data related to organic N-haloamines to inform judgments for individual members of the class. Chemicals in this class are known to exist in chlorinated and chloraminated drinking water, but concentrations have not been systematically determined.

➢ Non-linear relationships should be anticipated for yields of different classes of DBPs. The non-linearity will be exacerbated with the emergence of mixed use of disinfectants.

- Some of these by-products may be intermediates in the formation of some terminal by-products that have been associated with chlorinated drinking water (THMs and HAAs). Therefore, reduction of risk by alternate methods of disinfection may not parallel the reduction in the terminal by-products produced by excess chlorine. This has particular implication for the use of chloramines for disinfection of drinking water. The concentrations of these compounds at the tap, and thus the risk they pose, may differ if chloramine is used as the primary disinfectant or if ammonia is applied to water after it has been subjected to a period of treatment with free chlorine.

➢ A different conceptual framework for visualizing the potential health hazards associated with DBPs needs to be developed. A general research approach is suggested to address the safety issues raised by the ever evolving use of reactive chemicals for purposes of disinfecting drinking water. Research efforts in this area need to specifically incorporate:

- A more organized approach to developing a broad perspective of the types of by-products that can be formed with any specific disinfectant strategy.
- A much better understanding of the formation and degradation of intermediate products of reactions of disinfectants with NOM is needed at varying disinfectant/carbon and nitrogen ratios. Some intermediates are orders of magnitude more toxic than the end-products of disinfection. Therefore, efforts must be directed at the development of kinetic models of DBP formation that include these intermediates. Such information will be essential for understanding the probable risks of modifying disinfectant treatment (e.g. using a quenching agent, such as ammonia, following addition of a primary disinfectant may minimize regulated DBPs, but is likely to increase the concentrations of more highly toxic intermediates at the tap).
- Health assessments need to more aggressively pursue studies of the health effects of DBPs beyond the regulated compounds. Toxicological research in the past decade has cast significant doubt on the risk associated with the THMs and HAAs that are subject to regulation.
- Selection of surrogates for monitoring purposes must be shown to be meaningfully related to risk. This is best done by developing a model that

reliably reflects the relationship between the formation of different groups of DBPs.

- The Candidate Contaminant List (CCL) process is poorly suited for solving the problem of health risks associated with the disinfection of drinking water (or other complex processes of concern).

Several hundred DBPs identified in a number of surveys and reviews were examined with QSTR tools. Where occurrence data could be used or inferred and combined with information about the toxicological potency, none appeared capable of making a significant contribution to the risks associated with chlorinated drinking water in epidemiological studies. However, there are large gaps in information with respect to some classes of DBPs that are known to be formed in that their occurrence and toxicology are inadequate to make any judgment as to their importance as health hazards.

Modeling chemical reactions with precursor substructures found in NOM predicted the formation of a large number of novel by-products, including haloquinones, halocyclopentenoic acids, organic N-haloamines, nitrosamines and nitrosamides, halonitriles and haloamides. Approximately 40 of which were found to be interesting from the QSTR analyses. Several chemicals were identified among these novel by-products that were predicted to be very toxic, particularly halogenated quinones.

The metabolism of quinones bears some resemblance to that of aromatic amines that are bladder carcinogens, in that they are conjugated in the liver to relatively non-toxic compounds, but tend to be reactivated in the kidney and bladder by further enzymatic processing. Among the haloquinones, 2,6-dichloro-3-methyl-1,4-benzoquinone (DCMBQ) was identified as an idealized candidate of a chlorination by-product with properties that would favor bladder carcinogenesis. This putative DBP identifies key structural features within the class, but does not rule out a contribution of other members of the haloquinone class. In some cases the predicted chronic LOAEL was lower for the other haloquinones than for DCMBQ, but they lacked the stabilization of reactive intermediates that is conferred by a non-halogen ring substitution (i.e. the 3-methyl group) that favors targeting the kidney and bladder. Recent data demonstrate that chlorine dioxide readily forms o-quinone compounds with certain aromatic amino acids and that chloramines form halogenated 1,4-benzoquinones from phenols, so the production of such chemicals is not limited to chlorine.

Another novel group of predicted DBPs were halocyclopentenoic acids. The halocyclopentenoic acids bear at least a superficial resemblance to MX. In general, the training sets of the QTSR models were not adequate to provide confidence in the predictions made for MX-related chemicals. Verification that experimental data on MX met the criteria used in this study was one reason these compounds were retained as high priority. Also influential were recent surveys that suggest that MX occurs at higher concentrations than previously reported. While MX, in itself, is unlikely to be the cause of cancer risks associated with chlorinated drinking water, the occurrence of a large group of toxicologically similar compounds should be of concern.

Study of the occurrence of halonitriles has been limited to the haloacetonitriles even though the likelihood that higher molecular weight compounds exist is strongly supported by model studies of reactions of chlorine with naturally occurring amino acids. QSTR analyses indicate that propionitriles are likely to be both carcinogens and developmental toxicants. The predicted chronic LOAELs are significantly lower than those of regulated DBPs. Therefore, it is concluded that occurrence of these higher molecular weight halonitriles is a key research need.

Despite the fact that formation of organic N-haloamines has long been recognized with the use of chlorine, chloramine and ozone, these compounds have not been characterized for their occurrence nor has their toxicology received serious study. The QSTR program performed badly in this class, because the chloramine fragment was not in the training sets for the models. Characterizing this group of by-products is considered to be of the highest priority. There has been significant effort in biomedical studies to associate the endogenous formation of N-haloamines as the cause of a variety of chronic inflammatory diseases. The formation of N-haloamines with certain drugs has also been implicated in the development of severe toxicities not detected in pre-market testing. Therefore, the high-priority research need is to identify one or more members of this group for toxicological characterization to determine whether or not they are likely to be of importance from a health effects point of view.

There is the possibility of forming nitrosamine products with a variety of substructures within NOM. While TOPKAT® addresses low molecular weight nitrosamines in a satisfactory way, the higher molecular weight members of the class are poorly addressed because of inadequacies in training sets. Review of recent literature with regard to nitrosamine formation from nicotine and related alkaloids suggests that higher molecular weight compounds derived from natural sources should be of concern. Differential formation of such compounds can be anticipated if there is significant contact time with free chlorine residual before ammonia is added. For example, sufficient contact with free chlorine will degrade purine rings such that a wider variety of N-nitrosamine compounds may formed with the addition of ammonia.

A framework for conceptualizing the extent and diversity of DBPs that may be formed with different disinfection strategies is suggested. The essential feature of the framework suggested in this document is that some disinfection strategies participate heavily in both oxidation and halogenation reactions. Such disinfectants substantially alter the basic structure of NOM and produce a wide variety of DBPs. Other disinfection strategies are less active with respect to one of these types of reactions or the other. As a result, the reaction products are less diverse. Mixed use of disinfectants, such as a period of contact with free chlorine before the addition of ammonia, has the possibility of producing a more complex set of by-products because it may "preserve" toxic intermediates at relatively high concentrations that would be destroyed by further reaction with chlorine. Therefore, it is important to more explicitly consider the kinetics of by-product formation and degradation with varying ratios of disinfectant to organic carbon ratios. It is anticipated that significant non-linear relationships between DBP classes are likely. This is particularly important as the industry increasingly relies upon the use of mixed disinfectants (e.g. free chlorine followed by ammonia) to minimize the formation of regulated DBPs (i.e. THMs and HAAs)

The CCL process used for prioritizing compounds for research and regulation is poorly suited for the process of identifying chemicals that may be responsible for adverse health effects of various drinking water disinfection strategies. To evaluate alternate approaches to drinking water disinfection, a research program is needed that is based upon sets of alternative hypotheses to explain epidemiological data. A systematic evaluation of these alternatives is necessary to solve the problem. The CCL is too unwieldy to develop the type of organized information that is required to deal scientifically with problems associated with the use of an established process, nor will it accurately predict the potential hazards associated with alternative processes.

RESEARCH RECOMMENDATIONS

Prioritized Research to Predict and Confirm By-product Formation

1. Determine whether or not haloquinone and halocyclopentenoic acids can occur in drinking water.
   - Analytical methodology is required for measuring haloquinone and halocyclopentenoic acid structures that is sensitive into the low ng/L range.
   - The kinetics of the formation and destruction of these compounds should be determined with free chlorine and varying conditions of chloramine use. There is a particular concern for whether or not health risk may be increased with a period of free chlorine contact prior to the addition of ammonia.
2. Characterize the differences in organic N-chloramines that are formed with NOM treated above and below breakpoint concentrations of chlorine (e.g. chloramine vs. excess chlorine).
3. Further explore the types of DBPs that can be formed by:
   - Identifying additional substructures in NOM.
   - Developing approaches for predicting products of the interactions between the oxidative and halogenation chemistry of disinfectants with NOM.
4. Determine the potential for forming N-nitroso compounds with substructures in NOM.
5. Develop semi-quantitative models for DBP formation based on known NOM structures and known yields from those structures.
   - Consider the impact of development in watersheds and other drivers rooted in watershed management.
   - Consider geographical differences in NOM substructure.

Prioritized Areas for Health Effects Research

1. Study halogenated quinone and furan derivatives to determine structure-activity characteristics that will facilitate genotoxic and cytotoxic damage to the urinary bladder at low doses (i.e. do not simply extend Ames's mutagenic activity, include in vivo measures of these types of damage to the bladder). This effort should come subsequent to the actual identification of such compounds as chlorination by-products in actual drinking water (see below).
2. Experimentally identify the important determinants of N-haloamine toxicity and carcinogenicity. Studies should be targeted at key differences in by-products formed with chloramine and with breakpoint chlorination.
3. Experimentally verify characteristics of halogenated quinone DBPs predicted to contribute to the development of bladder cancer based upon metabolic profiles and projected structure-activity relationships.
4. Characterize the toxicological properties of selected members of higher molecular weight (i.e. > haloacetonitriles) dihalonitriles projected to be carcinogenic. Choice of halonitriles for study should be made based on occurrence measures as well as structure-activity predictions.
5. Develop a QSTR methodology for predicting health hazards associated with DBPs. Such methodologies may be based in part upon commercial computer programs, but it must be recognized that these programs were developed using on chemicals in commerce and that several important classes of DBPs are frequently poorly represented in their training sets.

Therefore, the lack of systematized information within these classes (e.g. N-haloamines, halonitriles, haloquinones, furan derivative, and higher molecular weight nitrosamines) makes it necessary to collect and organize data so that the critical structural features that contribute to toxicity can be identified.

Prioritized Areas for Occurrence Research for DBPs

1. Determine the extent to which haloquinones occur in drinking water. The toxic potential of these compounds suggest that the sensitivity of the methods should extend into the low ng/L range.
2. Extend halonitrile analyses to include halopropionitriles (saturated and unsaturated) and perhaps higher molecular weight compounds. Suggested sensitivity would be 100 ng/L.
3. Develop a means for measuring the extent of total nitrosamine/nitrosamide in drinking water with a sensitivity of 10 ng/L or lower.
4. Explore methods for characterizing variations in the geographical occurrence of DBPs that better characterize the variation in the natural composition of NOM.
5. Develop methods for detecting novel products that arise from the use of nitrogen in disinfection (e.g. chloramination).

CHAPTER 1. INTRODUCTION

ASSOCIATION OF DRINKING WATER DISINFECTION WITH ADVERSE HEALTH EFFECTS

A major impetus for this project is the fact that there has been very little effort to rectify the differing implications of epidemiological and toxicological studies for assessing risks from disinfection by-products (DBPs). Unfortunately, the focus of research has been upon justifying particular regulatory strategies. Protection of the public health in this area requires an effort to identify causal relationships for associations that suggest substantial risks. Only in this way can alternative disinfection strategies be evaluated and a more comprehensive approach be taken to reduce risks of adverse health effects. Repeated epidemiological associations between chlorinated drinking water and elevated risks of cancer have been made, both within the U.S. and around the world in the last 25 years (Mills et al. 1998). While toxicological studies of individual by-products have identified chemicals with carcinogenic effects, the target organs are different from those associated with chlorination and the projected risks from these studies are much smaller than those identified in epidemiological studies (Bull et al. 2001). There are a variety of possible explanations for these apparent differences, which largely revolve around an insufficient understanding of the chemical nature of by-products that are formed and the fact that among by-products that have been identified, few have had systematic investigations of their toxicological properties. Most fundamentally, epidemiological associations involve the entire mixture of by-products that are consumed, whereas toxicological studies have generally focused on individual by-products or simple mixtures. Twenty-five years of empirical research related to chlorination by-products has failed to resolve this issue, so there is little hope for clear direction in evaluating the use of other reactive chemicals as disinfectants (e.g. ozone, chlorine dioxide, and chloramines) without a more efficiently designed research program. Experimental approaches to confirm or deny epidemiological findings cannot be practically based on the empirical testing all DBPs and combinations of DBPs. Instead methods to focus the research in order to save resources and reduce the time to the resolution of the overall problem must be used. The question is how to organize a research program that addresses the toxicology of chemicals with the highest potential for harm, for which there is little or no empirical data?

The epidemiology studies have been subjected to a number of scholarly reviews (Mills et al. 1998, Reif et al. 1996) and have been used to support the development of regulations by national (USEPA 2003a, USEPA 2003b, Health Canada 2003) and international bodies (WHO 2000). Therefore, only those aspects of this body of literature that are essential background for this project are discussed. The studies have been of different design, power, and quality. In general, odds ratios for increased cancer risks are found in a range of 1-3. Case-control studies of the two more common of these cancers, bladder and colorectal, have been explored in meta analyses (Morris et al. 1992, USEPA 1998, King 1999, Villaneuae et al. 2002). These studies indicate that increased risk of bladder cancer appears to be found with a consistency that is hard to ignore, whereas findings with colon and rectal cancer have been less consistent. The question of whether or not the apparent variation in risks of colorectal cancer is a real reflection of site-specific variations in DBPs that are formed or if they simply reflect differences in the populations studied has not been resolved. If the former scenario is correct, then the risks from colorectal cancer in the geographical areas in which such an association has been found are roughly the same order of magnitude as the bladder cancer risk (Morris et al. 1992).

In its review of the cancer risks associated with chlorination of drinking water, the U.S. Environmental Protection Agency (USEPA)(2003a) concluded that the population attributable risk (PAR) was between 2 and 17% of the lifetime bladder cancer risk in the USA based upon the meta analyses mentioned above. As shown Table 1.1, PARs of this magnitude correspond to lifetime risks in males of between 7/10,000 to as high as 6/1000. The same PARs in females suggest lower risks, approximately 1/3 of that of males. Collectively, these data suggest a human lifetime risk of bladder cancer attributable to chlorination of drinking water to be between 5/10,000 to as high as 5/1000, with a central value of approximately 1/1000.

Table 1.1
Risks of bladder cancer attributed to chlorination of drinking water

	Males	Females
Bladder Cancer Incidence	39/100,000*	10.1/100,000
Lifetime probability for developing bladder cancer	0.0356†	0.0113†
Population Attributable Risks‡		
2%	0.0007	0.0002
17%	0.006	0.002

* Age adjusted incidence for years 1997-2001, Cancer Facts 2005, American Cancer Society (http://www.cancer.org)
† Years 1999-2001, Cancer Facts, 2005, American Cancer Society (http://www.cancer.org)
‡ USEPA, 2003a

The cancer risk associated with chlorinated drinking water has been attributed to by-products that result from the reaction of chlorine with the natural organic matter (NOM) and bromide. This statement is based largely on the observation that some of the by-products produced cancer in various sites when administered to experimental animals. Two groups of by-products have been selected for regulation and are routinely measured by water utilities, four trihalomethanes (THMs) and five of the haloacetic acids (HAA5). In addition to these by-products, several other DBPs of the halofuranone, halopropenal, haloketone, and haloacetonitrile classes have been shown to possess carcinogenic and mutagenic activity in experimental studies (Bull et al. 2001). The U.S. Environmental Protection Agency (USEPA) has suggested the THM and HAA5 by-products that are regulated are reasonable indicators of the carcinogenic risk associated with chlorination (USEPA 2003a).

The human risk estimate derived from the epidemiological data can be contrasted with the risks that are calculated based upon extrapolation of data on individual chlorination by-products to humans. Table 1.2 provides the unit risks for chlorination by-products that were identified in the Stage II DBP rule (USEPA, 2003a) in terms of the lower bound on the concentration in drinking water that would contribute an additional lifetime risk of 1/1,000,000. Unit risks for a number of unregulated chlorination by-products were obtained from Bull et al. (2001). These figures are coupled with median and 90th percentile concentrations (or maximum if 90th percentile could not be estimated) of chlorination by-products garnered from the Information Collection Rule (ICR) database (McGuire et al. 2002) and some subsequent surveys. A concentration and a unit risk level allowed calculation of the median and >90th percentile risk. Choosing among the chlorination by-products that are in the THM and HAA classes, a total risk of 21/1,000,000 is calculated as the median risk. At concentrations that approximate the 90th percentile, 55 cases/1,000,000 was calculated. If other carcinogenic chlorination by-products are included, the sums of the risks increase only to 24/1,000,000 and 73/1,000,000, respectively. Therefore, in terms of raw numbers, the risk of bladder attributed to chlorination of drinking water is approximately 20 times that of all types of cancer seen in experimental animals treated with individual chlorination by-products.

Table 1.2
Estimates of cancer risk attributable to individual disinfection by-products formed in the chlorination of drinking water

Disinfection by-product	Median conc. µg/L	LB on conc. with 10^{-6} Risk	Median risk (X 10^{-6})	90th percent. or highest conc. µg/L	High estimate of risk (X 10^{-6})
Chloroform	23.1	--*	0*	51	0*
BDCM	8.4	1	8.4	19	19
DBCM	4.2	--*	0*	11.8	0*
Bromoform	1.4	8	0.2	3.8	0.5
Dichloroacetic acid	8.9	0.7	12.7	24.3	35
Bromochloroacetic acid	4.3	--†	--†	19	--†
Dibromoacetic acid	1.3	--†	--†	18	--†
Trichloroacetic acid	6.8	--*	0*	24.9	0‡
Bromodichloroacetic acid	3.9	--†	--†	19	--†
Tribromoacetic acid	ND	--†	--†		--†
Chloral hydrate	2.8	--*	0	--*	0
Formaldehyde	ND	21	ND	21.8	1
Trichlorophenol	<0.4	2	<1		<1
MX	0.06	0.03	2	0.31	10
NDMA Cl_2	0.0005	0.0007	0.7	0.0011	1.1
NDMA $ClNH_2$	0.0029	0.0007	4	0.0045	6.3
Total risk			28**		73††

* Concentration below the MCLG
† Insufficient evidence to assess carcinogenic potential – probably carcinogenic
‡ Sufficient to determine non-carcinogenic and below MCLG based on toxicity. Actual MCL developed on absence of reproductive/developmental toxicity tests.
** Total risk includes NDMA formed with chlorine and excludes value for chloramine
†† Total risk includes NDMA formed with chloramine and excludes value for chlorine.

The difference in these risks is likely to be larger than smaller for a variety of reasons. First, the calculation of risk from experimental animals has assumed a linear relationship with the upper bound of risk with dose. With the regulated DBPs, it would appear that modes of action suggest less than linear relationship between dose and response. Second, assumptions used in calculation of risk (i.e. that of the active metabolite at its target site) assume that the parent chemical is toxic and metabolism leads to detoxification. Chlorination by-products such as the THMs are activated by metabolism, therefore, toxicity will increase with rate of metabolism rather than decrease. The convention of assuming metabolism terminates toxic action significantly adds to the estimated risk assigned to each THM.

Several of the DBPs included in Table 1.2 are indicated to contribute no risk. There are several reasons for this. In the case of chloroform, this is because it was determined that risk at low doses was not linearly related to risk. In the cases of dibromochloromethane, trichloroacetic acid, and chloral hydrate the ability to induce cancer was not supported by the weight of evidence despite adequate chronic bioassays. In both cases these compounds were assigned non-zero maximum contaminant limit goals (MCLGs). The concentrations of these chemicals in drinking water were below the MCLG or where the MCLG would be if one simply considered carcinogenic outcomes. Therefore, any risk associated with these compounds is considered to be negligible, if not zero. There are also several HAAs identified in Table 1.2 that are not assigned risk. In all probability the dihalo- and trihaloacetic acids except trichloroacetic acid are carcinogenic, but adequate studies have yet to be published. Therefore, they were assigned no risk. Characterization of the carcinogenic activity of some of these chemicals might increase the overall risk, but this will be balanced by recent experimental findings. For example, studies of bromodichloromethane in drinking water found that it did not increase cancer incidence at any site in rats or mice (NTP 2005b). The original bioassay was conducted with corn oil gavage and produced cancer in several organs in both sexes of both species. These data are reminiscent of findings that lead to reducing the apparent cancer risk from chloroform. There is a strong case to be made that risks from dichloroacetic acid are not linear with dose as well, as it appears to act principally by altering hormonal influences in the liver rather than as a mutagen (Bull 2000).

Bromodichloromethane and dichloroacetic acid account for virtually all of the risk attributed to the THMs and the HAAs. If a non-zero MCLG were to be developed based on their carcinogenicity, the risk attributable to the sum of individual chlorination by-products drops to between 1/1,000,000 and 1/100,000. If anything, research focused on these traditional chlorination by-products is resulting in further divergence of the risk that can be attributed to chlorinated drinking water.

Therefore, it is unlikely that members of the regulated classes of DBPs (THM and HAA5) are causally related to the cancers observed in epidemiological studies of chlorinated drinking water. The major reasons for doubting a causal relationship include: 1) no chlorination by-product has been shown capable of targeting the urinary bladder, 2) the potency and concentrations of the studied by-products are too low to account for the high degree of risk suggested by the epidemiological studies, 3) halogenated by-products do not account for the totality of by-products produced in chlorinated drinking water, and 4) the odds ratios for associations are not significantly increased when THMs are used as the exposure variable rather than chlorinated vs. non-chlorinated water. Epidemiological studies designed to more critically test the relationship do suggest that bladder cancer is primarily related to chlorination. A recent case control study, conducted in hospitals in France, found that drinking water supplies disinfected with ozone do not carry the same bladder cancer risk found with chlorinated supplies (Chevrier et al. 2004). An earlier case-control study of chloraminated supplies had lower risks of bladder cancer than chlorinated waters from the same source (Zierler et al. 1988). Thus, the probability that chlorination is uniquely responsible for increased bladder cancer risk has been supported by study of alternative disinfection strategies.

The problem is that DBPs associated with ozonation and chloramination introduce higher risks for cancers other than those of bladder (e.g. kidney and liver) and these were not addressed in the epidemiology studies. Therefore, this project asks whether there are highly carcinogenic DBPs that arise from chlorination or in the use of other disinfectants. To do this, it is necessary to focus on non-halogenated as well as halogenated chemicals that might be produced. This focus is necessary for determining changing disinfection strategy actually reduces the magnitude of overall cancer risk and not just that which has been associated with chlorination.

CARCINOGENIC ACTIVITY OF DISINFECTION BY-PRODUCTS IN EXPERIMENTAL SYSTEMS

The most thoroughly studied DBPs are the THMs, selected HAAs, 3-chloro-4-(dichloromethyl)-5-hydroxy-2(5H)-furanone (MX), bromate, and N-nitroso-N-dimethylamine (NDMA). The target organs identified in studies of those that produced cancer in experimental animals and the lowest dose that produces a 50% incidence of tumors in a lifetime (TD_{50}) are provided in Table 1.3. In this table we have provided the TD_{50} as a measure of each compound's potency as a carcinogen for two reasons. First, this value can be directly derived from an experiment without consideration of how that data should be extrapolated to low dose for purposes of risk assessment. Second, the quantitative structure toxicity (QSTR) program utilized makes predictions of chronic Lowest Observed Adverse Effect Levels (LOAELs). This figure is more comparable to a TD_{50} (frequently seen to be approximately equivalent to an ED_{10}) and facilitates comparisons of likely potency (the relationship between TD_{50} and the chronic LOAEL is dealt with in some detail in Chapter 6).

Some additional difficulties with the use of the THMs as markers for carcinogenic activity associated epidemiologically with chlorinated water need to be discussed further. First, chloroform was shown to be inactive as a carcinogen in mice when administered in drinking water as opposed to using stomach tube in a corn oil vehicle (Jorgenson et al. 1985). Although chloroform still did produce kidney tumors when administered to rats via the drinking water, these tumors have since been found to be secondary to cytotoxicity and reparative hyperplasia. As indicated earlier bromodichloromethane has also now been found to produce no tumors in mice or rats when given in drinking water (NTP 2005b). In fact there was a negative trend of cancer incidence with dose. This inhibition of spontaneous tumorigenesis was presaged by the observations that chloroform in drinking water inhibits the cell proliferative activity of chloroform given in corn oil (Pereira 1994, Pereira and Grothaus 1997). These data suggest that the risks calculated for the THMs, with the most robust of datasets to support its being considered a carcinogen, appear to be considerably overestimated. In fact, these findings raise the question of whether or not exposure to THMs in drinking water is a cancer hazard at all.

Aside from issues with the THMs, the table further illustrates the problem of depending upon THMs as surrogates of risk from DBPs. Not only have no compounds been identified in controlled experiments in animals that produce bladder cancer, but also the combined consideration of potency ($1/TD_{50}$) and concentrations of the DBPs that have been studied are insufficient to account for a risk of the magnitude that is suggested by epidemiology studies. The TD_{50} of MX is low. In general, the very low concentrations at which it has been found suggest a very small probability that it contributes significantly to risk. Unlike the THMs and HAAs, there has been little effort to study other DBPs related to MX. The carcinogenic potency of bromate is significant, but it is not produced as a by-product of the chlorination of drinking water (it does occur as a contaminant of chlorine used in disinfection). Moreover, there are very large species differences in the sensitivity to bromate. The TD_{50} for mice is 53.8-mg/kg day^{-1} and for hamsters is 533-mg/kg day^{-1}. The carcinogenic potency of NDMA is very high, but its concentrations generally range below 10 ng/L. It is generally higher in waters that have been chloraminated (Mitch and Sedlak 2002). The limited epidemiological data that are available suggest that chloramination represents less of a bladder cancer risk than chlorination (Zierler et al. 1988, McGeehin et al. 1993). Nevertheless, NDMA has been reported as high as 180 ng/L in chloraminated supplies (Charrois et al. 2004). At concentrations of this magnitude, its calculated risk would approach those found in epidemiology studies of chlorinated water. To account for the estimated 1 in 1000 additional risk for bladder cancer that has been associated with chlorinated drinking water, a chemical or group of chemicals at the concentrations of the THMs (i.e. 10-100 μg/L) would have to have TD_{50} of approximately 1 mg/kg body weight.

Table 1.3
By-products of disinfection that have been studied in chronic experiments for their ability to induce cancer in animals under controlled conditions

Disinfection by-product	Target organs rat	Target organs mouse	Lowest TD_{50} mg/kg	Reference
Chlorinated water	negative	negative		Gold 2005
Monochloramine	negative	negative		Gold 2005
Chloroform	kidney	liver	90.3	Gold 2005
Bromodichloromethane	kidney, colon, liver	kidney, liver	47.7	Gold 2005
Dibromochloromethane	negative	liver	93.9	Gold 2005
Bromoform	colon	negative	648	Gold 2005
Monochloroacetic acid	negative	negative		Gold 2005
Dichloroacetic acid	liver	liver	102	Gold 2005
Trichloroacetic acid	negative	liver	584	Gold 2005
3-Chloro-4-(dichloromethyl)-5-hydroxy-2(5H)-furanone (MX)	thyroid, liver, mammary	not tested	1*	Komulainen et al. 1997
2,4,6-Trichlorophenol	hematopoietic	liver	405	Gold 2005
Chloral hydrate	not tested	liver	106	Gold 2005
Chloropicrin	Inadequate	negative		Gold 2005
Sodium chlorite	negative	negative		Gold 2005
Sodium chlorate	thyroid	pancreas	Not established	NTP 2005a
Formaldehyde	hematopoietic, nasal,	nasal	2.19	Gold 2005
Acetaldehyde	nasal	not tested	153	Gold 2005
Benzylaldehyde	negative	stomach	1390	Gold 2005
Potassium Bromate	kidney, thyroid, mesothelioma of testes	kidney	9.82	Gold 2005
N-Nitroso-N-dimethylamine	liver, kidney, lung, testis, vasculature	liver, lung, nervous system	0.096	Gold 2005

* Estimated from the yield of thyroid follicular adenomas in the male rat (Komulainen et al., 1997.

RESOLVING THE ISSUE

It is necessary to very broadly review the literature to ascertain whether or not as yet uncharacterized by-products might present better explanations for cancer risks that have been associated with chlorinated drinking water epidemiologically than the THMs and the five regulated haloacetic acids (HAA5). To address this issue, the following steps were undertaken:

- the toxicology of DBPs that have been studied were reviewed,
- the possibility that other types of by-products could be formed was addressed on a theoretical basis,
- the toxicology of compounds in these classes was assessed, and
- in selected cases confirmation of chemistry envisioned will be tested in the laboratory.

As a result of the limited toxicological data available for most DBPs, QSTR have been used to aid in the identification of chemicals that might be of particular importance for study. It is anticipated that these analyses will form the basis of a more systematic experimental exploration of possible causes of cancer risk associated with chlorinated drinking water and the beginning of an approach to assessing whether or not similar reactions can ensue with alternative methods of disinfecting drinking water.

It was anticipated that this effort would fall short of the goal of identifying those by-products that are responsible for the epidemiological findings with chlorinated drinking water. Moreover, this study, in and of itself, cannot rule out the possibility that such risks exist. The most substantive goal of the project was to identify chemicals that may be potential causes of such cancers and to suggest a research agenda that can systematically address this important public health issue.

In the development of our approach, it became apparent that it would also have utility for developing and pursuing hypotheses related to other health effects that have been associated with consumption of chlorinated drinking water (e.g. reproductive and developmental toxicities). Our conclusions in these areas will be limited, reflecting the results of a single model for predicting potential developmental toxicity of DBPs. It is hoped that this approach can be used in the development of public policy that objectively minimizes by-product induced risk from drinking water, but still results in the absolute minimal risk for causing waterborne infectious disease.

CHAPTER 2. GOALS OF PROJECT

The overall goal is to develop a logical scheme for prioritizing research to identify disinfection by-products (DBPs) that have the potential of producing significant adverse outcomes in humans considering available information on by-product formation, along with toxicological and epidemiological data that bear on the issue. Thus, the intent is to provide guidance for research that may potentially resolve the issue of whether or not there is significant risk associated with the use of reactive chemicals in the disinfection of drinking water. If such research does, in fact, identify significant causes of risk more rational and focused regulatory strategies may be developed to ensure that the public health is protected

Quantitative structure-toxicity relationships (QSTR) are investigated in this project to predict whether or not individual DBPs are carcinogens or developmental toxicants with the intent of narrowing a research strategy that has the intent of accounting for the risks suggested by epidemiological studies. Therefore, DBPs that are predicted to be non-carcinogens or non-developmental toxicants are not of direct interest. In other words, QSTR is being used to select chemicals associated with a particular drinking water treatment from a large list of identified and postulated by-products that have a high probability for contributing to risks.

These predictions should not be taken as providing a critical level of evidence for regulation. The expectation is that predictions of probable general toxicity, carcinogenicity or developmental toxicity will need to be tested in appropriate experimental systems before any regulatory action is undertaken. It must also be pointed out that the predictive tool used was shown to fail at identifying some compounds that are produced by chlorination that have been demonstrated to be carcinogenic in animals. Therefore, the potential for false negatives cannot be ignored. In selected cases, more limited research efforts should be undertaken to determine that compounds discarded in the current effort do not pose significant risks.

EFFORTS TO IDENTIFY CRITICAL RESEARCH ON DBPs

The principle goal is to provide a basis for identifying DBPs that could be responsible for health hazards that have been associated with DBPs by epidemiological evidence. The effort consists of:

a. Reviews were conducted of occurrence and toxicological literature on DBPs.
b. In the absence of toxicological data on identified DBPs, chemicals were subjected to QSTR analysis for potential activity as carcinogens or as developmental toxicants.
c. Principles of organic chemistry were used to predict reactions with natural organic material as a means of identifying novel DBPs .
d. Chemicals that have been demonstrated to be bladder carcinogens in humans or experimental animals were examined to see if these structures could inform future chemical research on the formation of DBPs from natural precursors.
e. A decision process was developed that allows prioritization of research efforts on by-products of chlorination by chemical class and within class that can be simply extended to address issues related to alternative methods of disinfection.
f. A set of chemicals was identified that should be of high priority in studies of DBP occurrence.

g. A set of by-products was identified that should be of the highest priority for toxicological characterization.
h. From this effort, guidance can be derived to improve exposure assessment for characterizing risks associated with various disinfection strategies.

CONTRIBUTING TO THE DEVELOPMENT OF THE REGULATORY AGENDA

While the main purpose of this project is to assist in the prioritization of research efforts to clarify the DBP issue, this effort should also contribute to the activities of the Office of Groundwater and Drinking Water of the USEPA related to prioritization of chemicals for the candidate contaminant list (CCL). The CCL provides a medium for updating research needs and communicating those needs to other federal research agencies that have much to contribute, most notably, the National Toxicology Program, the National Institute of Environmental Health Sciences, and the Centers for Disease Control and Prevention. This report has developed recommendations for research on occurrence and health effects for those DBPs that appear to be the most likely to present a hazard.

CHAPTER 3. DESCRIPTION OF APPROACH - TOXICOLOGY

Two general approaches have been used in the development of the information in this report. The first was the approach identified in the initial proposal to the Awwa Research Foundation (AwwaRF). This is referred to as the feed-forward process. The feed-forward process examines the chemical nature of precursor materials found in source waters and predicts their interaction with disinfectants to identify potential by-products. Included in this group of compounds are chemicals that have been established as disinfection by-products (DBPs). The second approach, a feed-back process, involves the identification of chemical structures with toxicological properties similar to those that have been associated with disinfected water in epidemiological studies and determines whether or not precursors for analogous compounds can be identified in natural waters. The components of each process are outlined below:

FEED-FORWARD PROCESS

- Identify molecular substructures in natural organic matter (NOM) and through the use of model compounds to predict probable DBPs
- Utilize the principles of organic chemistry to predict by-product occurrence
- Utilize results to identify those classes of DBPs that require additional research. In part, these classes will be identified by a lack of empirical data to support reliable quantitative structure-toxicity relationship (QSTR) predictions.
- Utilize QSTR methodology to identify those compounds within a DBP class that are most likely to produce adverse health effects and thus have the highest priority for research
- Utilize results to provide some perspective on the risks projected in epidemiological studies.

FEED-BACK PROCESS

- Identify chemicals that produce toxicities associated with diseases observed in epidemiological studies of disinfected drinking water
- Look for precursor material that could give rise to analogous compounds
- Articulate principles used in selecting model precursors and analogies to potential NOM substructures.
- Utilize results to focus research needed to account for "causes" of particular diseases (e.g. bladder cancer).

The feed-back process was felt necessary because it became clear that there were both qualitative and quantitative shortcomings in the database associated with the use of the trihalomethanes (THMs) and five of the haloacetic acid (HAA5) as surrogates for DBPs capable of inducing the health effects imputed by the epidemiological studies of chlorinated drinking water. It is important to indicate that epidemiological data associating health effects with established DBPs or classes of DBPs fall short of the criteria for causality. Epidemiological studies have established causal relationships between exposures to specific chemicals and bladder cancer in other settings (i.e. industrial exposures or with certain chemotherapeutic agents). In these cases, the odds ratios have been sufficiently robust to systematically rule out other potential causes. In contrast, the odds ratios obtained in epidemiological studies of specific

DBPs have not improved the odds ratios significantly over simple comparisons with amount of chlorinated water consumed relative to non-chlorinated water. For the purposes of this study, this problem was pursued primarily in the context of bladder cancer. The progression of questions that guide the feedback process are summarized below and discussed in more detail in the following text:

- Have halogenated compounds been shown to induce bladder cancer in any experimental system? If so, what are their structural characteristics?
- Do the halogenated by-products that have been identified possess sufficient potency to account for the risk suggested by epidemiological studies?
- Are there other chemical classes that would be more attractive as being causal for the cancers associated with chlorinated drinking water?
- Is the potency of members of the alternative classes consistent with a causal relationship?
- Are there precursors in drinking water sources that would yield members of these alternative classes of by-products?
- Do structure-toxicity relationships suggest that the specific by-products formed are of sufficient potency to account for risk?
- Are similar structures likely to be formed with alternative methods of disinfection?

The first two questions address the issue of whether or not identified carcinogenic by-products are plausible causes of bladder cancer, both in terms of specificity and potency. Assuming that the answer to the first two questions is negative, the third and fourth questions broaden the horizons of the study, asking whether or not examples of chemicals that are more plausible causes of a cancer risk of the magnitude suggested by epidemiological studies of chlorinated drinking water exist. To narrow the study to manageable dimensions it was decided that a focus on chemical classes that have produced bladder cancer in humans or at least one experimental animal would be used as a point of departure. If there are such compounds, the fifth question asks whether there are precursors in drinking water sources that would generate such chemicals with chlorination. The chemical nature of precursors that are available broadly in natural waters provides some limits on the number of chemicals that will remain in the final analyses. If the formation of specific by-products can be reasonably hypothesized or measured, do QSTR analyses suggest that the by-products are of sufficient potency to account for the risk? Finally, while the project necessarily focuses on chlorine as the disinfectant, it is important to consider whether or not the chemistry of alternate methods of disinfecting drinking water would form the same by-product/s given the appropriate precursors.

Obviously, the present project will not completely answer these questions. Its primary purpose is the development of an analytical structure for a more efficient approach to guide research that will ultimately resolve the issue. The intent is to suggest how the feed-back approach can be utilized to determine whether or not chemical classes with the appropriate toxicological properties can be formed based on consideration of the chemistry of the disinfectant and the nature of the precursors that are likely to be present.

DEVELOPING PRIORITIES

This effort does not resolve the question of whether or not chlorinated water causes bladder cancer or other adverse health effects. That question can be answered only with a well-designed research program. To be efficient that research program must find a way to focus on those DBPs that have a high probability of producing the effect of concern and be present in sufficient concentrations to be plausible causes. An idealized prioritization structure is provided that should be useful on and beyond the set of compounds available for analysis at the time of this report. Although a particular QSTR program was selected for this effort, the structure is not dependent upon this specific program. It was a convenient, though not necessarily the best possible tool. The particulars of these criteria are:

1. Chemicals with a chronic lowest observed adverse effect level (LOAEL) < 1-mg/kg day^{-1} were considered to be of high interest, irrespective of whether or not they were predicted to be carcinogens or developmental toxicants. In general, the LOAEL submodel automatically selected by TOPKAT® was utilized. In cases where very low LOAELs were predicted, but fell outside the optimal prediction space (OPS) of the model, indices and structures in training sets of the selected or alternative models were examined in detail and the model value finally selected based on this evaluation.
2. Chemicals predicted to be carcinogens or developmental toxicants with a TD_{50} or chronic LOAEL of < 10-mg/kg day^{-1} were considered to be of high interest.
3. A probability of 0.7 (recommended in TOPKAT® instructions) was required to accept a positive prediction of carcinogenicity or developmental toxicity within any given QSTR model. Predictions that fell outside the OPS were discounted if they were outside the permissible limits of the OPS of the model.
4. Predictions of carcinogenic activity were considered of greater interest the more consistent the prediction was among species (rats and mice) and sex. Both the National Toxicology Program (NTP) and Food and Drug Administration (FDA) models for each species and sex were utilized. A positive result within a species or gender was considered a positive response even if the model based on the other database predicted a lower probability or was outside the OPS. This approach was taken because the databases (or training sets) for the FDA and NTP models differ significantly and were considered complimentary rather than competing models.
5. A final decision point for accepting a positive prediction of carcinogenicity or developmental toxicity by TOPKAT® was that the chemical fall into certain limits of similarity with the chemicals in the training set for the model in question. A similarity index of < 0.2 to its closest analog in the training set is the recommended cut-off in the instructions for the program. In the first application of this criterion, this range was expanded to < 0.3 to observe the range of data across different models. However, at least one prediction based on a similarity index of <0.2 was necessary for a chemical to be placed in the highest prioritization category.
6. Identification of natural precursors for a putative DBP was critical to its prioritization. To the extent possible, putative DBP precursors will be grouped into three categories: ubiquitous, common, and rare.
7. To the extent possible, the concentrations of the precursors were grouped in the context of probable yield of a putative DBP: sufficient to yield > 0.1 mg/L, > 0.01 mg/L, > 0.001

mg/L, or > 0.0001 mg/L where the by-products have been measured in drinking water or where there are sufficient data to infer a concentration range from a study of model substrates.

8. A weighted priority was developed between the observed or predicted potency and the probable yield of the putative DBP. For example, a carcinogen with a LOAEL of 1-10 mg/kg would be considered a high priority if it were predicted to occur at >0.1 mg/L, at a medium priority if it occurred at < 0.1 mg/L and >0.01 mg/L, and as a low priority if < 0.01 mg/L and not considered at lower concentration ranges. Parallel prioritization schemes will be used for compounds that are predicted developmental toxicants or simply based on a low LOAEL consistent with points 1 and 2. This permits comparison of the relative importance of regulated and non-regulated DBPs in terms of the probable health risk. For example, the ratios of dose/TD_{50} for regulated DBPs ranged from 10^{-7} to 10^{-5}. Therefore, putative DBPs with a ratio dose/TD_{50} or predicted chronic LOAEL > 10^{-5} was considered of increasing priority in direct proportion to the ratio. Chemicals with smaller ratios were considered to be of low priority.
9. Finally, a DBP predicted to be a carcinogen and to be positive with respect to Ames' mutagenicity is ranked higher in priority than a chemical not meeting the mutagenesis criteria, but having a similar LOAEL.

The Ames' mutagenicity criterion articulated in point 9 is consistent with the hypothesis that "genotoxic" carcinogens have linear dose-response in the low-dose region. A non-mutagenic carcinogen is likely to be treated by utilizing the lower 95% confidence limit of the benchmark dose predicted to give rise to a 10% response as a point of departure for applying uncertainty factors of 100-1000 to arrive at a virtually safe dose.

In practice, priorities were determined by the calculation of a ratio between the dose anticipated to cause health effects and that which might be anticipated from drinking water. This is analogous to the margin of exposure (MOE) approach that is suggested under the new cancer risk assessment guidelines (USEPA 2005). The occurrence and concentration data on most DBPs are so limited that no calculation of a ratio could be reasonably made. Those DBPs that were predicted to have chronic LOAELs below 1-mg/kg day^{-1} are suggested to be of a high priority for the development of occurrence data.

PROBABILITY OF OCCURRENCE

The probability of occurrence is a little more nuanced than implied in the points above. Especially with reference to point 7, the probability of occurrence could also qualify a chemical based on the following considerations:

- The probability of occurrence increases as the model compound or its analog is identified up the precursor chain.
- The ultimate determination would be to demonstrate formation in a source water treated with disinfectant.
- Absent direct demonstration of formation, one can estimate occurrence by ubiquity of the precursor substructure/s
- Concentrations have not been projected for chemicals that have not been observed in disinfected water. Occasionally a concentration range may be estimated from model

studies where experimental conditions are sufficiently defined and the reactivity of precursor substructure has been pursued sufficiently to estimate probable yields.

Again, these criteria may help to identify compounds that are likely to be produced. However, the predictions are only as good as one's knowledge of precursors that are present and one's understanding of the chemistry of a particular disinfectant. Therefore, this approach cannot rule out the possibility that a particular DBP will occur.

TOXICOLOGY DATA

Established DBPs for which there is a substantive database for toxicological characterization were identified by literature review. The general, reproductive, and developmental toxicities of those DBPs that have received appropriate toxicological testing are represented in the results section with the dose calculated to give 50% of the animals tumors in a lifetime (TD_{50}s), LOAEL, or no observed adverse effect level (NOAEL) of the key studies noted. These data are compiled in Table A.1 in Appendix A. The details related to carcinogenic responses are identified by target organ in each species and gender that was studied. The latter figures are derived largely from the Carcinogen Potency Database (CPDB, Gold 2005).

Among chlorination by-products, the THMs, selected HAAs, selected halophenols, chlorate, and 3-chloro-4-(dichloromethyl)-5-hydroxy-2(5H)-furanone (MX) have produced cancer in controlled experiments in animals. A number of halogenated by-products have been tested in animals and did not increase the incidence of cancer. These include iodoform and monochloroacetic acid. Non-halogenated by-products have also been tested, including acetaldehyde, formaldehyde, benzaldehyde and benzoic acid. Some chemicals reported to be by-products have been eliminated from the list because they are generally contaminants of the disinfectant rather than resulting from reactions of the disinfectants with precursors in the water (e.g. dichloromethane, carbon tetrachloride, and hexachloroethane). Another group were eliminated because they are very unlikely products of disinfection (e.g. 1,2-dichlorobenzene, 1,4-dichlorobenzene, chlorobenzene, 1,4-dioxane, benzene, and toluene). There are also a few compounds that have been shown to increase the incidence of cancer in short-term in vivo tests with cancer as an endpoint (initiation/promotion studies in skin, liver or other organ, lung adenomas, etc), but are not generally accepted as the bases for assessing cancer risk. Within this category are found the haloacetonitriles (Bull et al. 1985) and selected halopropenals and haloacetones (Robinson et al. 1989).

By-products produced by alternative disinfectants to chlorine that have been studied are also identified in Table A.1 in Appendix A. Some of these are produced by chlorine as well, but are more closely associated with the alternative. Bromate, N-nitroso-N-dimethylamine (NDMA), and chlorite associated with ozone, chloramine and chlorine dioxide, respectively, are cases in point.

LOAELs reported for reproductive and developmental toxicants were abstracted from the review of Tyl (2000). At this time, these chemicals do not appear to have sufficient biological potency to warrant regulation and have not been included in the main body of the report. However, the data from Tyl (2000) can be found in Table A.2, Appendix A. The maternal LOAELs generally reflect decreased weight gain and allow one to segregate true developmental toxicants from chemicals that simply reduce appetite (Carney et al. 2004). Effects on spermatogenesis were the most frequent endpoint for male reproduction. The LOAEL for developmental toxicities relate to a variety of measurements of malformations in more or less traditional teratogenicity studies.

DESCRIPTION OF A QUANTITATIVE STRUCTURE- AND TOXICITY-RELATIONSHIP (QSTR) APPROACH TO BY-PRODUCTS

General Nature of QSTR

QSTR is an easily developed and powerful analytical tool where the effects of a chemical are mediated through a specific receptor. Interactions of drugs with receptors have been extensively exploited in the pharmaceutical industry with such tools. Activation or inhibition of a physiological function at a specific receptor molecule (generally a protein involved in transmitting regulatory signals within or between cells) necessarily reflects the specificity of that receptor for its natural ligands. As a result steric and chemical properties of chemicals that influence the activity of these receptors generally fall into a narrow range. For this reason, structure-activity relationships have a long history in drug development.

Chemically induced toxicities can also involve very specific interaction of a chemical with a receptor. As long as the interaction is straightforward, very good QSTR can be developed. A good example is the group of toxicants that are classified as endocrine disruptors. Some of these chemicals have very specific interactions with receptors for particular hormones. While the reversible activation of such receptors is integral to normal biological function, a chronic activation or inhibition of the receptor can lead to adverse health outcomes (NRC 1999). However, it is equally possible to disrupt endocrine function without interacting with the identified receptor. This can arise by less specific interactions with post-receptor components of the hormonal signaling system. As many different sites are potentially involved and the nature of chemicals that can produce the post-receptor effects can be diverse, it is much more difficult to develop all-inclusive QSTR models. Therefore, QSTR is most easily developed for a group of chemicals that interact with a single specific receptor (e.g. an estrogen receptor).

Predicting whether or not a chemical is a carcinogen is much more complex because there are multiple mechanisms that can lead to cancer (Rosenkranz 2004). Some chemicals induce cancer through interactions with specific receptors. Compounds that fall into this group are frequently tumor promoters. They are a subset of those carcinogens that are referred to as 'non-genotoxic' carcinogens. On the other hand, genotoxic carcinogens produce cancer either by interacting covalently and/or interfering with the faithful replication of DNA at cell division. The reaction of some carcinogens or their metabolites with DNA can be specific for a particular DNA base, but the result ends up depending on a reaction with that base in very specific locations within a gene. There are many gene targets for mutation. Thus, modeling of these carcinogens tends to focus on their ability to interact with DNA, but the outcome of those interactions depends on the gene mutated and the function of its product. Moreover, many host, environmental, and lifestyle variables frequently modify the disease outcome. The enzyme that activates the carcinogen may be expressed differently in different organs and this, in turn, may vary among species. Enzymes that detoxify the active form of a carcinogen may also vary between organs and among species. In addition, the ability to repair the damaged DNA may vary by organ and species. Finally, the organ in which the cancer develops will be influenced by factors embedded in the physiology, environment, and dietary habits of the species being examined and even among individuals within that species. As a consequence, QSTR is much more limited in its ability to capture all the variables necessary predict specific carcinogenic outcomes.

Despite these shortcomings structure-activity relationships have utility when attempting to address the carcinogenic potential of chemicals. While QSTR models focus on specific chemical properties, their ability to identify toxicants depends upon a database that includes chemicals that have been established as toxicants and non-toxicants (or carcinogens and non-carcinogens). In some cases, there are sufficient data to predict carcinogenic potency with some accuracy. In general, this depends heavily on there being a sizeable database for chemicals that have analogous structural and chemical properties. Given a substantial database for such a grouping of chemicals, the ability to predict the potency of a carcinogen has been much more successful than the ability to predict qualitative aspects of the carcinogenic effects, such as the organs in which the cancer is likely to occur in humans (Benigni 2003).

As will be discussed in greater detail below, bladder carcinogens have been recognized as chemically induced cancers for many years. There is a substantial database in humans, as well as in experimental animals, that reflect this experience. It is important to recognize that the chemicals that are identified as bladder carcinogens in this paper frequently have other target organs (IARC 1999). However, there are some specific metabolic processes in which bladder carcinogens in the aromatic and heterocyclic amine classes are involved that illustrate this fact and are discussed in some detail Chapter 6. It is also important to recognize that chemicals that produce bladder cancer in one species may not produce bladder cancer in another. There are a variety of reasons for this, including the possibility that one tumor may be lethal and preclude the development of a bladder cancer within the shortened lifespan of the animal. Another reason is that very subtle differences in the metabolism of a chemical can affect the target organ. For example, variations in enzymes that oxidize and acetylate aromatic and heterocyclic amines have been shown to determine whether an amine produces bladder or colon cancer (Ambrosone and Kadlubar 1998).

It is important to recognize that a DBP which can be shown to produce bladder cancer reflects only one part of the controversy over whether or not DBPs produce increased bladder cancer risk. Most important is the question of whether or not the potency of the chemical as a carcinogen is sufficient to produce cancer at the doses that are likely to be encountered in the environment. Thus, the limited capabilities of QSTR tools to identify target organs made estimates of potency of paramount importance in this project.

Specific Properties of TOPKAT®

Several commercial programs were available that approach structure-activity relationships in somewhat different ways. Because of the diverse nature of DBPs, it was necessary to select an established tool for making predictions of harmful health outcomes. TOPKAT®, marketed by Accelrys, was selected primarily because of the diversity of endpoints that can be addressed with the series of models that are used in this software. In particular, the model in TOPKAT® that predicts chronic LOAELs in rats allowed an approximation of a chemical's potency to be made. The relationship of the chronic LOAEL of a chemical to a specific effect has its limits. For example, there is little conceptual basis to support the idea that the potency of a chemical as a carcinogen approximates its ability to cause other toxicological effects that are usually the basis of a chronic LOAEL. Empirically, however, it has been found that there is a proportionality of the carcinogenic potency of chemicals to the maximum tolerated dose (Gaylor and Gold 1995). Based upon the empirical evidence it was decided to use the chronic LOAEL as an approximation of potency for the purposes of this project. Only TOPKAT® included a model for predicting the chronic LOAEL of a chemical.

Within the TOPKAT® model formulations, certain descriptors are utilized that relate to the rates of absorption of a chemical such as molecular bulk, shape, and symmetry. These are combined with descriptors that quantify the chemical properties of a compound. The chemical properties utilized by TOPKAT® relate to the information-rich electrotopological state (E-state) values developed by Kier and Hall (Accelrys 2001) to quantify the electronic attributes of molecular structure to account for interactions at a molecular site of action within the body. Each model within TOPKAT® has a training set of compounds that are used as comparators. For example, there are separate models for mice and rats and for males and females within each species for predictions of carcinogenicity. Therefore, diverse predictions occur because of species and gender differences that are reflected in responses to chemicals within the training set of each model. It addition, there are two databases that can be used, those compounds that have been tested under a consistent protocol by the NTP and those that are included in the FDA database. The latter group consists of data utilized in the registration of certain products (e.g. new pharmaceuticals) by manufacturers. Thus, the studies may involve somewhat different protocols, are conducted under a different set of guidelines, involve chemicals that may be under represented in the NTP database (e.g. involve a much higher proportion of nitrogen containing compounds), and frequently involve administration of a chemical by a wider variety of routes. Moreover, the data are frequently considered confidential business information. As a consequence, it can be difficult to really explore important details of supporting studies. Therefore, there are differences in predictions based on the training set that was used to develop each model and some potential limits for exploring why predictions vary between models. In the current project, both the NTP and FDA models are considered to be largely independent and complimentary in their ability to predict carcinogenicity. As this project focuses on positive predictions of carcinogenicity, a positive prediction that is well-founded in the training set of one model was not discounted by a negative prediction of carcinogenicity with another model.

In addition to the computed probability that a chemical is a carcinogen or a developmental toxicant, TOPKAT® provides a number of warnings to alert the analyst to the confidence that should be placed in a prediction. TOPKAT® provides three types of warnings if a chemical is poorly represented in the training set. The first is an indication of whether or not the prediction is outside the OPS of the model. The program distinguishes between predictions that are outside the OPS, but within permissible ranges and those that exceed the permissible range. In the tables constructed depicting the model outputs a prediction that falls in the first category is marked by an asterisk in the column adjacent to the probability score for that model. Those that exceed the permissible range are identified in the tables by a cross. Second, within each model's database, the program provides a measure of the similarity of the query structure with the chemicals that were included in the equation on which the prediction is based. Matches in similarity are reflected in a range from 0.000 (identity) to 1.0 (no similarity). In general, the query structure needs to match at least one chemical in the database with a similarity of < 0.2 to have confidence in the prediction. Finally, the program indicates when a particular molecular fragment is not represented in the model's database. These were also identified in the assembled database.

Carcinogen Potency Database

The CPDB is an online database that summarizes most of the cancer bioassays that have been conducted on chemicals in experimental animals. The website is maintained by Gold and coworkers at the University of California, Berkeley (http://potency.berkeley.edu/cpdb.html). The database represents a total of 1451 compounds that have been subject to test in at least one species and the data are of sufficient quality to make an estimate of potency. The potency of carcinogens is expressed as the TD_{50}, which is a value calculated from the data to estimate the dose that would cause a 50% increase in the incidence of tumors during the lifetime of a test animal. The advantage of the TD_{50}, as opposed to the figures utilized to characterize risk in support of regulation, is that the derivation of the TD_{50} is generally within the range of responses actually measured in the studies. Therefore, there is minimal extrapolation of the data involved and the TD_{50}s are not dependent upon untested and untestable assumptions that are frequently used to estimate risks at low environmental doses. Since there are few assumptions made at this level, chemicals can be much more reliably compared for relative potency by utilizing the TD_{50}. An approximation of the daily dose that would give rise to a 10^{-6} lifetime risk can be obtained by dividing the figure given by 500,000. This would be a rough approximation of the one-hit model, but would generally reflect a lower dose than would be derived by alternative models (e.g. the linearized multistage model).

The possibility of converting model-predicted chronic LOAELs to TD_{50}s was considered. It was decided that using the LOAEL, which would generally approximate a dose that produces a 10 or 20% response (ED_{10} or ED_{20}, respectively) depending upon the statistical power of the study that was performed, would infer more precision in the prediction than was warranted. In general, the confidence intervals surrounding the prediction would include ED_{50}. The decision was made that it is better to maintain a clear distinction between measured and estimated values.

Structures of Compounds that Produce Bladder Cancer

As indicated in the introductory section, the most difficult problem in predicting carcinogenic activity is identifying target organs for the query structure. In part, this is because there are strong genetic elements involved in determining how each species/strain responds to a carcinogenic chemical. Coupled with this are differences in the nature of the controlled diet and environment provided to experimental animals vs. that of humans who do not live in the same consistent conditions. In part, the difficulty arises because the focus of much toxicological testing is to ensure that a potentially dangerous compound is not introduced uncontrolled into the environment. Therefore, there is an emphasis on consistent conditions and reproducible experiments. Despite the differences between experimental conditions and those in which humans live, there is a good general relationship of carcinogenic potency among species, although there are many clear exceptions. However, neglecting the carcinogenic activity of a chemical in an organ not found in humans is not considered a prudent public health policy (USEPA 2005).

It is important to recognize that the present project is not based on the idea of preventing the introduction of harmful chemicals into the environment. Rather, it is an effort to determine whether or not a risk that has been projected in epidemiological studies can be supported with experimentation under controlled conditions. Bladder cancer has not been identified in animal studies as being caused by identified chlorination by-products. More importantly, the potency of the studied compounds suggests that these occur at much too low concentrations in drinking water to account for the magnitude of the risk that has been projected. The question is whether or not there are other by-products potentially formed by disinfection of drinking water that are more probable causes of a risk of this magnitude. This necessitates a critical reexamination of some of the basic assumptions that have been made about which by-products of disinfection are important. From a practical point of view, such an effort is essential before it can be determined if methods of mitigating the apparent risk observed with chlorinated drinking water will be effective.

A secondary database was developed from the CPDB that lists those chemicals that are known bladder carcinogens in humans or experimental animals (Appendix B, Table B.1). It is used in this effort to identify those classes of compounds that do include bladder carcinogens.

Potent Carcinogens

Carcinogens were identified in the CPDB that had TD_{50}s below 10-mg/kg per day^{-1} and tabulated. This provided a set of chemicals that are approximately 10 times the potency of the by-products that are regulated under the DBP rule (i.e. the THM and HAA classes). Potency on this order or greater would be necessary to reasonably account for the degree of risk implied by epidemiology studies of chlorinated drinking water. The chemical classes that were represented in the group of potent carcinogens can be found in Appendix C, Table C.1.

A total of 331 carcinogens were identified in the CPDB that had a TD_{50} in at least one species below 10-mg/kg day^{-1}. As the carcinogenic potency of DBPs is examined in subsequent pages they can be compared to chemicals that have been shown to be potent carcinogens by our definition. The < 10-mg/kg day^{-1} cutoff as the boundary for potent carcinogens was selected primarily because a chemical with a higher TD_{50} is unlikely to contribute significantly to the risk of developing cancer from drinking chlorinated water unless its concentrations were above 100 µg/L.

ORGANIZATION OF DBPS BY CLASS AND GROUP

Attempts were made to prioritize both classes of DBPs and individual chemicals within a class. Halogenated DBPs were initially placed into 30 groups. To facilitate tracking of a particular DBP in the database, each was assigned a serial number that also identified the group in which it was included. These are displayed in tables as a code (e.g. K01C01 = the first chemical in the KO1 group). The definition of these chemical groups depended upon functional groups (other than halogen) that are part of the structure and may be identified as a chemical class. However, the groupings are also somewhat arbitrary. Certain by-products could be placed into more than one chemical class because they have more than one functional group. Generally, well-recognized DBP groupings were generally separated from other less well-established DBPs with the same functional groups. Therefore, the THMs are grouped separately from other haloalkanes, haloacetic acid groups were grouped separately from the general class of saturated haloacids, and the haloacetonitriles were grouped separately from other halonitriles. These more or less arbitrary groupings of DBPs have the advantage of allowing assessment of whether or not non-regulated DBPs within and outside of these classes are likely to be of greater or lesser concern than the regulated DBPs. The criteria for whether or not a particular class of DBPs requires attention was judged primarily by comparison of the distribution of its potencies relative to the distribution of potencies of halogenated compounds in general.

Additional groups were added to include non-halogenated by-products and as more novel classes of halogenated DBPs were identified. Groups K31 to K49 include non-halogenated DBPs. Novel DBPs predicted to be formed by reaction of disinfectants with substructures of NOM were included in group K50. This latter group contains chemicals from a variety of chemical classes. Not all of these groups were populated, in large part because of the limited information that was available on alternative forms of disinfection. In addition, some groups were simply not addressed because they are recognized, non-toxic constituents of the diet. By-products for which no QSTR analyses were performed were not assigned a serial number within the class. The most obvious of these is the myriad of fatty acids that are formed in water with oxidants.

CHAPTER 4. DESCRIPTION OF APPROACH - CHEMISTRY

DESCRIPTION OF APPROACH

In Chapter 3, the feed-forward and feed-back approaches were presented. In this chapter, the two approaches are described in greater detail as they pertain to the chemistry of disinfection by-product (DBP) formation.

Feed-forward Process

The fundamental steps in a feed-forward method as envisioned here are:

- Identification of significant molecular substructures in natural organic matter (NOM)
- Prediction of probable disinfection byproducts from those structures through a combination of:
  1. Laboratory model compound studies and
  2. Mechanistic organic chemistry

The authors envision two types of feed-forward approaches; one based more heavily on NOM-focused model compound testing, and having a more empirical flavor. This employs a scheme that is quite accessible and can be easily implemented for a wide range of by-products at the current time. This will be referred to as the first generation feed-forward protocol. The other approach is more heavily embedded in mechanistic and computational organic chemistry. This is potentially much more powerful and holds greater promise for discovery of new by-product structures. However, this second generation feed-forward protocol will require a substantial development effort for it to be the comprehensive tool that the authors envision.

First Generation Feed-forward Process

This first approach takes advantage of our current understanding of NOM at a biochemical class level. Using the latest concepts of NOM composition and genesis (e.g., per Leenheer, MacCarthy, Hatcher, Christman, etc.), one can propose a set of biochemical classes that represent NOM structure in a general sense. The exact breakdown among these classes will depend on all of the biological, chemical, climatic and geographic factors that go into NOM genesis. However, the belief is that such a biochemical-based classification can be made based on targeted characterization of any given NOM type.

In parallel with the NOM classification would be a focused effort to acquire and compile data on the reactions of chlorine and other disinfectants with key members of the biochemical classes. This should result in a targeted model compound – DPB database that is mined to produce overall DBP profiles expected under typical disinfection conditions.

An illustration of how this would work is presented below. Table 4.1 contains a listing of eight major classes of plant products that are likely to be solubilized and incorporated into aquatic NOM. For each of these eight classes, there exists an estimate of abundance as determined for a particular site-specific NOM. One can also propose other environmentally important data such as general biodegradation rates. Here numbers representing the expected rank order are presented for illustrative purposes. For each of these classes, a class-specific trihalomethane formation potential (THMFP) was included. These are roughly based on typical structures and known trihalomethane (THM) yields from model compound studies. Combining these data, one deduces that the particular NOM represented by the relative abundance numbers would have a carbon specific THMFP of about 25 ug/mg-organic carbon (-C).

To further this approach, one can postulate the impacts of aging on this material in the form of biodegradation. Such degradation may occur in a river or reservoir or even in a biological filter (albeit at an accelerated rate). To simulate such a process, the various classes of plant products were allowed to decay based on the fictitious rates and a first order rate law. Along with these the associated THMFP decayed as well (Figure 4.1). Also shown in Figure 4.1 is the modeled ratio of THMFP to residual total organic carbon (TOC) (i.e., the specific-THMFP or Sp-THMFP). This value climbs with time, due primarily to the quick loss of the relatively non-reactive carbohydrates, and the persistence of the highly reactive tannins.

Table 4.1
Model parameters used to create Figure 4.1

Constituent	Initial abundance (% of TOC)	Biodegradation half-life (d)	Sp-THMFP (µg/mg-C)
Fats & Waxes	2	60	0
Terpenoids	12	400	10
Tannins	4	200	400
Lignins	10	150	40
Cellulose	22	20	3
Hemicellulose	25	10	3
Protein	10	5	30
Sugars & Starches	15	1	3

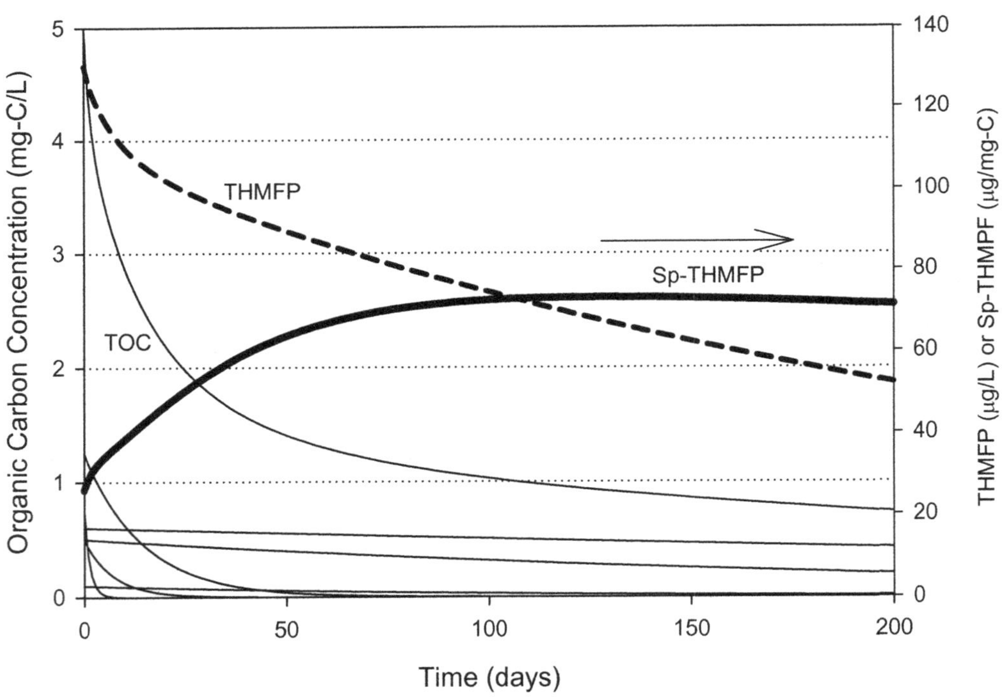

Figure 4.1 Fictitious example of selective degradation of NOM constituents and resulting specific THMFP (individual constituents are shown at bottom)

The above analysis is most applicable to young NOM, particularly the organic matter in flowing surface waters. This is also a gross simplification of the natural processes, which do not result in direct mineralization of the NOM classes. Biodegradation rates will certainly depend on the immediate environment, the standing biomass, the presence of light and availability of terminal electron acceptors, etc. Nevertheless, the authors view this as a useful framework upon which uncertainties and additional data can be assembled.

While understanding THM formation is a useful goal, it is not the focus of this research. The recommendation is that major biochemical constituents be examined carefully for formation of other, non-regulated DBPs and that this information be incorporated into the framework. While empirical in approach, we believe this would help to focus existing empirical DBP detection efforts, and increase the likelihood of finding new and relevant by-products. Ultimately, this would lead to a first generation feed-forward model that may be used to make semi-quantitative predictions of DBP formation for a much wider class of compounds. The predictions would then feed into DBP risk models.

Second Generation Feed-forward Process

The second feed-forward approach will take more time to develop, but it should ultimately be much more powerful. This method would begin with a similar breakdown of NOM constituents as represented in Table 4.1, but with the possibility of going into much greater chemical detail. For example, geography, and native flora may dictate the type of tannins that are most likely to be incorporated into NOM. In this way, one can draw heavily from the vast literature on natural product chemistry.

This second-generation approach is conceived as relying much more on fundamental chemistry and quantitative structure-activity relationships (QSARs). It also takes a reductionist approach. The focus will be on substructures and reactive centers in constituent molecules. We propose that a library be compiled of natural product structures, using existing commercial software (ChemDraw® or equivalent). These structures can then be used to populate a searchable database (e.g., ChemFinder® or equivalent), which can be queried by structural features of one's choosing. This same type of database will be needed for the feed-back approach, described later in this chapter.

Associated with each small structural feature will be information on relative reactivity and likely by-product structures. This information must be built into the database and it is, itself, built upon the basic chemical literature. Computational chemistry will play a substantial role in this approach. Combining the reaction pathways and by-product structures of the reactive centers with the surrounding molecule (which may include other reactive centers) will allow the "construction" of larger, multifunctional hypothetical by-products.

Feed-back Process

The feed-back approach begins with a proposed carcinogen or developmental toxicant of high potency (Figure 4.2). This may be a single compound of defined structure or a class of compounds with some critical structural feature. In either case, the compound or compound class is viewed as an end product of an oxidant-induced sequence of reactions. In most cases the researchers will be considering a sequence of chlorine reactions (oxidation, addition, substitution, hydrolysis, etc.) The feed-back approach will also require the development of a searchable database of natural products, much like the one recommended for the second generation feed-forward approach.

Working backward from the proposed carcinogen in a stepwise fashion, one can propose immediate precursors. These will be used to query the natural product database. At each step in the process as one moves backward down the reaction sequence, possible NOM substructures will be mined from the database along with estimates of their likely abundance (broad range). This analysis will provide information useful on two fronts. First, it will provide an initial assessment of the likelihood that the proposed carcinogen is a DBP and the type of plant products that would serve as a precursor. Second, it will identify model compound substructures that should be tested to verify or refute the proposed oxidative links to potent carcinogens.

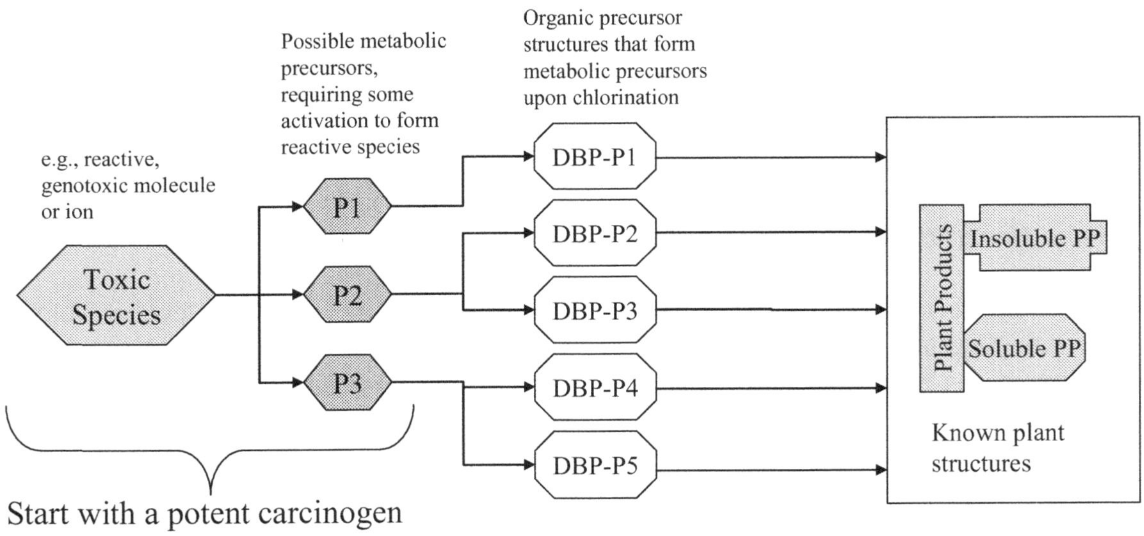

Figure 4.2 Schematic of feed-back approach

Underlying Chemical and Reactivity Data

The first step in either type of feed-forward approach is to critically assess the current state of knowledge regarding NOM substructures. For this purpose, published data from degradative and non-degradative studies of NOM are used along with the vast body of knowledge on natural plant products.

The most appropriate literature for understanding NOM substructures is the diffuse written work on whole surface waters and surface water extracts. Successively more abstract are the studies on aquatic microbial exudates, terrestrial organic leachates, and finally direct studies of plant components. Each of these studies can encompass various levels of specific structural information. The general categories are:

Confirmed molecular structures (usually by MS)
Structures of monomers (e.g., after acid hydrolysis)
Specific structural information on products from harsh degradation methods (e.g., after pyrolysis).
Non-specific characterization of structural elements (e.g., NMR, IR)

As summarized by MacCarthy (2001), investigators are now coming to an agreement that humic substances are a complex mixture of plant and bacterial products that have been partly degraded and then recombined by nearly random, non-enzymatic reactions. Largely because of the non-enzymatic processes, these humic substances are quite refractory. They also tend to be the single most important reservoirs of DBP precursors.

In contrast, the non-humics are largely natural plant products that may have undergone some degree of biodegradation or biosynthesis. One newly recognized and prominent group of non-humics is the acylheteropolysaccharides (APS)(Benner et al. 1992). These are high-molecular weight (>1000 daltons), algal-derived polymers of simple sugars, acetate and lipids. They are recalcitrant, much like the humic substances. Recent reports have placed this material at 10%-35% of the dissolved organic carbon (DOC) in lakes and rivers (Repeta et al. 2002). The APS pool probably constitutes the majority of what has often been characterized as the hydrophilic neutral fraction of NOM. This fraction is quite abundant in large reservoirs, and it is known to have a substantial THM and haloacetic acid (HAA) precursor content, albeit somewhat lower than the humic fractions (Reckhow et al. 1999a).

Much of the remaining NOM is thought to be among the simple identifiable plant products. Included among these compounds are carbohydrates, nucleic acids, proteins, lipids, and their various building blocks (e.g., sugars, nucleotides, amino acids, fatty acids), as well as some other prominent compounds (e.g., phenolics, terpeniods, alkaloids). Recent work by Leenheer and colleagues (2003) suggests that the terpenoids are especially prevalent.

Numerous studies of organic carbon in surface waters have demonstrated the importance of algal sources (autochthonous) as well as those from terrestrial plants (e.g., Wetzel 2001). Walker (1983) illustrated the strong empirical relationship between reservoir phosphorus level and TOC. He proposed that THM precursors would behave similarly. However, others have disputed the wisdom of directly linking TOC with DBP precursors (Stepczuk et al. 1998). Nevertheless, there exists a body of evidence that supports algae as important sources of precursors in some reservoirs (e.g., Karimi and Singer 1991; Plummer and Edzwald 2001).

For the purposes of this assessment project, the authors propose the following conceptual model for NOM structure selection. As illustrated in Figure 4.3, the recommendation is to start with some knowledge of specific soluble plant products (SPP) and anthropogenic substances in aquatic systems. Added to this should be information on laboratory degradation studies (both mild and severe) which gives rise to identifiable fragments. This fragment information can then be compared to known substructures in terrestrial and aquatic plant cells. If there is an acceptable level of agreement, the particular structure is deemed likely to exist in aquatic systems and therefore a potential DBP precursor. It is now our opinion that all major plant products should be considered in the various feed-forward and feed-back approaches. Nevertheless, strategic use of resources is imperative when deciding on structures most critical for laboratory degradation tests.

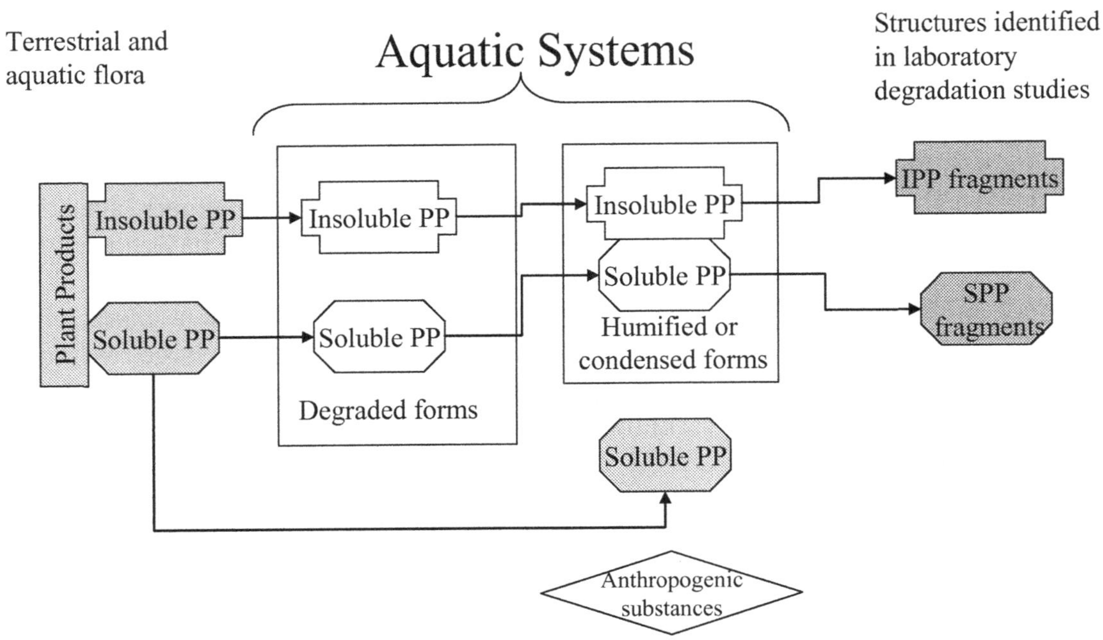

Figure 4.3 Schematic representing identification and confirmation of NOM structures

Once a significant plant substructure is identified as being of interest, there are several questions that must be asked before it should be selected for further study. These questions pertain to (1) abundance in nature; (2) evidence for their presence in aquatic organic matter, as already discussed; and (3) likelihood that the structure will be reactive with disinfectants (Figure 4.4). Chemical evaluation (e.g., model compound testing with postulated disinfectant reaction pathways) should proceed only after each decision point has been satisfactorily bridged.

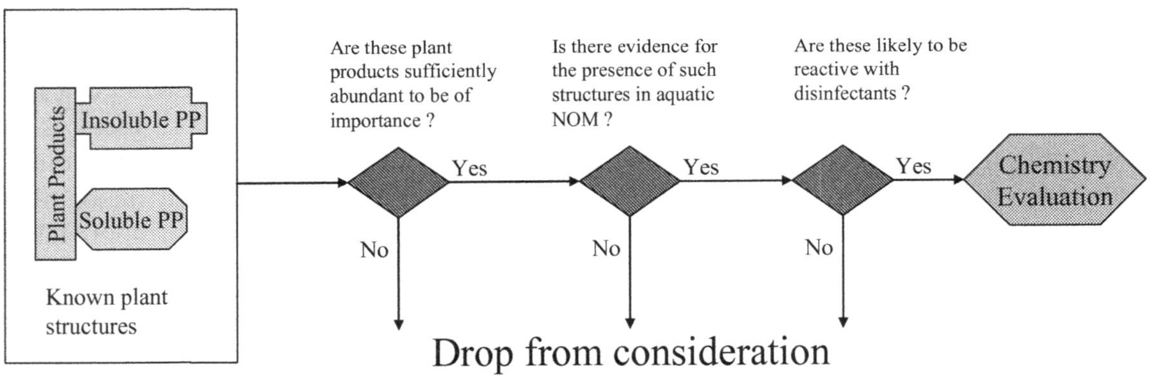

Figure 4.4 Schematic representing the confirmation and selection process for test precursors originating from plants

For the current work, we propose that a few key precursor structures be examined as case studies. The following two generic substances represent contrasting AOM categories. Each of them also presents a high probability of being among an important precursor group.

- Insoluble or High MW PP: Lignin, isolated protein
- Soluble PP: Nucleic Acids, terpenoids, tannins

DEVELOPMENT OF MODEL STUDIES AND ILLUSTRATIVE EXAMPLES

Review of Literature on Identified DBPs

There is a broad literature on DBP identification and occurrence. Some of these data were collected using approved methodologies, and some with innovative, yet unvalidated methods. There is also a broad range of abstraction in the DBP research literature. The most empirical and defensible studies were conducted on actual treated drinking waters. Some were performed on highly-concentrated drinking water extracts. While still others have used isolated NOM fractions and even plant or soil extracts. It is important that these data be carefully assessed for relevance and validity. After a critical review, they should be summarized for use in toxicological assessments.

Review of Literature on Model Compound Studies

The feed-forward approach relies heavily on predictive chemistry, which itself rests on fundamental chemical knowledge (originally gleaned from laboratory experiments with compounds that may or may not have any relevance to drinking water), and on empirical chlorination experiments with compounds that are likely to be in raw drinking waters (i.e., model compound studies). For this reason, it will be necessary to extract as much specific or empirical information on pathways, products and kinetics as possible from the literature, and blend this with sound organic chemistry fundamentals.

For illustrative purposes, the researchers elected to focus this literature review on a few groups of nitrogenous organic compounds. These include amino acids, selected nucleic acid constituents and some simple aromatic amines. The relevant compound-specific literature encompasses:

Environmental studies on the problem of drinking water DBPs
Chemical studies on use or role of chlorine in synthesis reactions
Medical studies on inflammation-induced reactions in vivo
Miscellaneous studies (e.g., biosynthesis of organochlorine products)

It is important that all of these sources be tapped. It is also necessary that the vastly different conditions used in these studies be acknowledged and interpretations made accordingly.

Laboratory Screening

Because of the paucity of mechanistic studies that are specifically directed at drinking water, it is helpful to collect some general screening data from the laboratory. This study has demonstrated the utility of such data by running a series of bench-scale chlorination tests with selected model compounds. Most of those chosen are known natural products, or monomeric constituents of common biopolymers.

DBP Assessment

The final step in the chemistry assessment is to integrate the literature data on by-product formation from model compounds, with new data on simple laboratory assessments and information on the likely abundance and characteristics of the precursors compounds in nature.

This requires careful attention to reaction conditions typical in water treatment plants and distribution systems. This also requires that some judgment be used in assessing the likelihood of substantial export and persistence of the proposed precursor structures.

CHAPTER 5. OVERALL RESULTS - CHEMISTRY

INTRODUCTION

In this chapter some key laboratory results from model compound chlorination tests conducted as a part of this research will be presented. These data will be first presented in brief without analysis. This will be followed by a detailed synthesis of the literature including the new laboratory data as it becomes relevant to the discussion. Finally, a set of examples will be presented for the feed-forward and feed-back approaches.

LABORATORY CHLORINATION EXPERIMENTS

New Model Compound Data

As part of this research, laboratory studies on the chlorination of 55 model compounds were conducted at the University of Massachusetts Amherst (UMass) (Table 5.1). Most of these fall into the following categories:

Purines & pyrimidines
Simple aromatic amines
Pyridine derivatives
Structural amino sugars (sialic acids)
Lignin monomers and derivatives

Many of these structures contain an acetylamino group or are likely to degrade to one. The Sialic acids have been recently recognized as major components of natural organic matter (NOM), especially the colloidal fractions. Most probably originate from algae and bacterial cell walls. They have been identified in marine and fresh water systems, and have sometimes been classified as part of the "hydrophilic neutral" and "colloidal" fractions.

Table 5.1
Nitrogenous model compounds

No.	Parameter	Molecular Weight	Molecular Formula
1	Anthranilic acid	137.14	C7H7NO2
2	3 Aminobenzoic acid	137.14	C7H7NO2
3	4 Aminobenzoic acid	137.14	C7H7NO2
4	Aniline	93.128	C6H7N
5	Acetanilide	135.2	C8H9NO
6	2-Aminophenol	109.13	C6H7NO
7	1,4-Phenylenediamine	108.14	C6H8N2
8	(+)-N-Acetylmuramic acid	293.28	C11H19NO8
9	(-)-N-Acetylneuraminic acid	309.28	C11H18NO9
10	Coniferyl Alcohol	180.21	C10H12O3
11	Pyridine	79.1	C5H5N
12	4-Hydroxypyridine	95.1	C5H5NO

(continued)

Table 5.1 (Continued)

No.	Parameter	Molecular Weight	Molecular Formula
13	4-Pyridinealdoxime	122.13	C6H6N2O
14	3-Pyridinepropanol	137.18	C8H11NO
15	Nicotinic acid	123.11	C6H5NO2
16	2-Hydroxynicotinic acid	139.11	C6H5NO3
17	6-hydroxynicotinic acid	139.11	C6H5NO3
18	Nicotinamide	122.13	C6H6N2O
19	Adenine	135.13	C5H5N5
20	Thymine	126.12	C5H6N2O2
21	Guanine	130.13	C5H5N5O
22	Cytosine	111.1	C4H5N3O
23	Uracil	112.09	C4H4N2O2
24	Indole	117.14	C8H7N
25	Caffeine	194.19	C8H10N4O2
26	4-Hydroxy benzoic acid	138.12	C7H6O3
27	4-Hydroxy benzaldehyde	122.12	C7H6O2
28	4-Hydroxy acetophenone	136.15	C8H8O2
29	Vanillic acid	168.15	C8H8O4
30	Vanillin	152.15	C8H8O3
31	4-Hydroxy benzoic acid	138.12	C7H6O3
32	4-Hydroxy benzaldehyde	122.12	C7H6O2
33	4-Hydroxy acetophenone	136.15	C8H8O2
34	Vanillic acid	168.15	C8H8O4
35	Vanillin	152.15	C8H8O3
36	4-Hydroxy benzoic acid	138.12	C7H6O3
37	4-Hydroxy benzaldehyde	122.12	C7H6O2
38	4-Hydroxy acetophenone	136.15	C8H8O2
39	Vanillic acid	168.15	C8H8O4
40	Vanillin	152.15	C8H8O3
41	4-Hydroxy benzoic acid	138.12	C7H6O3
42	4-Hydroxy benzaldehyde	122.12	C7H6O2
43	4-Hydroxy acetophenone	136.15	C8H8O2
44	Vanillic acid	168.15	C8H8O4
45	Vanillin	152.15	C8H8O3
46	4-Hydroxy benzoic acid	138.12	C7H6O3
47	4-Hydroxy benzaldehyde	122.12	C7H6O2
48	4-Hydroxy acetophenone	136.15	C8H8O2
49	Vanillic acid	168.15	C8H8O4
50	Vanillin	152.15	C8H8O3
51	4-Hydroxy benzoic acid	138.12	C7H6O3
52	4-Hydroxy benzaldehyde	122.12	C7H6O2
53	4-Hydroxy acetophenone	136.15	C8H8O2
54	Vanillic acid	168.15	C8H8O4
55	Vanillin	152.15	C8H8O3

Test Methodology

Solutions of the pure compound were prepared at the target concentration of 2 mg-C/L. These were prepared in high purity lab water (Super-Q), and subsequently buffered at pH 7. Model compound test solutions were analyzed for total organic carbon (TOC), ultraviolet absorbance at 245 nm (UV254) at the time of preparation.

Samples were chlorinated in the laboratory under standard conditions (20 mg/L dose, 48 hours, 20° C; Figure 5.1) as used with prior model compound work. These conditions were selected to mimic the chemical environment that is typical of treatment and distribution systems. At the end of the 48 hour incubation period, these were partitioned into five aliquots for analysis of the following:

Residual chlorine

Total organic carbon (TOC)

- Total organic halide (TOX)
- Trihalomethanes (THMs) and other neutral extractables
- Haloacetic acids (HAAs)

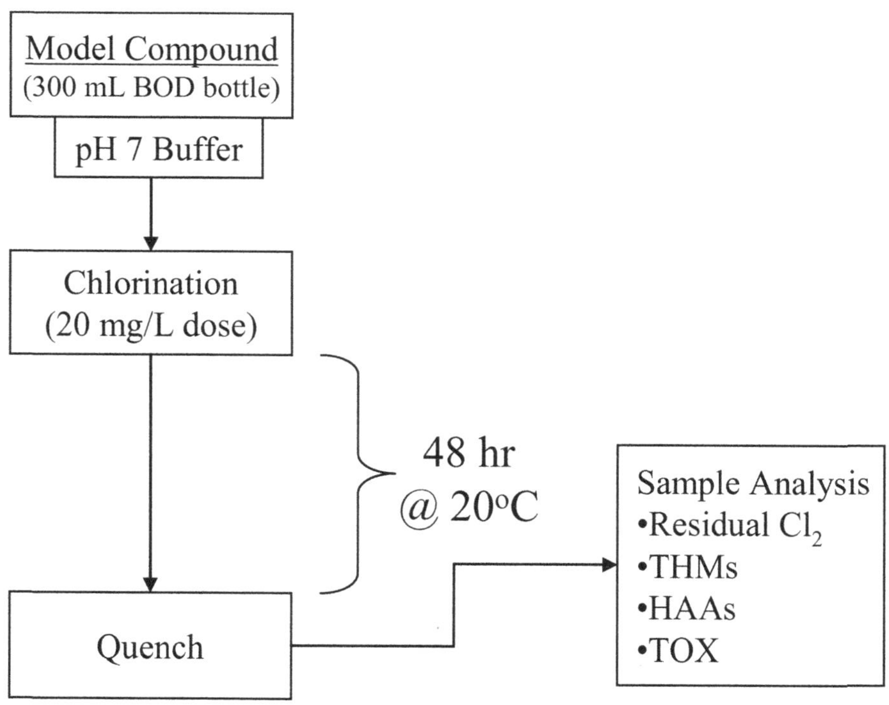

Figure 5.1 Design of laboratory chlorination experiments

With the exceptions of the N-acetylmuramic acid and a few of the pyridine derivatives, all compounds were quite reactive as measured by chlorine consumption (Table 5.2). However the amount of halogenated by-product (TOX) observed varied greatly from one compound to the next. Uracil produced 3 moles of TOX per mole of compound, which accounted for about half of the chlorine demand (Table 5.3). On the other hand adenine reacted almost exclusively by oxidative reactions with little or no chlorine incorporation.

Table 5.3 shows the molar yields* for TOX, chlorine demand and major by-product classes. The unknown TOX was determined by subtraction of the analyzed disinfection by-products (DBPs) from the TOX. Between 16% and 100% of the TOX was observed to be in the form of non-regulated or "unknown" compounds.

Trihalomethanes were detected in significant concentrations at neutral pHs with the aminobenzoic acids and aniline (Table 5.3). In most other cases, they were less important. Because no bromide was added to these model solutions, no brominated DBPs were observed.

Table 5.2
Disinfection by-product concentrations from model compounds treated with chlorine*

(unit: μg/L)

Compound	Chlorine demand ($mg\ Cl_2/L$)	THMs	1XAA	2XAA	3XAA	HAA6	HANs	HKs	CP	TOX
Anthranilic acid	10.20	162.91	0.39	42.40	223.37	266.15	6.32	0.00	1.08	378.61
3 Aminobenzoic acid	13.07	103.96	0.00	32.70	58.70	91.41	11.93	0.55	0.58	373.38
4 Aminobenzoic acid	13.33	171.89	0.00	45.62	75.81	121.44	18.87	0.00	0.52	486.38
2-aminophenol	9.25	78.65	0.00	78.70	76.56	155.26	6.34	0.00	0.69	304.66
Aniline pH 7	16.90	229.84	0.00	68.64	111.20	179.84	10.00	0.00	0.74	392.68
Aniline pH 9	17.80	441.09	0.00	125.52	228.64	354.16	3.00	0.00	0.93	874.84
Aniline pH 11	14.60	649.74	7.72	137.98	9.60	155.30	6.45	0.00	0.45	1183.10
Acetanilide pH 7	1.70	0.00	0.00	0.74	0.28	1.01	0.00	0.00	0.09	1.60
Acetanilide pH 9	3.70	65.62	0.83	18.91	43.35	63.10	1.41	0.00	0.27	133.86
Acetanilide pH 11	6.90	361.02	6.38	58.45	12.96	77.79	3.33	0.00	0.39	641.96
1,4-phenylenediamine	6.95	4.63	0.00	0.00	0.60	0.60	0.00	0.00	0.00	32.52
(+)-N-Acetylmuramic acid	0.40	0.00	0.00	0.00	0.00	0.00	0.00	0.64	0.00	71.10
(-)-N-Acetylneuraminic acid	2.45	0.91	0.00	0.00	0.00	0.00	0.00	0.00	0.00	109.33
Coniferyl Alcohol	7.87	27.03	0.00	19.20	36.39	55.59	2.96	0.32	2.96	1222.80
Pyridine	0.30	0.52	1.47	2.16	0.34	3.97	0.00	0.00	0.00	20.16
4-hydroxypyridine	1.45	7.97	0.00	0.00	29.37	29.37	0.00	0.00	0.00	352.53
4-pyridinealdoxime	9.60	2.42	0.00	16.92	10.05	26.97	0.00	0.00	47.95	82.64
3-pyridinepropanol	0.60	1.47	0.06	2.73	0.78	3.58	0.00	0.00	0.00	17.03
Nicotinic acid	0.25	0.76	0.00	2.18	0.39	2.57	0.00	0.00	0.00	23.80
2-hydroxynicotinic acid	3.55	0.74	0.17	0.00	18.12	18.29	0.47	0.00	0.00	144.29
6-hydroxynicotinic acid	0.65	0.27	0.00	0.00	0.00	0.00	0.72	0.00	0.00	34.50

* 1XAA (total monohaloacetic acids); 2XAA (total dihaloacetic acids); 3XAA (total trihaloacetic acids); HAA6 (sum of monochloro, monobromo, dichloro, trichloro, bromochloro and dibromo acetic acids); HANs (haloacetonitriles); HKs (haloketones0; CP (chloropicrin); TOX (total organic halide)

(continued)

Table 5.2 (Continued)

(unit: μg/L)

Compound	Chlorine demand (mg Cl_2/L)	THMs	1XAA	2XAA	3XAA	HAA6	HANs	HKs	CP	TOX
Nicotinamide	5.40	0.00	0.19	3.47	0.81	4.48	0.00	0.00	0.00	30.26
Adenine	18.90	1.20	0.00	5.40	0.34	5.74	0.00	0.00	0.00	16.58
Guanine	19.55	1.68	0.00	0.35	0.00	0.35	0.00	0.00	0.00	21.53
Caffeine	4.20	3.38	0.00	0.00	0.00	0.00	0.00	0.00	0.60	3.89
Indole	10.15	66.46	1.23	32.59	98.92	132.74	10.47	0.00	3.64	198.42
Thymine	12.80	0.00	0.00	2.01	2.07	4.08	0.00	0.00	0.00	1119.02
Cytosine	20.00	0.00	0.00	5.87	0.39	6.26	9.71	0.00	0.00	147.15
Uracil	18.40	28.86	0.00	310.01	54.44	364.45	3.68	0.00	0.11	5013.35
4-hydroxy benzoic acid	16.00	85.22	0.00	52.02	1063.79	1115.81	0.00	0.00	0.00	988.10
4-hydroxy benzaldehyde	15.40	82.07	0.00	45.83	1024.51	1070.33	0.00	0.00	0.00	1097.86
4-hydroxy acetophenone	13.20	256.99	0.00	59.58	882.27	941.84	0.00	5.81	0.00	1153.99
Vanillic acid	12.70	98.85	0.00	53.25	985.99	1039.24	0.00	0.00	0.00	932.18
Vanillin	11.70	99.57	0.00	77.40	1039.49	1116.89	0.00	0.00	0.00	924.42
Aceto vanillione	11.40	215.64	25.94	89.15	167.56	282.65	0.00	196.20	0.00	627.80
Syringic acid	9.10	12.48	0.00	56.23	314.32	370.55	0.00	0.00	0.00	459.67
Syring aldehyde	11.00	45.61	0.00	130.83	1439.67	1570.49	0.00	0.00	0.00	1358.63
Aceto syringone	9.20	610.49	112.52	226.93	59.55	399.00	0.00	559.20	0.00	1195.94
p-coumaric acid	12.20	20.42	76.95	317.70	108.61	503.25	0.00	0.00	0.00	847.11
Ferulic acid	12.20	57.44	0.00	1070.26	707.62	1777.88	0.00	0.00	0.00	1173.10
3,4,5-Trimethoxybenzaldehyde	0.00	2.92	0.00	19.61	55.62	75.24	0.00	0.08	0.05	66.75

(continued)

Table 5.2 (Continued)

(unit: μg/L)

Compound	Chlorine demand (mg Cl_2/L)	THMs	1XAA	2XAA	3XAA	HAA6	HANs	HKs	CP	TOX
3,4,5-Trimethoxybenzoic acid	1.30	3.73	0.00	16.25	25.75	42.00	0.00	0.18	0.07	573.25
3,4,5-Trimethoxyacetophenone	1.00	219.40	0.00	31.03	14.74	45.77	0.00	24.58	0.06	320.79
3,4,5-Trimethoxyphenylacetic acid	0.90	3.48	0.00	16.75	11.95	28.70	0.00	0.11	0.05	576.09
3,4,5-Trimethoxybenzamide	6.30	7.18	0.00	153.63	74.66	228.29	3.52	0.43	0.12	174.17
3,(3,4,5-Trimethoxyphenyl) Propionic acid	1.30	33.11	0.00	13.10	14.72	27.82	0.00	0.06	0.06	584.78
3,4,5-Trimethoxybenzyl Alcohol	1.80	5.67	0.00	21.14	78.80	99.94	0.00	0.64	0.08	582.12
3,4,5-Trimethoxy phenylacetonitrile	4.70	11.55	0.00	113.88	120.63	234.51	13.37	0.26	0.08	453.86
4-(3,4,5-Trimethoxybenzoyl) Butyric acid	1.70	39.25	0.00	24.44	39.22	63.66	0.00	0.94	0.05	124.55
Ethyl 3,4,5-Trimethoxybenzoylacetate	3.80	3.83	0.00	592.21	8.91	601.12	0.00	0.23	0.00	507.27
3,4,5-Triethoxybenzoic acid	0.90	3.44	0.00	12.67	13.40	26.06	0.00	0.21	0.13	300.86
3,4,5-Triethoxybenzyl alcohol	0.50	5.16	0.00	19.73	70.17	89.90	0.00	0.09	0.06	493.25
4-Methyl-2,6-Dimethoxyphenol	6.40	44.16	0.00	48.98	200.72	249.70	0.00	55.85	0.04	236.55
Sinapic Acid	9.70	29.39	0.00	70.42	194.90	265.32	0.00	0.04	0.06	259.23
Sinapyl Alcohol	6.60	26.01	0.00	43.76	274.85	318.61	0.00	0.16	0.06	275.89
Trans-3,5-Dimethoxy-4-Hydroxycinnam	10.30	31.28	0.00	74.80	500.72	575.52	0.00	0.00	0.14	512.58
Methyl Syringate	8.40	0.76	0.00	849.99	26.35	876.35	0.00	0.72	0.03	88.22
4-Allyl-2,6-Dimethoxyphenol	7.30	56.21	0.00	48.55	253.38	301.92	0.00	0.66	0.00	386.07
Urea	16.20	0.00	0.00	0.00	0.00	0.00	0.08	0.00	0.09	4.63

Table 5.3
Disinfection by-product yields with compounds treated with chlorine

(unit: mM/mM-compound)

Compound	Chlorine demand ($mMCl_2$/mM)	THMs	1XAA	2XAA	3XAA	HAA6	HANs	HKs	CP	TOX (mM Cl/mM)	Unknown TOX (mM Cl/mM)
Anthranilic acid	6.042	0.057	0.000	0.014	0.057	0.071	0.002	0.00	0.000	0.449	0.071
3 Aminobenzoic acid	7.740	0.037	0.000	0.011	0.015	0.026	0.005	0.00	0.000	0.442	0.256
4 Aminobenzoic acid	7.898	0.060	0.000	0.015	0.019	0.034	0.007	0.00	0.000	0.576	0.292
2-aminophenol	4.697	0.024	0.000	0.022	0.017	0.039	0.002	0.00	0.000	0.309	0.139
Aniline pH 7	8.581	0.069	0.000	0.019	0.025	0.044	0.003	0.00	0.000	0.399	0.072
Aniline pH 9	9.038	0.133	0.000	0.035	0.050	0.085	0.001	0.00	0.000	0.888	0.266
Aniline pH 11	7.413	0.196	0.003	0.039	0.002	0.044	0.002	0.00	0.000	1.201	0.523
Acetanilide pH 7	1.151	0.000	0.000	0.000	0.000	0.000	0.000	0.00	0.000	0.002	0.001
Acetanilide pH 9	2.505	0.026	0.000	0.007	0.013	0.020	0.001	0.00	0.000	0.181	0.048
Acetanilide pH 11	4.671	0.145	0.003	0.022	0.004	0.029	0.001	0.00	0.000	0.869	0.372
1,4-phenylenediamine	3.529	0.001	0.000	0.000	0.000	0.000	0.000	0.00	0.000	0.033	0.028
(+)-N-Acetylmuramic acid	0.372	0.000	0.000	0.000	0.000	0.000	0.000	0.00	0.000	0.132	0.132
(-)-N-Acetylneuraminic acid	2.281	0.001	0.000	0.000	0.000	0.000	0.000	0.00	0.000	0.204	0.202
Conifenly Alcohol	6.660	0.014	0.000	0.009	0.013	0.022	0.002	0.00	0.001	2.070	1.964
Pyridine	0.127	0.000	0.000	0.001	0.000	0.001	0.000	0.00	0.000	0.017	0.015
4-Hydroxypyridine	0.614	0.002	0.000	0.000	0.005	0.005	0.000	0.00	0.000	0.298	0.276
4-Pyridinealdoxime	4.874	0.001	0.000	0.005	0.002	0.007	0.000	0.00	0.011	0.084	0.034
3-Pyridinepropanol	0.406	0.001	0.000	0.001	0.000	0.001	0.000	0.00	0.000	0.023	0.019
Nicotinic acid	0.127	0.000	0.000	0.001	0.000	0.001	0.000	0.00	0.000	0.024	0.022
2-Hydroxynicotinic acid	1.803	0.000	0.000	0.000	0.004	0.004	0.000	0.00	0.000	0.147	0.134
6-Hydroxynicotinic acid	0.330	0.000	0.000	0.000	0.000	0.000	0.000	0.00	0.000	0.035	0.034
Nicotinamide	2.742	0.000	0.000	0.001	0.000	0.001	0.000	0.00	0.000	0.031	0.028

(continued)

Table 5.3 (Continued)

(unit: mM/mM-compound)

Compound	Chlorine demand	THMs	1XAA	2XAA	3XAA	HAA6	HANs	HKs	CP	TOX	Unknown TOX
Adenine	7.997	0.000	0.000	0.001	0.000	0.001	0.000	0.00	0.000	0.014	0.010
Guanine	8.272	0.000	0.000	0.000	0.000	0.000	0.000	0.00	0.000	0.018	0.017
Caffeine	2.843	0.001	0.000	0.000	0.000	0.000	0.000	0.00	0.000	0.005	0.001
Indole	6.872	0.027	0.001	0.012	0.029	0.042	0.005	0.00	0.001	0.269	0.064
Thymine	5.416	0.000	0.000	0.000	0.000	0.001	0.000	0.00	0.000	0.947	0.945
Cytosine	6.770	0.000	0.000	0.001	0.000	0.001	0.002	0.00	0.000	0.100	0.093
Uracil	6.228	0.006	0.000	0.058	0.008	0.066	0.001	0.00	0.000	3.394	3.236
4-hydroxy benzoic acid (PAD)	9.478	0.030	0.000	0.017	0.273	0.290	0.000	0.00	0.000	1.171	0.226
4-hydroxy benzaldehyde (PAL)	9.123	0.029	0.000	0.015	0.263	0.278	0.000	0.00	0.000	1.301	0.394
4-hydroxy acetophenone (PON)	8.937	0.103	0.000	0.022	0.259	0.281	0.000	0.00	0.000	1.563	0.424
Vanillic acid (VA)	8.598	0.040	0.000	0.020	0.290	0.309	0.000	0.00	0.000	1.262	0.234
Vanillin (Vn)	7.921	0.040	0.000	0.029	0.305	0.334	0.000	0.00	0.000	1.252	0.158
Aceto vanillione (AcV)	8.683	0.098	0.015	0.037	0.055	0.108	0.000	0.08	0.000	0.956	0.174
Syringic acid (SA)	6.931	0.006	0.000	0.024	0.104	0.127	0.000	0.00	0.000	0.700	0.325
Syring aldehyde (Sn)	8.378	0.021	0.000	0.055	0.476	0.531	0.000	0.00	0.000	2.070	0.471
Aceto syringone (AcS)	7.786	0.307	0.071	0.106	0.022	0.199	0.000	0.25	0.000	2.024	0.004
p-Coumaric acid (CA)	9.292	0.009	0.044	0.133	0.036	0.213	0.000	0.00	0.000	1.290	0.845
Ferulic acid (FA)	10.324	0.029	0.000	0.498	0.260	0.758	0.000	0.00	0.000	1.985	0.123
3,4,5-Trimethoxybenzaldehyde	0.000	0.001	0.000	0.009	0.020	0.030	0.000	0.00	0.000	0.113	0.029
3,4,5-Trimethoxybenzoic acid	1.100	0.002	0.000	0.008	0.009	0.017	0.000	0.00	0.000	0.970	0.921
3,4,5-Trimethoxyacetophenone	0.931	0.121	0.000	0.016	0.006	0.022	0.000	0.01	0.000	0.597	0.148
3,4,5-Trimethoxyphenylacetic acid	0.838	0.002	0.000	0.009	0.005	0.013	0.000	0.00	0.000	1.073	1.035
3,4,5-Trimethoxybenzamide	5.331	0.004	0.000	0.071	0.027	0.099	0.002	0.00	0.000	0.295	0.054

(continued)

Table 5.3 (Continued)

(unit: mM/mM-compound)

Compound	Chlorine demand	THMs	1XAA	2XAA	3XAA	HAA 6	HANs	HKs	CP	TOX	Unknown TOX
3,(3,4,5-Trimethoxyphenyl) propionic acid	1.320	0.020	0.000	0.007	0.006	0.014	0.000	0.00	0.000	1.188	1.094
3,4,5-Trimethoxybenzyl alcohol	1.523	0.003	0.000	0.010	0.029	0.039	0.000	0.00	0.000	0.985	0.870
3,4,5-Trimethoxy phenylacetonitrile	4.375	0.006	0.000	0.058	0.049	0.107	0.008	0.00	0.000	0.845	0.546
4-(3,4,5-Trimethoxybenzoyl) butyric acid	2.014	0.028	0.000	0.016	0.020	0.036	0.000	0.00	0.000	0.295	0.118
Ethyl 3,4,5-Trimethoxybenzoylacetate	4.502	0.003	0.000	0.386	0.005	0.390	0.000	0.00	0.000	1.202	0.408
3,4,5-Triethoxybenzoic acid	0.990	0.002	0.000	0.008	0.006	0.014	0.000	0.00	0.000	0.662	0.620
3,4,5-Triethoxybenzyl alcohol	0.550	0.003	0.000	0.012	0.033	0.045	0.000	0.00	0.000	1.085	0.951
4-Methyl-2,6-dimethoxyphenol	4.874	0.020	0.000	0.021	0.066	0.087	0.000	0.02	0.000	0.360	-0.007
Sinapic acid	9.030	0.016	0.000	0.036	0.079	0.115	0.000	0.00	0.000	0.483	0.125
Sinapyl alcohol	6.144	0.014	0.000	0.022	0.111	0.133	0.000	0.00	0.000	0.514	0.092
Trans-3,5-Dimethoxy-4-hydroxycinnam	9.588	0.017	0.000	0.038	0.202	0.241	0.000	0.00	0.000	0.954	0.219
Methyl syringate	7.109	0.000	0.000	0.396	0.010	0.405	0.000	0.00	0.000	0.149	-0.673
4-Allyl-2,6-dimethoxyphenol	6.795	0.031	0.000	0.025	0.102	0.127	0.000	0.00	0.000	0.719	0.268
Urea	1.371	0.000	0.000	0.000	0.000	0.000	0.000	0.00	0.000	0.001	0.001

Table 5.4 shows the results of TOC and total nitrogen (TN) following chlorination. Along with TOX, these measurements allow the determination of an elemental balance on carbon, nitrogen and halide. The amount of carbon ("C") lost is strictly determined from the difference between the measured starting non-volatile total organic carbon (NVTOC) and the final NVTOC measured after chlorination. In the case of guanine, the starting NVTOC and TN suggest that this compound was recovered by about 40% in both analyses. All other NVTOC recoveries were 102 ±7%. All remaining TN analyses showed recoveries of 87 ±8%. In the few instances where recovery was poor, the "C lost" and "N lost" were adjusted assuming that the % recovery was the same for staring compound and products.

Table 5.4
Elemental balance and inorganic chloramine formation from nitrogenous model compounds

#	Compound	Cl_2 demand (M/M)	Residual chloramines (M/M)	C Lost (M/M)	N Lost (M/M)
1	Anthranilic acid	6.40	0.85	0.82	0.09
2	3-Aminobenzoic acid	7.57	0.78	0.46	0.04
3	4-Aminobenzoic acid	8.92	0.68	1.68	0.04
4	2-Aminophenol	4.04	0.62	0.00	0.08
6	Acetanilide pH 7	1.31	1.86	0.00	0.00
	Acetanilide pH 9	2.06	1.65	0.86	0.00
	Acetanilide pH 10	3.99	1.17	0.12	0.00
7	1,4-phenylenediamine	3.30	0.66	0.00	0.09
12	4-hydroxypyridine	0.98	0.61	0.00	0.02
13	4-pyridinealdoxime	4.82	0.61	1.02	0.42
15	Nicotinic acid	0.42	0.56	0.00	0.00
16	2-hydroxynicotinic acid	0.63	0.63	0.00	0.00
17	6-hydroxynicotinic acid	0.30	0.45	0.75	0.03
18	Nicotinamide	2.90	0.89	0.76	0.77
19	Adenine	7.34	0.15	ND	2.27
20	Guanine*	7.73	0.51	4.10	1.63
21	Caffeine	2.44	1.10	2.23	0.13
23	Thymine	5.27	0.33	2.71	2.00
24	Cytosine	5.97	0.18	2.79	1.69
25	Uracil	6.06	0.16	2.66	1.63
41	3,4,5-Trimethoxybenzamide	5.56	0.90	3.22	0.81
44	3,4,5-Trimethoxy phenylacetonitrile	5.05	1.15	2.78	0.31
55	Urea	1.24	0.05	0.58	0.64

* Data may be unreliable due to poor C and N recovery

SYNTHESIS OF LAB RESULTS AND LITERATURE ON CHLORINE REACTIONS WITH MODEL COMPOUNDS OF ENVIRONMENTAL SIGNIFICANCE

The feed-forward approach relies heavily on predictive chemistry, which itself rests on fundamental chemical knowledge (originally gleaned from laboratory experiments with compounds that may or may not have any relevance to drinking water), and on empirical chlorination experiments with compounds that are likely to be in raw drinking waters (i.e., model compound studies). For this reason, we have embarked on a detailed literature review of the reactions of chlorine with selected classes of compounds. Much of this work has focused on nitrogenous model compounds. Some of the text and analysis is shown below for illustrative purposes.

Researchers have often worked on these problems well within their disciplinary boundaries. Engineers have looked at DBP yields from treatment systems. Some have conducted laboratory studies of NOM, NOM isolates and model NOM compounds and measured formation of the simple regulated DBPs. Analytical chemists have explored the identity of some non-regulated DBPs using modern instrumentation. However, these are rarely translated to quantitative or mechanistic studies. In addition, proposed or "tentative" structural identifications are often confused with "confirmed" identifications requiring authentic standards. Organic chemists have primarily looked at chlorine reactions from the standpoint of chemical synthesis and chemical mechanism. Some have actually focused on the DBP problem. Each has its own shortcomings from the perspective of the DBP problem. Synthetic chemists generally work under conditions atypical of drinking water treatment. Mechanistic studies are often focused on a single reaction, and do not consider the full pathway. Even those focusing on DBP issues often suffer from a lack of quantitative data. Finally the biochemists have studied chlorination reactions with an interest in inflammation in vivo. Little of this information has been brought to bear on the DBP issue, because it is still relatively new and the reaction conditions may depart significantly from those of treated drinking water.

In this section the goal is to show that information thoughtfully transferred across the disciplinary boundaries can shed new light on the DBP issue. To accomplish this some of cross disciplinary research is highlighted and an attempt is made to provide some integration. One of the major objectives is to propose new, yet plausible structures for as-yet unidentified DBPs. This section is intended as an illustration of how this work could be done. It also represents a first step in that direction. It is not intended to be a complete analysis in itself.

The areas targeted in this analysis are primarily the reactions of free chlorine with naturally-occurring nitrogenous organic compounds. Because some of these have activated benzene rings that serve as the site of attack, we will also examine hydroxyl-activated aromatic structures. Finally, some discussion of lignin monomers will be presented.

Nitrogenous Organic Compounds Found in Nature

Dissolved organic nitrogen in US surface waters is in the range of 4%-12% of the dissolved organic carbon (DOC) (10th – 90th percentage), averaging about 7% (Figure 5.2; Westerhoff and Mash 2002). This spans much of the range reported for living plant tissues: 3%-25% for higher plants to algae (Caraco and Cole 2002). As primary productivity increases, so does the N/C ratio. Oligotrophic waters may be as low as 3% (N/C), whereas eutrophic waters may be as highas 25% (Wetzel 1975; Buffle 1988).

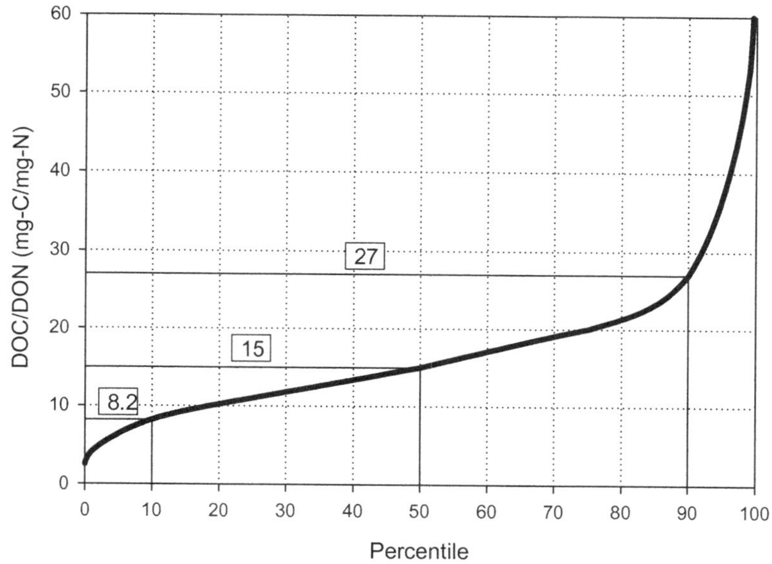

Redrawn from Westerhoff & Mash, 2002

Figure 5.2 Cumulative frequency distribution of organic C/N Ratios in US surface water samples

Table 5.5 presents a summary of dissolved organic nitrogen (DON) levels in US surface waters, along with estimates from the literature as to the composition of that DON. The overall DON levels are taken from Westerhoff and Mash's (2002) statistical analysis of the USGS NAQWA data. This is probably the least biased estimator we have of total DON. These values fit well with other summaries based on smaller data sets.

A large fraction of NOM is thought to be derived from amino acids and proteins. Thurman has estimated that 15-35% of the DON is in the form of amino acids. The 20 standard α-amino acids include forms with non-polar (mostly hydrocarbon) side chains, uncharged polar side chains (oxygenated or containing sulfhydryl), polar anionic (acids), or polar cationic (amines). A few of the structures are shown below. Aspartic acid, glycine and glutamic acid are prominent in surface waters.

Estimates of DON components are presented in Table 5.5. All of these are drawn from the literature, especially published surveys. Most reported parallel DON measurements, and could therefore be compared to the national distribution. When frequency distributions of specific nitrogenous organic components were uncertain, they were assumed to reflect the overall DON distribution from Westerhoff and Mash (2002). Like the data in Figure 5.2, all frequency distributions showed a non-normal distribution, skewed toward higher concentrations.

As a result the 90th and 99th percentiles are high compared to the median.

The estimates of free amino acids in Table 5.5 are based the freshwater dataset of Dossier-Berne et al. (1994 and 1996), considering also the literature summaries in Thomas 1997; and Bronk 2002. The combined amino acid values are based on several data sets and generalized in terms of the ratio of combined to free amino acids. These estimates range from about 1:1 (Bronk 2002) up to 10:1 (Peake et al. 1972). We selected a more typical value, closer to 2:1 as used by Wetzel (2001). This issue is complicated by the difficulty in recovering bound amino acids, and uncertainty concerning the original structure. The amino sugar estimate is based on semi-quantitative pyrolysis-GC/MS data from Hwang et al. (2001), and hydrolysis GC/MS data from Repeta et al. (2002), and fractionation/ characterization data from Rostad et al. (1997).

The estimates for humic nitrogen in Table 5.5 are based on the comprehensive review of Perdue and Ritchie (2004). These authors summarized our best current estimates of the median and standard deviation in N/C ratio in isolated aquatic humic materials (0.016±0.017 M/M for fulvic acid). This was combined with estimates of the mean and frequency distribution for raw water DOC from the Information Collection Rule (ICR) data (McGuire et al. 2002), and the portion of DOC isolated into the hydrophobic acid fraction (also from Perdue and Ritchie 2004).

Aside from the amino acids and urea, there is very little good quantitative data on specific nitrogenous organic compounds in natural waters. As a result, we have used basic plant biochemical assays to extrapolate to what the aquatic concentrations might be. For example, Imafidon and Sosulski (1990) tested 13 categories of food plants for nucleic acid content. They found that 1.3-9.6% of the total nitrogen could be accounted for in nucleic acids. Assuming that this ratio is preserved in aquatic organic nitrogen, and taking an equal distribution of the bases, we can postulate that the purines and pyrimidines are present at levels shown in Table 5.5, below.

Table 5.5
Order of magnitude estimates for organic nitrogen in surface waters (all values in µg-N/L)

Classification	50%ile	90%ile	99%ile
DON	350	800	2000
Free AA	20	50	200
Combined AA	40	100	400
Nucleic acids	20	50	200
Amino Sugars	40	100	400
Humic-N	25	200	1000
Others		See text	

Alkaloids are common plant and animal derived-compounds that all contain nitrogen. They are generally produced from an amino acid and usually have preserved a recognizable segment of that amino acid. About 900 of the 10,000 plant genera are known to produce alkaloids (Mann 1994). However, Robinson (1991) cites evidence that 15-30% of plant species produce alkaloids exceeding 0.05% of their dry weight. Concentrations of alkaloids and their remnants in surface waters are still relatively unknown. Among the alkaloids, heterocyclic nitrogens are quite common, and many are derivatives of pyridine, piperidine, indole, isoquinoline and tropane (Figure 5.3). These compounds are bases and would be isolated in the hydrophilic and hydrophobic base fractions.

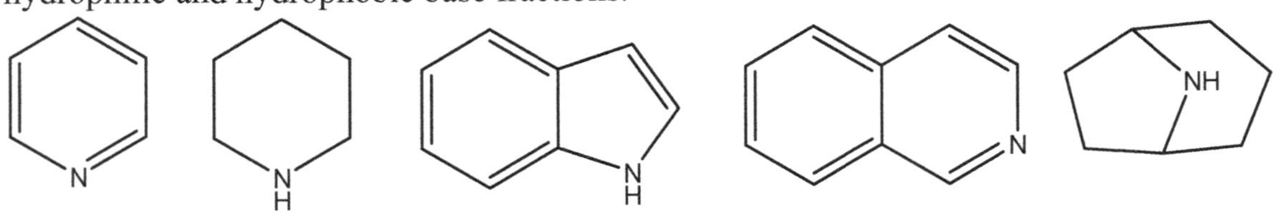

Pyridine Piperidine Indole Isoquinoline Tropane

Figure 5.3 Major nitrogen heterocyclic structures in alkaloids

Porphyrins are found across the animal and plant kingdoms. The basic structure is a tetrapyrrole ring with extensive resonance stabilization (Figure 5.4). The substituents are limited to the following:

- $-CH_3$, or $-CHO$, or $-CH_2COOH$
- $-CH_2CH_3$, or $-CH{=}CH_2$, or $-CH_2CH_2COOH$

And the positions around the nucleus almost always correspond with the arrangement shown in Figure 5.4.

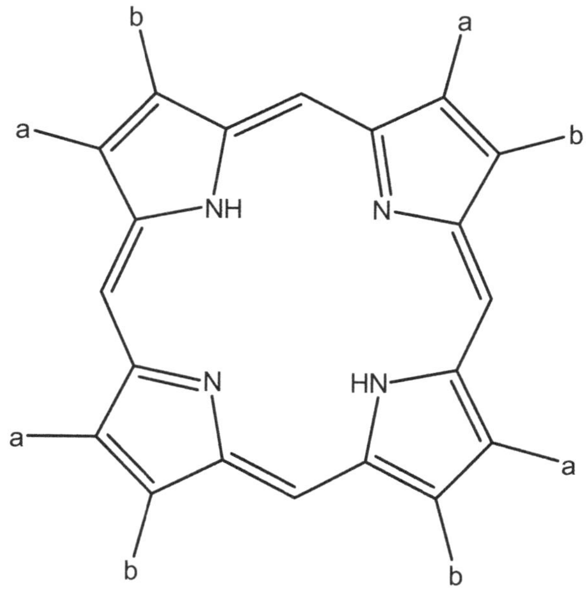

Figure 5.4 The cyclic tetrapyrrole porphyrin nucleus

General Role of Nitrogen in Chlorine Reactions

In general, the presence of nitrogen in NOM would be expected to stimulate reactions of chlorine with that organic material. Most chlorination reactions are electrophilic in nature. Electron withdrawing substituents such as nitro groups slow the reaction of chlorine with phenolic compounds (Soper and Smith 1926). However, most natural organic nitrogen is in the –III oxidation state. Amines, which are electron donating will tend to speed up chlorination reactions. Hydrolysis of amines, imines and nitriles will likely result in the formation of new C-O bonds that could further elevate the reactivity of the carbon backbone.

Formation and Decomposition of N-halo-organics

Formation

It has long been recognized that primary and secondary amines will react with aqueous chlorine (e.g., HOCl or OCl^-) and inorganic chloramine (e.g., NH_2Cl or $NHCl_2$) to form organic chloramines.

$$R-NH_2 \xrightarrow{HOCl} R-NHCl \xrightarrow{HOCl} R-NCl_2 \quad (5.1)$$

$$R_2-NH \xrightarrow{HOCl} R_2-NCl \quad (5.2)$$

Equilibrium considerations are closely tied to the basicity of the amine. As the amine becomes a stronger base, it will tend to form more stable chloramines (Pitman et al. 1969). The primary amine groups of the common amino acids are quite basic and therefore readily donate their electrons to electrophilic reactants such as chlorine. Amidines are also quite basic (pKa $\geq$ 11.2), especially the guanidine derivatives (pKa ~ 13.5). The primary amines on nucleic acids undergo amino-imino tautomerization which renders them far less basic. In this case the proton can be represented as residing on the ring nitrogens (position #1 for adenine; #7 for guanine #3 for cytosine). Table 5.6 shows some relevant pKa values for the common amino acids and three nucleosides.

Table 5.6
pKa Values* for some nitrogenous organic compounds

Compound	R_3 Group†	pKa
Cysteine	$-CH_2SH$	10.7
Proline	$-CH_2CH_2CH_2-$	10.64
Aspartic Acid	$-CH_2COOH$	9.9
Alanine	$-CH_3$	9.87
Glycine	-H	9.78
Isoleucine	$-CH(CH_3)Et$	9.76
Valine	$-CH(CH_3)_2$	9.74
Leucine	$-CH_2CH(CH_3)_2$	9.74
Glutamic Acid	$-CH_2CH_2COOH$	9.47
Tryptophan	$-CH_2$-indole	9.41
Histidine	$-CH_2$-imidazole	9.33
Phenylalanine	$-CH_2$-Ph	9.31
Methionine	$-CH_2CH_2SCH_3$	9.28
Serine	$-CH_2OH$	9.21
Tryrosine	$-CH_2$-Ph-OH	9.21
Glutamine	$-CH_2CH_2CONH_2$	9.13
Threonine	$-CHOHCH_3$	9.1
Lysine	$-(CH_2)_4NH_2$	9.06
Arginine	$-(CH_2)_3NHC(NH_2)_2$	8.99
Asparagine	$-CH_2C(=O)NH_2$	8.72
Cytidine		4.5
Adenosine		3.8
Guanosine		2.4

* These PKa values are for protonation of nitrogen. When more than one nitrogen is present, the highest pKa is shown (most basic)
† For amino acids

Kinetics often parallel thermodynamics, and this appears to be true for the formation of many N-chloramines. Kinetic studies have shown that the reaction is between the uncharged species (i.e., HOCl and $R-NH_2$), which results in maximum rates under moderate alkaline conditions, mid-way between the respective pK_as (e.g., see: Abia et al. 1998). First order rate constants for the formation of N-chloro-α-amino acids were reported to range from 0.002 to 0.7 (Awad et al. 1990). Simple amines will react very quickly, at near diffusion controlled rates, unless the nitrogen is sterically hindered (e.g., Kirmeyer et al. 1993).

Aromatic amines, such as aniline, are relatively weak bases, and therefore slow to form N-halo derivatives. Since amine groups are also electron donating, they tend to increase reactivity on aromatic rings. For this reason, p-chloroanaline is preferred over N-chloroanaline (Hwang et al. 1990).

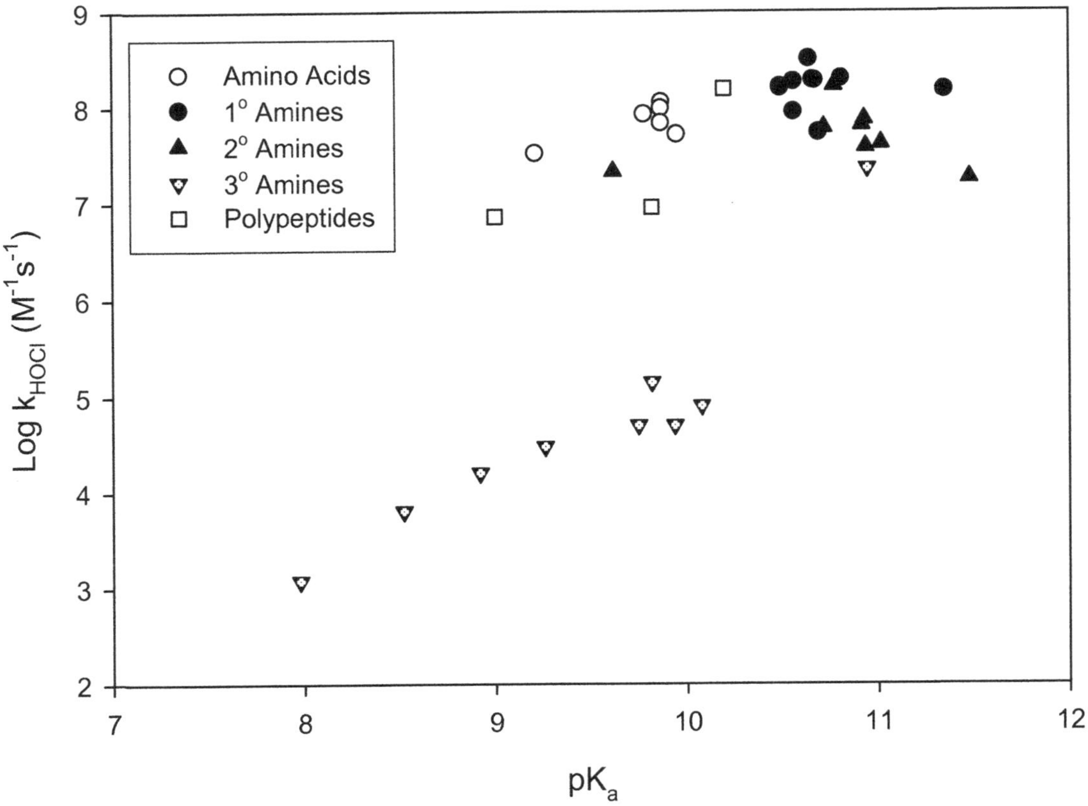

Data Sources: Friend 1956; Hussain et al. 1972, Isaac et al. 1983, Armesto et al. 1993, Armesto et al. 1994, Antelo et al. 1995, Abia et al. 1998

Figure 5.5 Relationship between basicity and 2nd order rate constants for reaction of HOCl with N-compounds

Morris (1967) showed that the rates of N-halogenation were strongly correlated with basicity over several orders of magnitude. Most of the primary and secondary amines react at rates that are nearly diffusion-controlled, but still exhibit some chemical control. Figure 5.5 shows much of the available data for simple amines, amino acids and some polypeptides. The tertiary amines are far less basic than the others and their lower reactivity is evident. Although not shown on this figure, monohalogenated N-chloroamines are also less basic and less reactive than their free counterparts (Armesto et al. 1998).

Armesto and coworkers (1998) have summarized the relationship between known 2nd order rate constants for the amine-hypochlorous acid reaction and the basicity of the amines, and found a curvature at low pKa's. They have analyzed this curvature using Marcus' theory (More O'Ferrall-Lewis approach), and found an intrinsic barrier ($\Delta^{\ddagger}G_0^{0}$) for the transfer of chlorine between HOCl and the unprotonated amine as 15 kJ/mole. Taft plots (with $\rho>0$) indicate that development of a negative charge on the nitrogen for the activated complex in primary and secondary amines.

These authors also conclude that rate constants for chlorination of tertiary amines are about three orders of magnitude lower than those for primary or secondary amines. In this case, the Taft plots (with $\rho<0$) indicate development of a positive charge on the nitrogen for the activated complex in tertiary amines. Halogenated tertiary amines may undergo elimination reactions ultimately leading to cleavage of a side-chain, and production of a secondary amine (Ellis and Soper 1954).

Subsequent reaction with chlorine leads to the formation of N-dihalo species. This reaction is much slower than the initial halogenation, due to the much lower basicity of the N-chloro compounds. Disproportionation of two monohalamines can lead to the formation of a dihalamine, however, this is less likely in highly dilute drinking waters containing a free chlorine residual.

Degradation Pathways Preserving Active Chlorine

Organic chloramines are inherently unstable, with half lives from minutes to weeks. Since they exist in equilibrium with aqueous chlorine, removal of the hypochlorite, hypochlorous acid, or inorganic chloramine will result in a shift of the equilibrium back to the starting amine or amino acid. Alternatively, organic chloramines may transfer their active chlorine to other amine groups. Bedner and colleagues (2004) presented evidence for the slow movement of active chlorine from small hydrophilic DON molecules to larger, less reactive and more hydrophobic forms. All of these reactions preserve the oxidized form of chlorine. Little kinetic data exist on the hydrolysis or halogen transfer of organic chloramines under conditions typical of drinking water treatment.

Degradation Pathways with Loss of Active Chlorine

Organic chloramines may also go on to form new oxidized products; a set of reactions which is commonly referred to as the "decomposition" pathway. When they decompose they may form carboxyl products, ammonia, carbon dioxide and chloride (Awad et al. 1990a).

Isaac and Morris (1983) used non-specific methods to study the rate of degradation of a broad range of organic chloramines (Table 5.7). Most experiments were done by titrating residual oxidant (using phenylarsenine oxide), and a few were corroborated using monitoring of UV absorbance. These experiments were conducted between 20 and 25 C, and at pHs ranging from 4.1 to 8.7. In one case, faster degradation was noted at lower pH (glycine) and in one case the reverse was true

Table 5.7
Degradation rates of N-chlorinated amines

Parent Amine	k_{obs} (s^{-1})	$t_{1/2}$ (min)
Alanine	1.3E-04	86
Glycine	1.4E-06	8400
Histidine	2.7E-04	43
Leucine	1.6E-04	72
Phenylalanine	2.2E-04	52
Serine	2.4E-04	49
Creatinine	3.5E-06	3300
Glycine N acetyl	6.0E-07	19000
Glycine ethyl ester	2.3E-04	50
Glycylglycine	1.0E-05	1100
Sarcosine	5.3E-05	210

Source: Data from Isaac and Morris 1983

(glycine ethyl ester). In this work, no attempt was made to determine products or pathways of degradation. However, by using the phenylarsine oxide (PAO) titration, the authors were selecting for pathways that lead to loss of active chlorine.

Bedner and coworkers (2004) argue that for macromolecular N-chloramines one must consider other factors. They have presented evidence for "hydrophobic stabilization" where the more hydrophobic N-chloro group is protected by the more hydrophilic portions of the molecule. Donnan exclusion of potential reducing species could also help to increase stability.

Figure 5.6 summarizes the two major pathways proposed for degradation of mono and di-N-haloamines. Residual chlorine species (i.e., hypochlorite, hypochlorous acid, monochloramine; represented as X^+) can readily transfer a chlorine atom to an amine nitrogen forming mono and di halamines. Following monohalogenation, amines undergo an elimination, whereas amino acids undergo Grob fragmentation (decarboxylation). In both cases, an H^+ and an X^- are lost. Amino acids also lose carbon dioxide.

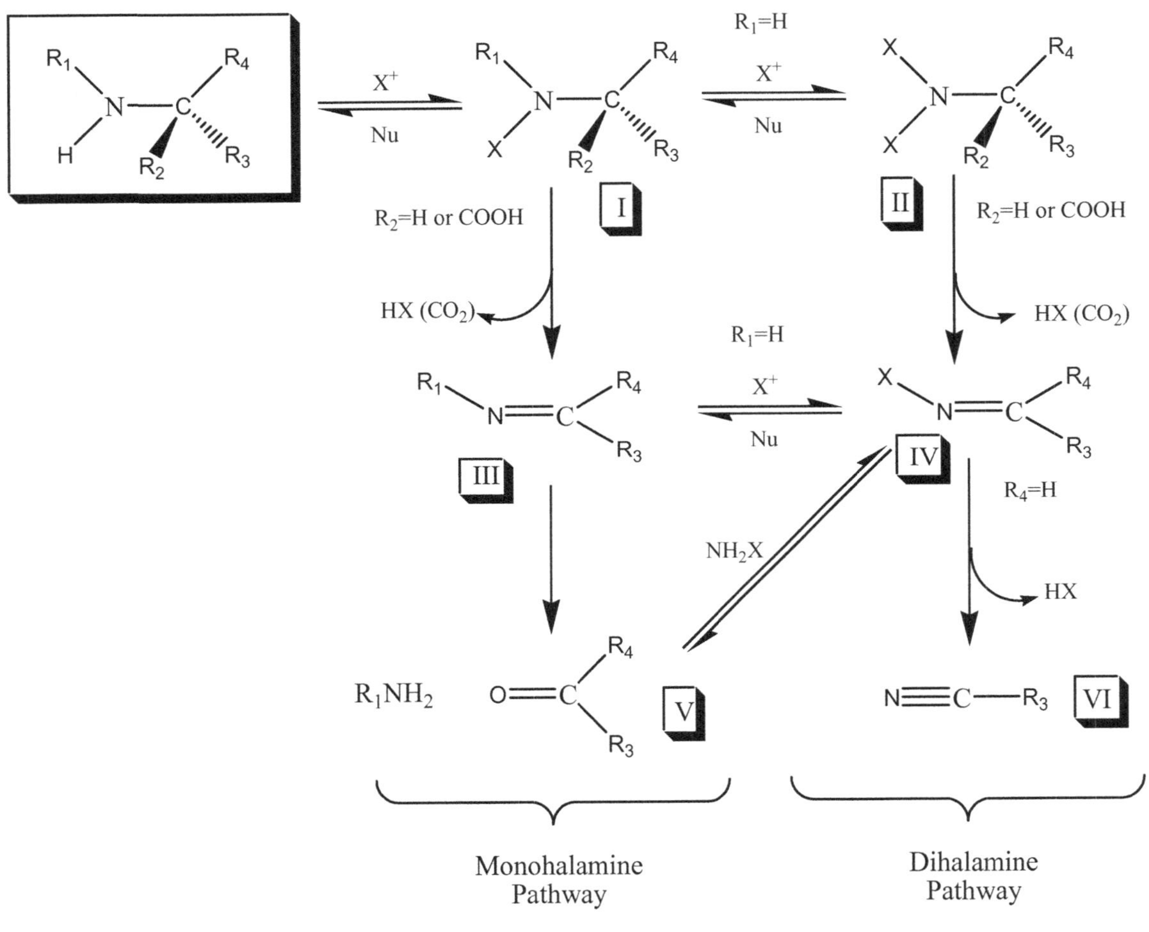

Source: Adapted from Nweke and Scully 1989 and Armesto et al. 1998

Figure 5.6 General scheme for carbonyl and cyano formation from chlorination of amines and amino acids.

For the dihalamine pathway, a second HX elimination can occur. Laboratory experiments indicate that this conversion of IV to VI may be sufficiently slow with some compounds that the N-haloaldimine (IV) persists as a major reaction product (Nweke & Scully 1989). The same may not be true of the unhalogenated aldimine (III).

Of course, with many amines and amino acids, there will be side groups (e.g., R_3) that will themselves react with active halogens. This can lead to products that are further oxidized, substituted or fragmented versions of the carbonyl or nitrile shown in Figure 5.6.

Grob fragmentation. Grob fragmentation begins with an anionic form of an amino acid (compound 10 in Figure 5.7) which first undergoes a unimolecular decomposition to an imine, giving off carbon dioxide and the halide anion. This is the rate-determining step, and the overall process is therefore independent of pH and ionic strength. The imine (11) then quickly hydrolyzes forming an amine (8) and an aldehyde or ketone (6). This mechanism is also referred to as the decarboxylative dechlorination of N-chloro amino acids.

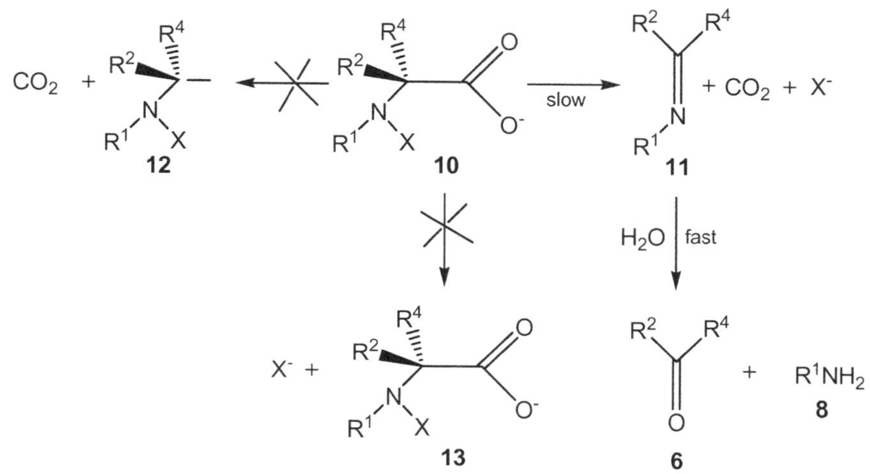

Source: redrawn from Armesto et al. 1998

Figure 5.7 Grob fragmentation scheme

The reaction in Figure 5.7 is a concerted fragmentation that is the preferred decomposition pathway when both an electron withdrawing group (e.g., halide) and an electron donating group (e.g., carboxylate) exist in an antiperiplanar configuration. In this configuration, the overlap of the p orbitals is maximized.

Chlorination of Selected Nitrogenous Organic Compounds

Amino Acids

Most amino acids undergo initial rapid formation of mono and di-N-chloramines. Due to the presence of an adjacent carboxyl group, N-halo substituted α-amino acids undergo a Grob fragmentation to an aldehyde or ketone and a nitrile.

Many amino acids are known to yield halogenated aldehydes and nitriles (e.g., chloral hydrate and dichloroacetonitrile) following reaction with chlorine. The degree to which this happens depends on the presence of an activating side chain (e.g., R3). After reaction, the amine nitrogen leaves a functional group (carbonyl or nitrile) that is electron withdrawing. When combined with a similarly electron withdrawing side chain, the α-carbon can become an active site for halogen substitution.

For example, aspartic acid reacts as expected in Figure 5.6, and in addition there is substantial substitution on a side chain. Aspartic acid has the following substituents:

R_2 = COOH
R_3 = CH_2COOH
R_4 = H

As stated previously, the carbonyl and nitrile are electron withdrawing groups. Combining this with the electron withdrawing nature of R_3, makes the protons on the methylene group in R_3 quite acidic. This will lead to substitution with a halogen at that carbon. Once the first substitution occurs, the remaining proton is even more acidic, and a second substitution follows rapidly. Finally decarboxylation occurs, giving dihaloacetaldehyde (on the monohalamine pathway) and dihaloacetonitrile (on the dihalamine pathway). The dihaloacetaldehyde reacts rapidly with free chlorine to form trihaloacetaldehyde (e.g., chloral hydrate), which is a relatively stable end product. The dihaloacetonitrile reacts more slowly (days) with chlorine eventually forming dihaloacetamide and dihaloacetic acid (Reckhow et al., 2001). It should also be noted that dihaloacetamides can form relatively stable N-haloderivatives (Rapp and Reckhow 2005).

Several researchers have studied the chlorination of amino acids as model compounds in drinking waters. Perhaps the most extensive of these studies is the PhD work of Hureiki (Hureiki et al. 1994). A re-analysis of her data (Table 5.8) shows that many of the 20 amino acids likely undergo extensive oxidation. Many of these also lead to high levels of TOX that cannot be accounted for among the measured THMs.

Significant levels of TOX were noted for amino acids having R_3 groups that could contribute to the molecule's reactivity. For example, those having the following properties were in this category:

- Amino acids with electron withdrawing groups (CO, COOH, SH) capable of acting directly on the β-carbon (aspargine, aspartic acid, cysteine)
- Amino acids with groups that can become oxidized to yield a methyl ketone (threonine)

Extensive oxidation was noted for:

- Amino acids with no carbonaceous R group (glycine)
- Amino acids with electron donating groups (-OH) on the β-carbon (serine, threonine)

General molecular fragmentation with both oxidation and halogen incorporation occurs for:

- Amino acids having activated aromatic and heterocyclic groups (histidine, tryptophan, and tyrosine).

Table 5.8
Mass balance on amino acid chlorination

Amino Acid Common Name	R_3 Group	Chlorine Demand (M/M) Measured (M/M)	Carbonaceous only (M/M)	Carbonaceous only ($mgCl_2/mgC$)	DBPs measured TOX (M/M)	DBPs measured $CHCl_3$ (M/M)	% Oxid. of Carbon
Glycine	H	5.6	3.6	10.7	0.00	0.00	120%
Alanine	CH_3	2.8	0.8	1.6	0.01	0.00	13%
Valine	$CH(CH_3)_2$	2.7	0.7	0.8	0.01	0.00	6%
Isoleucine	$CH(CH_3)Et$	2.6	0.6	0.6	0.02	0.00	4%
Leucine	$CH_2CH(CH_3)_2$	2.6	0.6	0.6	0.00	0.00	4%
Serine	CH_2OH	5.3	3.3	6.5	0.01	0.00	66%
Threonine	$CHOHCH_3$	5.6	3.6	5.3	0.29	0.01	45%
Methionine	$CH_2CH_2SCH_3$	6.0	0	0.0	0.14	0.00	0%
Cysteine	CH_2SH	6.2	0.2	0.4	0.46	0.00	4%
Aspartic Acid	CH_2COOH	5.5	3.5	5.2	1.58	0.01	58%
Glutamic Acid	CH_2CH_2COOH	2.4	0.4	0.5	0.03	0.00	4%
Lysine	$(CH_2)_4NH_2$	3.8	-0.2	-0.2	0.04	0.01	-1%
Ornithine	$(CH_2)_3NH_2$	4.6	0.6	0.7	0.06	0.00	5%
Arginine	$(CH_2)_3NHC(NH_2)_2$	8.2	0.2	0.2	0.11	0.00	2%
Histidine	CH_2-imidazole	12.0	6.0	5.9	1.23	0.02	60%
Asparagine	$CH_2C(=O)NH_2$	5.6	1.6	2.4	1.63	0.01	27%
Glutamine	$CH_2CH_2CONH_2$	3.8	-0.2	-0.2			-2%
Proline	$CH_2CH_2CH_2$-	5.4	3.4	4.0	0.32	0.00	31%
Tryptophan	CH_2-indole	16.0	12.0	6.5	2.32	0.69	52%
Phenylalanine	CH_2-Ph	2.7	0.7	0.5	0.02	0.00	4%
Tyrosine	CH_2-Ph-OH	13.4	11.4	7.5	1.50	0.35	60%
AVG				**2.8**			**27%**

Source: Data from Hureiki, Croue & Legube 1994[6]

Compounds containing a thiol group are all quite reactive with chlorine. Folkes et al. (1995) found that glutathione (a tripeptide containing glycine, cysteine, and glutamic acid) reacts with free chlorine at a rate greater than $10^7\ M^{-1}s^{-1}$. The site of reaction is almost certainly the thiol group. It is proposed that an S-chloro derivative is first formed, and followed by hydrolysis to a sulfenic acid:

$$R\text{-}SH + HOCl \rightarrow RSCl + H_2O \rightarrow R\text{-}SOH + HCl \qquad (5.3)$$

Others have shown that the end products of the reaction between chlorine and amino acids are frequently aldehydes and nitriles (Stanbro & Smith 1979; Le Cloirec & Martin 1985, Hrudey et al. 1988; Bruchet et al. 1992). This can account for some of the chlorine reactivity noted by Hureiki, but not all of it. Furthermore, it does not account for the large amount of TOX.

Predicted DBPs Based on Estimated Natural Abundance. Table 5.9 shows estimated levels of common amino acids as they exist in the free form. These values were calculated using some of the most reliable recent data on amino acid abundance (as summarized in Thomas 1997 and Bronk 2002). For the purpose of developing exposure estimates, it would be helpful to have some measure of the high-end extreme in amino acid concentration. Although such distributions in concentration are not available, upper percentiles (90th, 99th) were estimated from the range of amino acid nitrogen to DOC values in the literature summaries and the existing DOC frequency distribution in the ICR data. These abundances were then multiplied by the respective DBP yields or chlorine demands. The results may be interpreted as expected DBP formation and chlorine consumption due to the free amino acids only.

Combined amino acids (combined = polypetides and proteins) are likely to be present at higher levels, probably about twice the levels shown in Table 5.9 (based on various literature estimates of combined and free amino acids). Some of these may be embedded into humic structures, and thereby isolated with the humic fraction. Despite their higher abundance, the combined amino acids may play a lesser role in fast DBP formation. The polypeptide bonds render bound amino acids (i.e., 2° amide-N) far less reactive than their free counterparts. Reckhow et al (2001) have argued that the combined amino acids may be responsible for a large amount of the slow reactivity observed in distribution systems.

From Table 5.9, it might be concluded that tryptophan residues are responsible for substantial amounts of THMs in most waters (2 μg/L as a median, and up to 20 μg/L in the extreme). However, the TOX formation indicates that there are few, if any other halogenated by-products from this structure besides the THMs. Given our poor understanding of the most toxic by-products, we may be well advised to pay more attention to amino acids producing substantial amounts of "other" TOX. The yield data in Table 5.9 come from Hureiki et al. (1994), who only measured TOX and THMs. Therefore, the "unknown TOX" as defined here could not be calculated. In its place is the "non-THM TOX". Based on this parameter, Aspartic Acid should be of greatest concern (6 μg/L up to 60 μg/L calculated). It may be that much of the unmeasured halogenated organic material is in the form of TCAA and DCAA, but the potentially high levels of unknown TOX invite further investigation.

Table 5.9
Calculated reactivity and DBP yields for free amino acids*

Amino acid	Abundance[†]	Cl2 demand ($mg\text{-}Cl_2/L$)			THM ($ug\text{-}CHCl_3/L$)			TOX (ug-Cl/L)			Non-THM TOX (ug-Cl/L)		
		50%ile	90%ile	99%ile	50%ile	90%ile	99%ile	50%ile	90%ile	99%ile	50%ile	90%ile	99%ile
Glycine	11.7%	0.02	0.06	0.23	0.00	0.00	0.00	0.00	0.00	0.00	0.00	0.00	0.00
Alanine	11.7%	0.01	0.03	0.12	0.01	0.04	0.14	0.05	0.13	0.50	0.04	0.09	0.38
Valine	5.2%	0.01	0.01	0.05	0.00	0.00	0.00	0.01	0.02	0.09	0.01	0.02	0.09
Isoleucine	4.4%	0.00	0.01	0.04	0.01	0.01	0.05	0.02	0.06	0.24	0.02	0.05	0.19
Leucine	5.6%	0.01	0.01	0.05	0.01	0.02	0.07	0.01	0.02	0.08	0.00	0.01	0.02
Serine	10.8%	0.02	0.05	0.21	0.03	0.07	0.26	0.05	0.12	0.47	0.02	0.06	0.23
Threonine	7.0%	0.01	0.04	0.14	0.06	0.15	0.59	0.73	1.83	7.31	0.68	1.70	6.78
Methionine	1.0%	0.00	0.01	0.02	0.00	0.00	0.00	0.05	0.12	0.49	0.05	0.12	0.49
Aspartic Acid	10.2%	0.02	0.05	0.20	0.10	0.25	0.99	5.79	14.48	57.91	5.70	14.26	57.03
Glutamic Acid	10.0%	0.01	0.02	0.09	0.01	0.03	0.12	0.09	0.22	0.90	0.08	0.20	0.79
Lysine	1.6%	0.00	0.01	0.02	0.01	0.02	0.10	0.02	0.06	0.23	0.01	0.03	0.14
Ornithine	2.0%	0.00	0.01	0.03	0.01	0.02	0.07	0.04	0.10	0.41	0.03	0.09	0.35
Arginine	7.4%	0.02	0.05	0.22	0.03	0.07	0.27	0.28	0.71	2.83	0.26	0.65	2.59
Histidine	2.5%	0.01	0.03	0.11	0.07	0.18	0.70	1.11	2.78	11.13	1.05	2.63	10.50
Asparagine	0.4%	0.00	0.00	0.01	0.00	0.01	0.02	0.21	0.53	2.10	0.21	0.52	2.08
Glutamine	0.5%	0.00	0.00	0.01	0.00	0.00	0.00	0.01	0.02	0.07	0.01	0.02	0.07
Tryptophan	2.5%	0.01	0.04	0.14	2.06	5.16	20.64	2.06	5.14	20.55	0.22	0.54	2.16
Phenylalanine	3.4%	0.00	0.01	0.03	0.00	0.00	0.00	0.03	0.07	0.29	0.03	0.07	0.29
Tryrosine	2.2%	0.01	0.03	0.10	0.91	2.28	9.14	1.17	2.93	11.73	0.36	0.90	3.59
SUM	**100.0 %**	**0.18**	**0.46**	**1.83**	**3.32**	**8.29**	**33.16**	**11.73**	**29.33**	**117.33**	**8.78**	**21.95**	**87.78**

* Total amino acid values are 20, 50, and 200 μg-N/L for the three percentiles. These numbers are based on amino acid nitrogen/DOC ranges from Thomas, 1997, and Bronk, 2002 literature summaries, and DOC ranges from the ICR. All yield data are based on Hureiki et al. 1994

[†] Based on an average of Isle Bay water (Yamashita & Tanoue 2003), Lake Biwa (Wu et al. 2003), and water column values, summer, from Thomas 1997

Protein.

Although free amino acids can be quite reactive with chlorine as previously discussed, polypeptides react more slowly. Much of the early reaction with proteins probably occurs on the heterocyclic or activated aromatic side chains (histidine, tryptophan, and tyrosine). This will lead to rapid formation of oxidation and substitution products, but may have little overall affect on the polypeptide backbone. Many polypeptide bonds are not degraded at all under drinking water treatment conditions (e.g., Pereira et al. 1973), or they are very slowly degraded in a stepwise fashion (Goldschmidt et al. 1927, Fox et al. 1997). Chlorine reactions with proteinaceous material are therefore, slow, and do not reach completion under drinking water treatment conditions. This may make N-DBP formation appear slow and nearly constant over the timeframe of interest. It may also make the kinetic dependence on chlorine concentration much more evident.

Heterocylic Compounds

Purines and Pyrimidines. The nucleic acid bases are quite reactive with chlorine, exhibiting chlorine demands of about 7 M/M. Among this group of heterocycles, pyrimidines are generally reported to be more reactive than purines (Jersey and Johnson 1992). Some of this demand may represent the formation of N-chloramines of low reactivity (i.e., not measured as a chlorine residual). Hoyano et al. (1973) found N-chloro derivatives of the pyrimidines, uracil and thymine, but not for the purines, adenine and guanine.

Dennis and coworkers (1979) explored the rate of loss of chlorine in the presence the four nucleotides at mole ratios of 1:3 (chlorine limited). They showed that at neutral pH, cytosine was most reactive followed by adenine, guanine and uracil. Their kinetic data suggest the following 2nd order rate constants: 53 $M^{-1}s^{-1}$, 14 $M^{-1}s^{-1}$, 8 $M^{-1}s^{-1}$, and <1 $M^{-1}s^{-1}$, for the monophosphate nucleotides of cytosine, adenine, guanine and uracil, respectively. These rates may be interpreted as describing the formation of an initial halide derivative (e.g., 8-chloroguanine, 5-chlorocytosine). Reaction rates for cytosine and adenine both increased with decreasing pH, whereas guanine showed the reverse behavior. These researchers did not examine thymine.

Uracil. Current knowledge of the chlorination of uracil is heavily influenced by initial product studies (often conducted by chemists) and final product analysis (more common among engineers). Early uracil degradation products include 5-chlorouracil and 3,5-dichlorouracil and trichlorouracils (Figure 5.8). All of these products are likely to be long-lived intermediates under chlorine limited conditions. With chlorine in excess, the reaction pathway leads to chlorohydrins

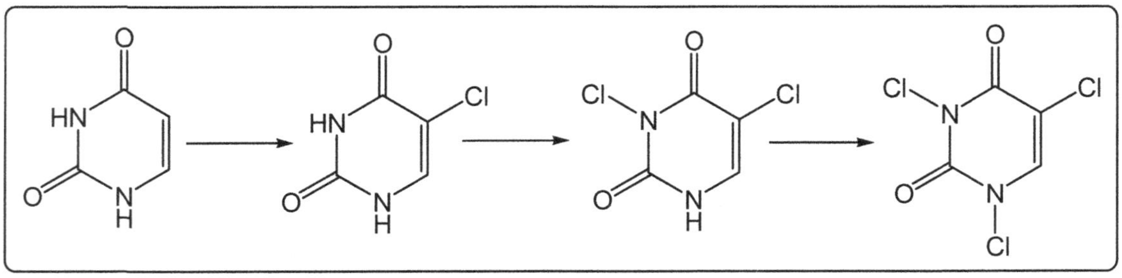

Figure 5.8 Initial attack of chlorine on uracil

and eventually, TCAA and CO_2 (Hayatsu et al. 1971; Hoyano et al. 1973; Gould et al. 1984). Substantial exposure to chlorine leads to the formation of ring products that have lost their aromaticity, which then leads to ring cleavage (Gould et al. 1984; Hoyano et al. 1973; Dennis et al. 1979). Gould and coworkers (1984) found quantitative formation of 5-chlorouracil at a chlorine molar dose ratio of 1:1. At a 5:1 ratio they found substantial amounts of 3,5-dichloruracil, smaller amounts of 5-chlorouracil and nearly quantitative amounts of NCl_3 as determined by the DPD-FAS method (Figure 5.9)

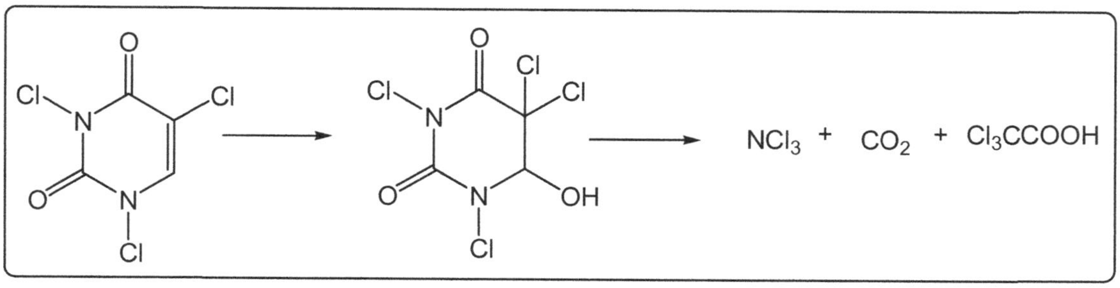

Source: Redrawn from Gould et al. 1984

Figure 5.9 Some uracil degradation products as proposed by Gould et al. (1984)

Young and Uden (1994) studied the aqueous chlorination of labeled uracil. They found a 19% molar yield of chloral hydrate from the 24 hour chlorination of uracil at an 8:1 molar ratio. The halomethyl and carbonyl carbons in this by-product (as well as for the less abundant TCAA) were found to originate from the 5 and 6 carbons as shown in Figure 5.10. This work clearly highlights the critical role of the 5 carbon site. As the only ring carbon in a position to be activated by each of the two carbonyl groups, it is doubly reactive. The carbonyl groups in nucleic acid bases can be strongly activating due to keto-enol tautomerization with the neighboring ring nitrogens (Figure 5.11).

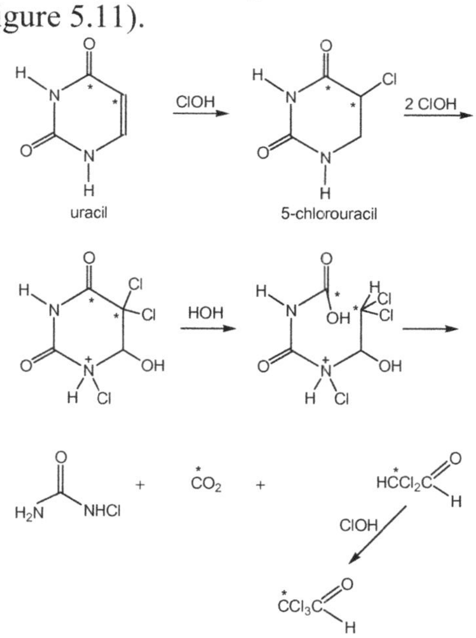

Source: Redrawn from Young and Uden 1994

Figure 5.10 Proposed degradation of uracil

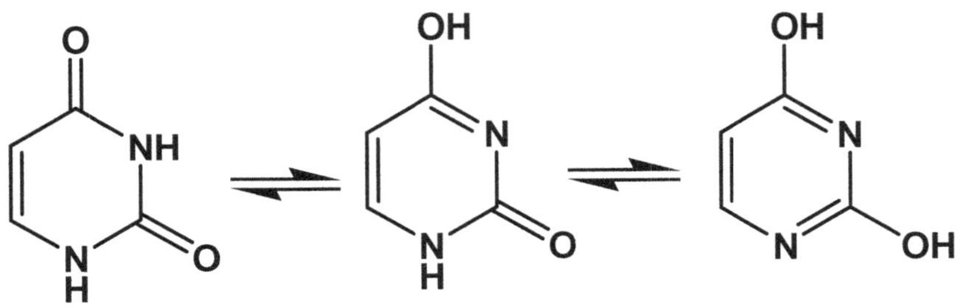

Figure 5.11 Keto-enol tautomers for uracil

Laboratory chlorination of uracil was conducted as part of this research. Both chlorine demand and TOX formation were quite high for this compound (Figure 5.12). Despite about 3.4 M/M of TOX formed, very little could be attributed to the measured by-products (6% yield of DCAA and traces of TCAA and THMs). However, chloral hydrate was not measured. Presuming that chloral hydrate was formed at the level (yield) reported by Young and Uden, there is still about 75% of the TOX that remains unaccounted for. Further, the carbon and nitrogen analysis (Table 5.4) points to extensive loss of both during acidification and gas purge. This suggests either extensive formation of CO_2 & N_2 or some other highly volatile forms (probably low charge, low polarity products). Some of this could be NCl_3 as measured by Gould and coworkers (1984) using the non-specific DPD-FAS method. However, there remain some inconsistencies among the various studies that do not readily accommodate a simple explanation such as quantitative NCl_3 formation.

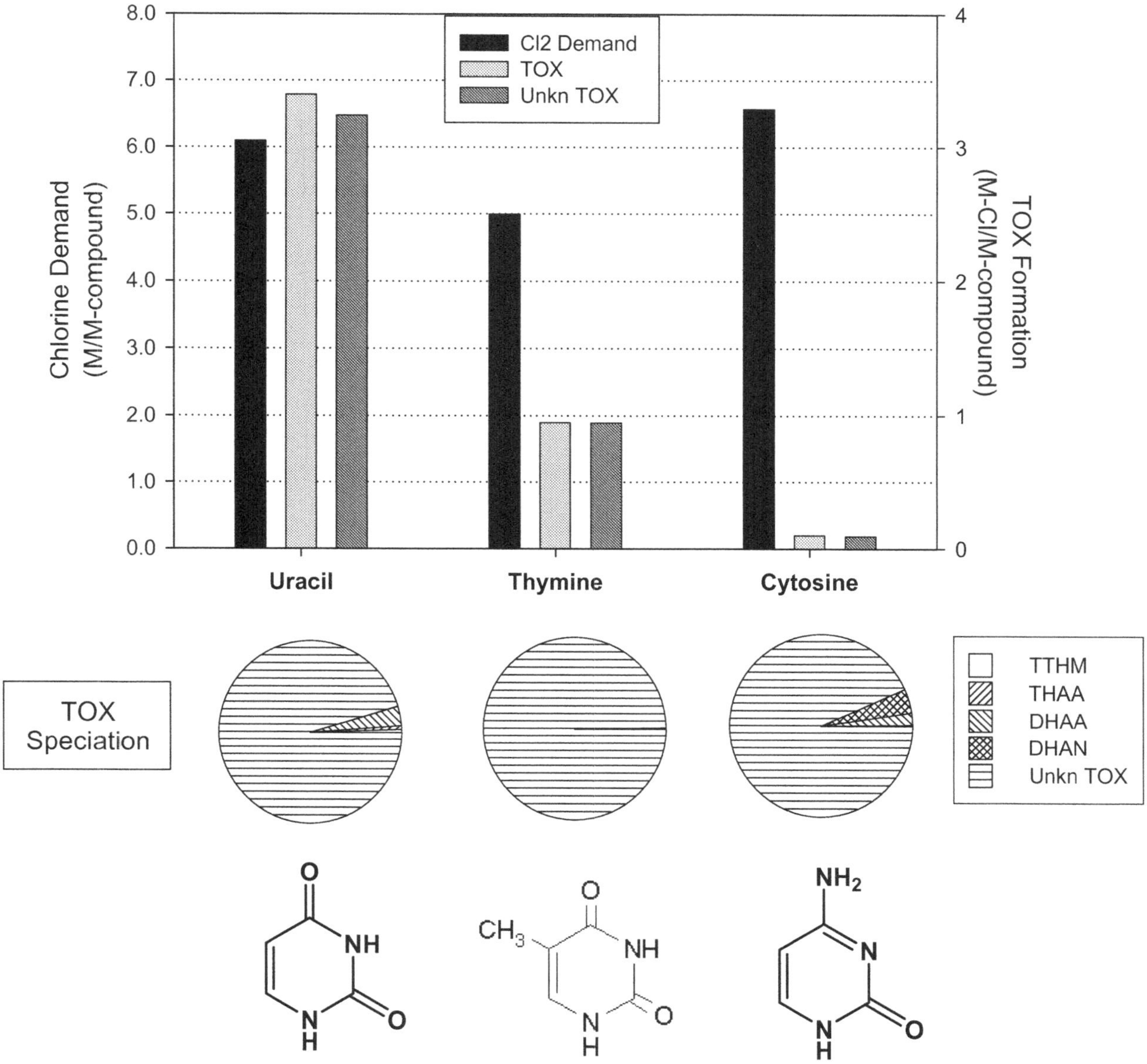

Figure 5.12 Free chlorine reaction data with pyrimidines

Thymine. Thymine, a methyl-substituted uracil, is also known to be quite reactive with chlorine. Yagi and Itokawa (1979) found that it was lost at a rate corresponding to a two -hour half-life when in the presence of an equimolar amount of chlorine. In laboratory studies, thymine produced a moderate level of TOX, but very little of the standard DBPs (Figure 5.12). Population of the 5 carbon with a methyl group would prevent dual chlorine substitution as proposed for this site in Uracil. This could explain why thymine produced about 2M/M less TOX than uracil.

Cytosine. Cytosine, along with the purines, adenine and guanine, has a primary amine which can undergo chlorine substitution forming an N-chloramine. As discussed earlier, this chlorine substitution could also occur on the neighboring heterocyclic nitrogen due to the amine-imine tautomerization. Nevertheless, Patton and coworkers (1972) reported a 76% yield of compound II (Figure 5.13) at a 1:1 mole ratio. Based on kinetic studies of Gould and coworkers (Gould and Hay 1982; Gould et al. 1984) the initial N-chloramine (compound II) is an especially stable one with an apparent half life of about 4 days. Formation of about 2M combined chlorine per M-cytosine was complete in about 10 minutes at pH 5. The first order decay rate of this apparent combined chlorine was 2×10^{-6} s^{-1} in the presence of a substantial free chlorine residual (~ 35 mg/L). The simplest explanation for these observations would be preferential formation of compound VI under these conditions of excess chlorine. However the possibility of additional N-chlorination (e.g., compound X) cannot be ruled out as some organic chloramines are detected as free chlorine by the conventional diphenylene diamine (DPD) method. Reynolds and coworkers (1988) also observed about 2M combined chlorine per M-cytosine, but they were unable to detect any chlorination products of the primary amine.

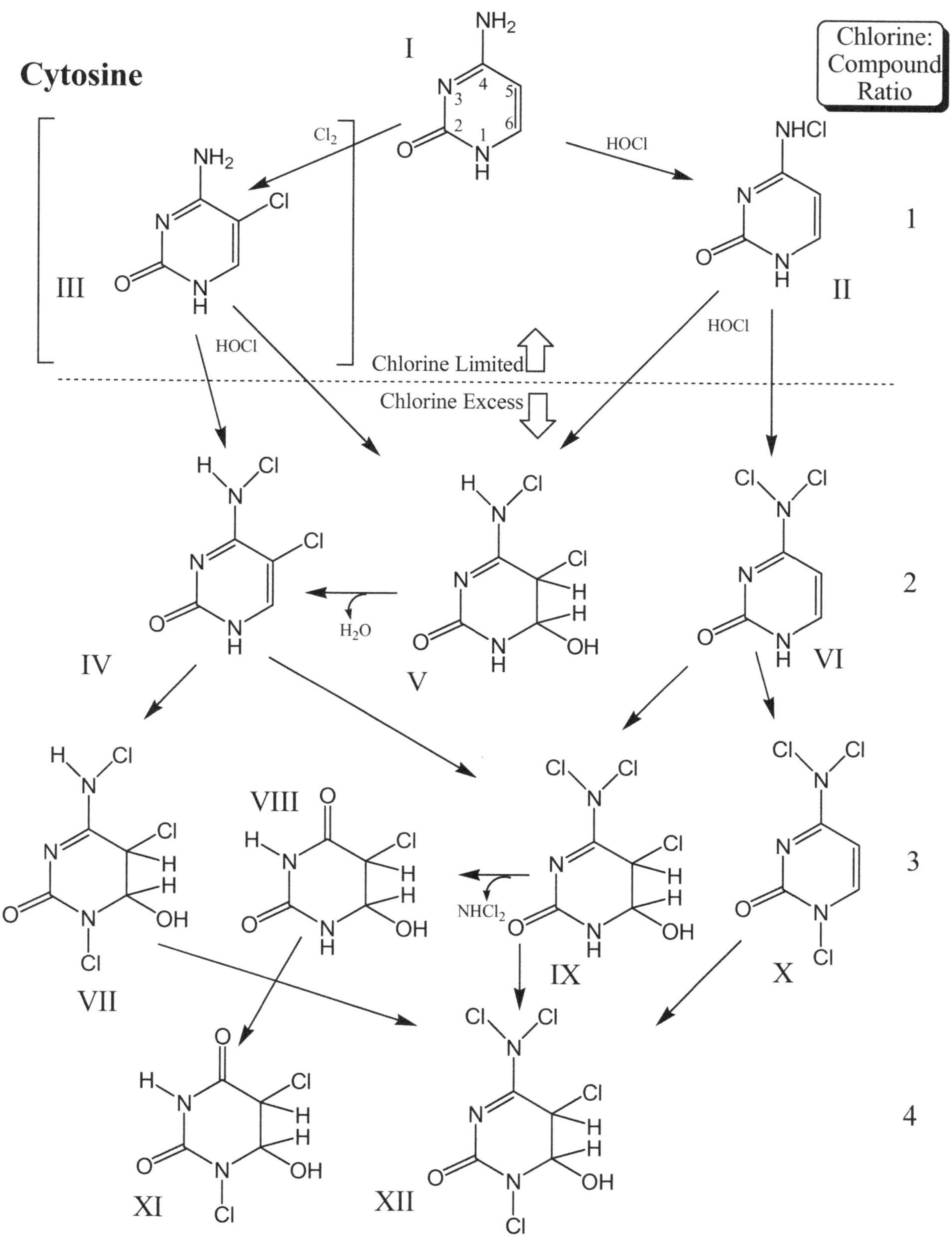

Figure 5.13 Proposed degradation pathways for the chlorination of cytosine

There is also considerable evidence for the formation of C-Cl bonds from the chlorination of cytosine. Patton et al. (1972) reported the isolation of compound II, IV, V and VII at 2 M/M chlorine dose and at 5 M/M they reported compound IX and a tetrachloro derivative (possibly XII). Patton used high concentrations of reagents over short time periods, and this might explain why their observations appear to differ from those of Gould.

Several researchers have noted that cytosine and its nucleoside, cytidine, can form the 5-chloro derivative (compound III), much like uracil does (Hayatsu et al. 1971; Henderson et al. 1999). Reynolds and co-workers (1988) have also proposed that the 1-chloro, 3-chloro, and 3,5-dichloro derivatives form. However, they acknowledge that their results differ from those of Patton et al., and that neither analytical method could be viewed as definitive.

Henderson and coworkers (1999) presented compelling evidence for molecular chlorine (Cl_2) as the chlorinating agent in the formation of 5-chlorocytosine (compound III). Yields* did not exceed 10% until the pH was below 5 at 100 mM chloride or below 4.5 at 30 mM chloride. These conditions are too acidic and saline as to be of great importance in drinking water treatment. While addition of HCl can convert II into III (Patton et al., 1972), this is also quite unlikely to happen in drinking water systems (see also Henderson et al. 1999). It should be noted that plasma levels of chloride (~100 mM) and localized low pHs are likely to result in production of 5-chlorocytosine derivatives in vivo, as shown by Kawai et al. (2004).

Kang and coworkers (2004) studied the reaction of chlorine with the DNA nucleoside form of cytosine (deoxycytidine). They proposed an initial reaction between cytosine and the chlorinating species forming a dihydro intermediate prone to deamination. Dehydration of the corresponding cytosine and uracil intermediates would yield 5-chlorocytosine without having to invoke the molecular chlorine pathway. Deamination might occur after hypochlorite addition to the ring, giving compound VIII.

In laboratory studies, cytosine produced very little TOX, despite a large chlorine demand (Figure 5.12). Almost no identifiable DBPs were seen. This supports the formation of N-chloro derivatives, but little or no stable C-Cl structures. Given the level of chlorine demand, degradation of cytosine must be substantially greater than shown in Figure 5.13. Elemental balance data (Table 5.4) showed loss of almost 3 moles of carbon and 1.7 M of nitrogen.
Of course cytosine in nucleosides and nucleic acids will be bound to ribose sugars at the N1 position (Figure 5.14). Without hydrolysis this will limit its reactivity at this position.

* Recall that yield refer to M/M ratios of product to starting compound, often presented as percentages. This particular yield is based on chlorine dose which was the limiting reactant.

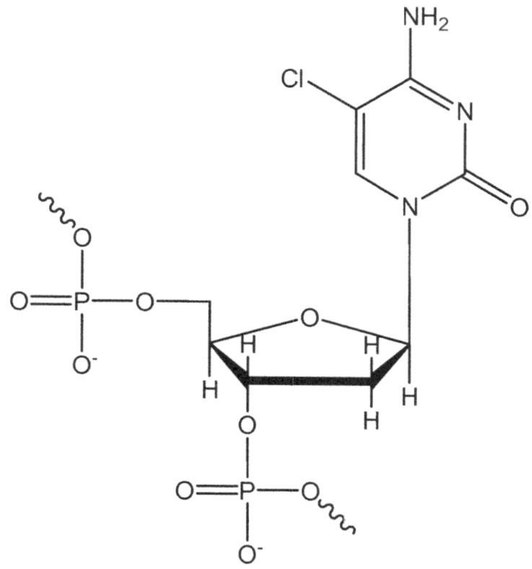

Figure 5.14 Structure of cytosine within DNA

Guanine. As already stated, purines are viewed as less reactive than pyrimidines (about 7M/M chlorine demand after 24 hours, which is less on a per carbon basis). Hoyano and coworkers (1973) found little reaction between chlorine and the purines, guanine, adenine and xanthine. Eventually, a cyclic fragmentation product, parabanic acid was detected. Neither guanine nor adenine produced more than 1% yield of CNCl (Young and Uden, 1994). Screening by GC failed to show any other degradation products from Guanine (Shang et al. 2000).

The research team of Suzuki has studied guanine:chlorine reactions under chlorine limited conditions. In most of their work they used the nucleoside form (guanosine, or Guo) or a acylated nucleoside (AcdGuo). Masuda and co-workers (2001) found the 8-chloro derivative as well as an 8-oxo product (Figure 5.15). They proposed that the N-7 position would be the primary site of attack as the most nucleophilic position on the molecule (this is the primary site of protonation; as discussed earlier in this chapter). The intermediate 7-chloro cation would ultimately lead to the 8-chloro compound by chlorine migration. This migration may be accelerated under basic conditions by deprotonation of the C8 hydrogen.

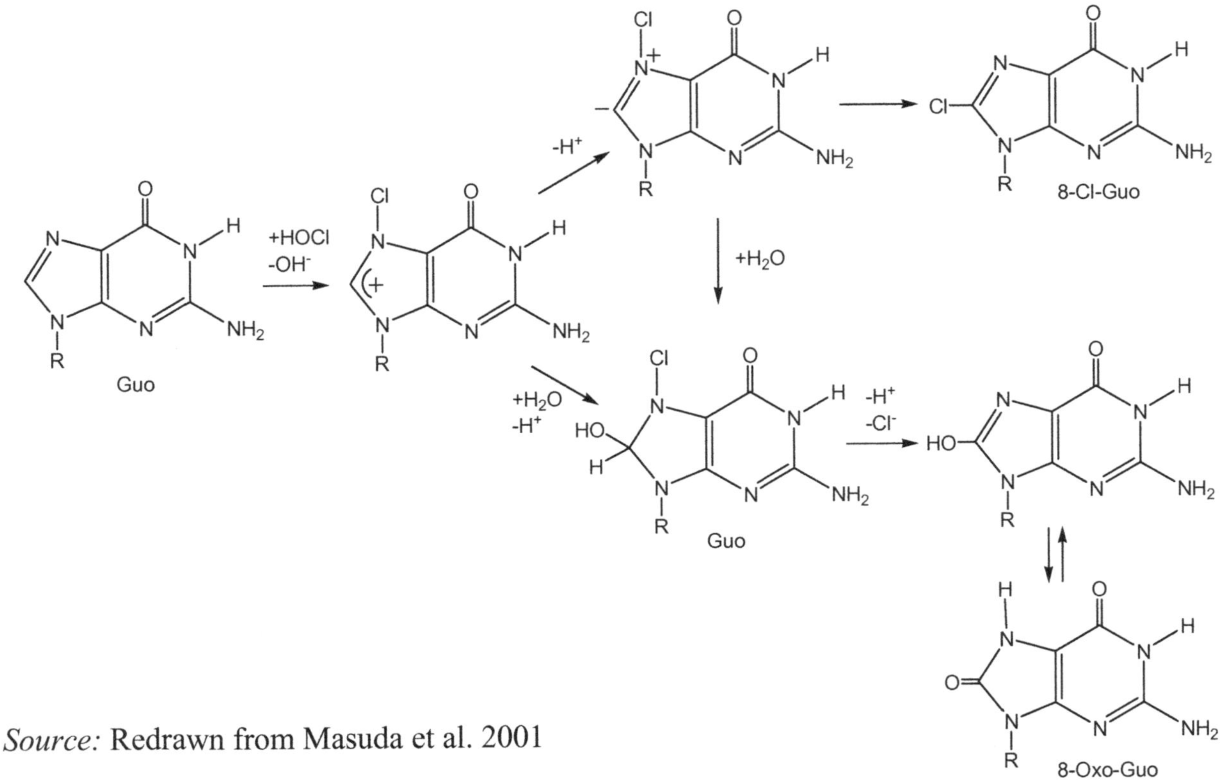

Source: Redrawn from Masuda et al. 2001

Figure 5.15 Chlorination of guanosine (guanine linked to a pentose sugar) under neutral conditions as proposed by Masuda et al. 2001

These authors found the 8-chloro derivative to be produced at a molar yield of about 1.6% and 30% guanosine destroyed, at equimolar doses of chlorine and guanosine (0.5 mM). This corresponds to a 5% M/M yield based on the amount of guanosine lost. The 8-oxo derivative was found at about 0.3% yield. Both products increased as the chlorine dose increased until a 2:1 ratio and then it decreased at higher levels. Under these conditions, the reactions were very rapid (complete within 1 minute), and it seems quite likely that some of the reaction products were as reactive or more reactive than the guanosine. Further degradation products are evidently formed under these conditions. The pH behavior of this reaction is complex, and it suggests an important role of all three major chlorine species (Cl_2, HOCl and OCl^-).

Following halogenation at the 8-carbon position, the molecule continues to react by halogenation of the amine attached to the 2-carbon. This ultimately leads to hydrolysis of the C-N bond and formation of a carbonyl derivative.

Hoyano and coworkers (1973) looked at reaction times of a week in the presence of excess chlorine (1M HOCl to 1 mM guanine) (Figure 5.16)*. They were able to show a 90% yield of parabanic acid. As with adenine, guanine produced less than a 1% yield of CNCl (Young and Uden 1994). Screening by GC failed to show any other degradation products from Guanine (Shang et al. 2000).

---

* Room temperature, unbuffered

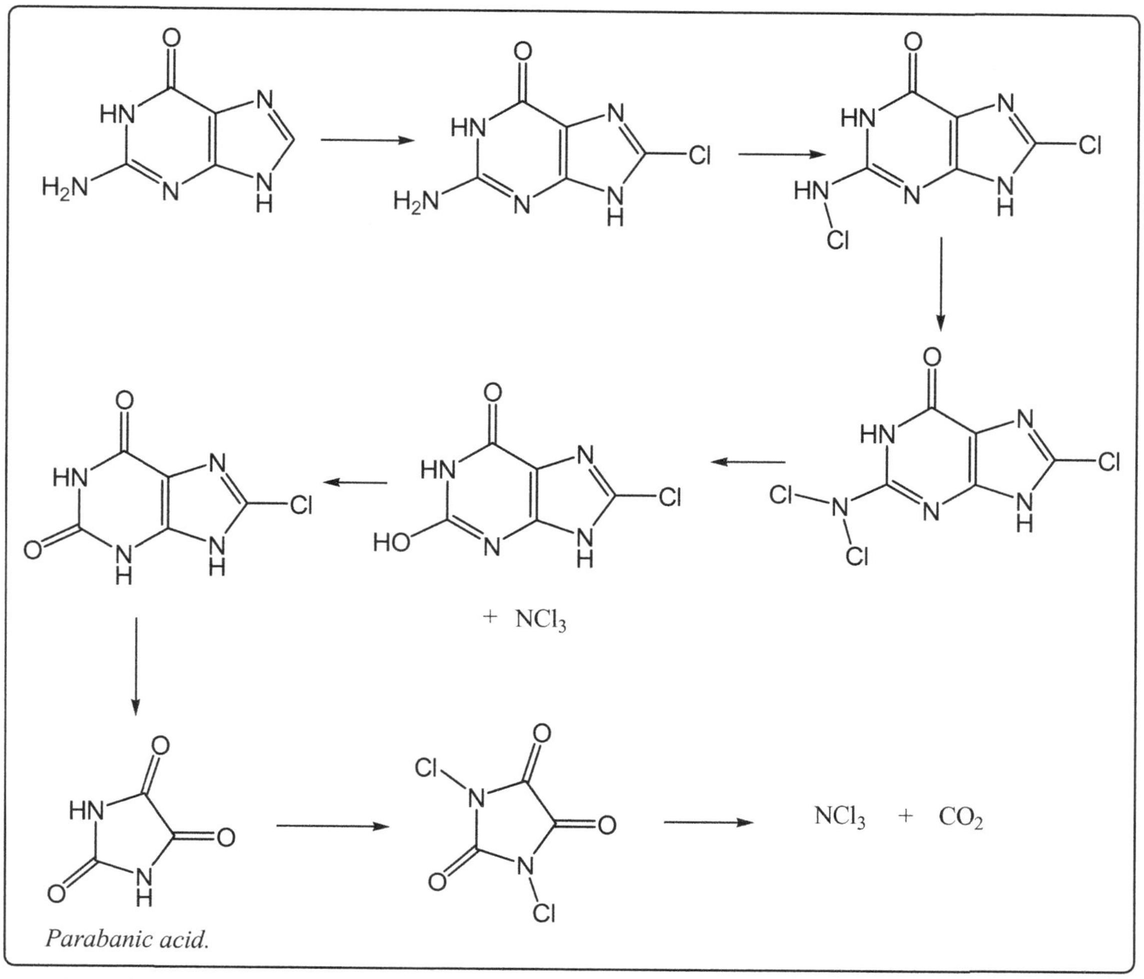

Source: Based on Hoyano et al 1973

Figure 5.16 Proposed pathway for chlorination of guanine

Suzuki and coworkers (2003) used acylation to further examine the major by-products at chlorine:guanosine stoichiometries between 0.25:1 and 2:1. They measured the acylated deoxyribose bound products (Acd) by HPLC. Throughout the chlorine dose range the spiroiminodihydantoin derivative (AcdSph) was the major product, followed by 8-chloro-guanosine (8-Cl-AcdGuo), diimino-imidazole (AcdDiz) and finally amino-imidazolone (AcdIz) (Figure 5.17). Both diamino-oxazolone (AcdZ) and the guanidinohydantoin/iminoallantoin mix (AcdGh/AcdIa) were minor by-products at low chlorine doses, but became nearly as abundant as AcdSph at high levels, suggesting that these are more slowly produced from an intermediate. The total yield of products was 10.3% at a 1:1 stoichiometry. Since only 14.3% of the guanosine had been lost, these products accounted for fully 72% of the lost compound.

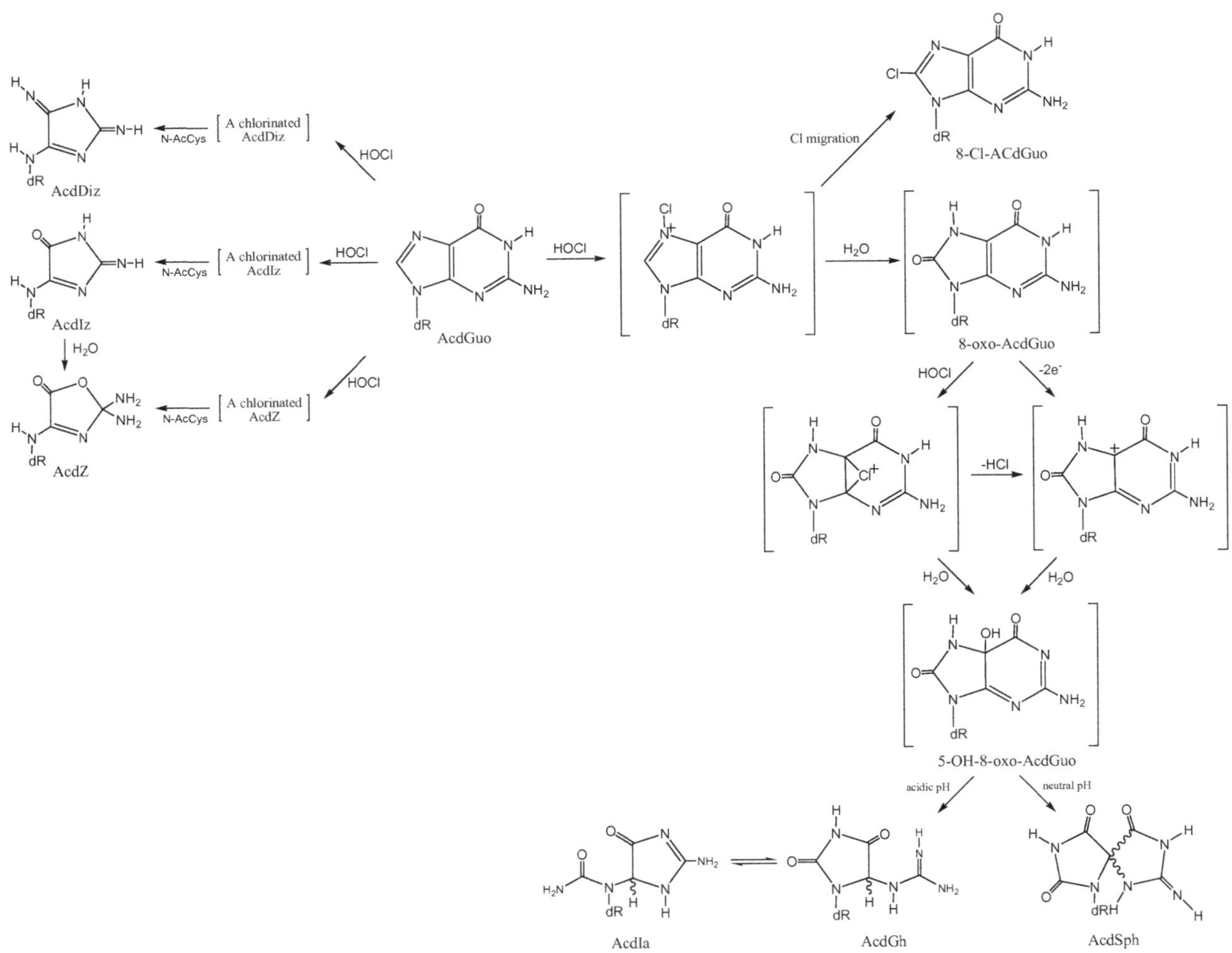

Source: Redrawn from Suzuki et al. 2003

Figure 5.17 Proposed mechanism for the reaction of AcdGuo with HOCl (dR stands for 3,5-di-*O*-acetyl-2-deoxyribose)

This guanine literature can be summarized in a more comprehensive way, and placed into the context of chlorine limited and chlorine excess reactions (Figure 5.19). The former might represent the common chloramination scenario (with a short free chlorine contact time), while the latter is more like simple chlorination. The first step is the chlorine induced oxidation of the 8-carbon as proposed by Masuda et al. (2001). This may be the rate limiting step (see Suzuki et al. 2004) (Figure 5.18). Hypochlorous acid then adds across the double bond of the fused rings. The next step could be loss of HCl, followed by carbonyl formation and ring rearrangement. This could be aided by the presence of chlorine leading to an N-chloro intermediate. In the absence of chorine, the final product would be spiroiminodihydantoin (Sph) as reported by Suzuki and coworkers (2003). In excess chlorine we propose the N-chloro intermediate would form, accelerating the degradation of that 5-membered ring. The ring cleavage product would undergo decarboxylation followed by chlorine-induced degradation of the amines/imines, ultimately producing the classic breakpoint products (mostly nitrogen gas and chloride) and carbon dioxide. The remaining 5-membered ring (parabanic acid) would be more stable to chlorination. Hoyano and coworkers (1973) found this as a final product in 90% yield. It's also possible that parabanic acid readily halogenates to the N,N-dichloro derivative, which becomes reduced to parabanic acid upon addition of a reducing agent.

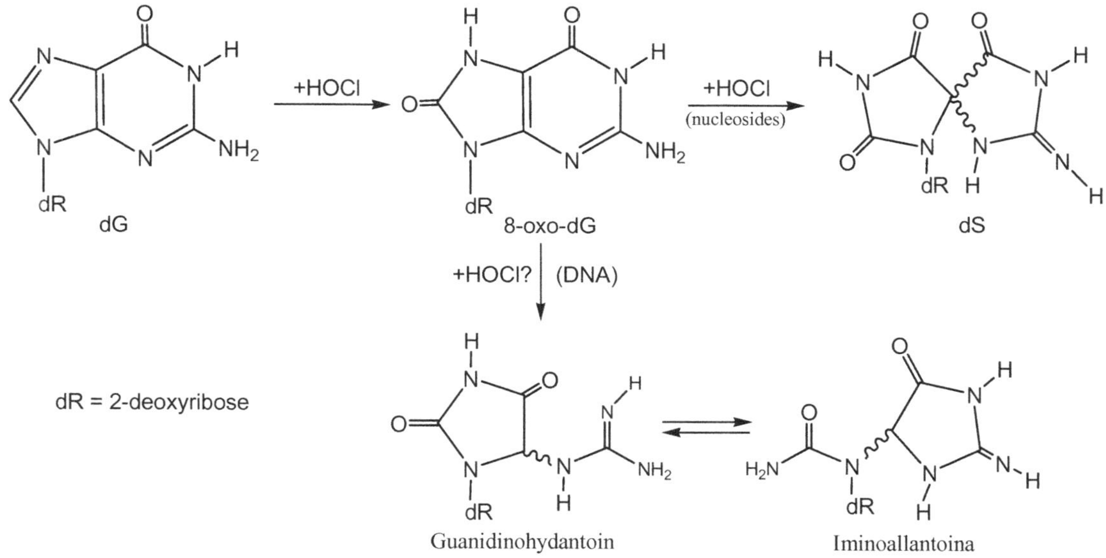

Source: Redrawn from Suzuki et al 2004

Figure 5.18 Proposed reaction path for the reaction of dG with HOCl and further oxidation products of 8-oxo-dG

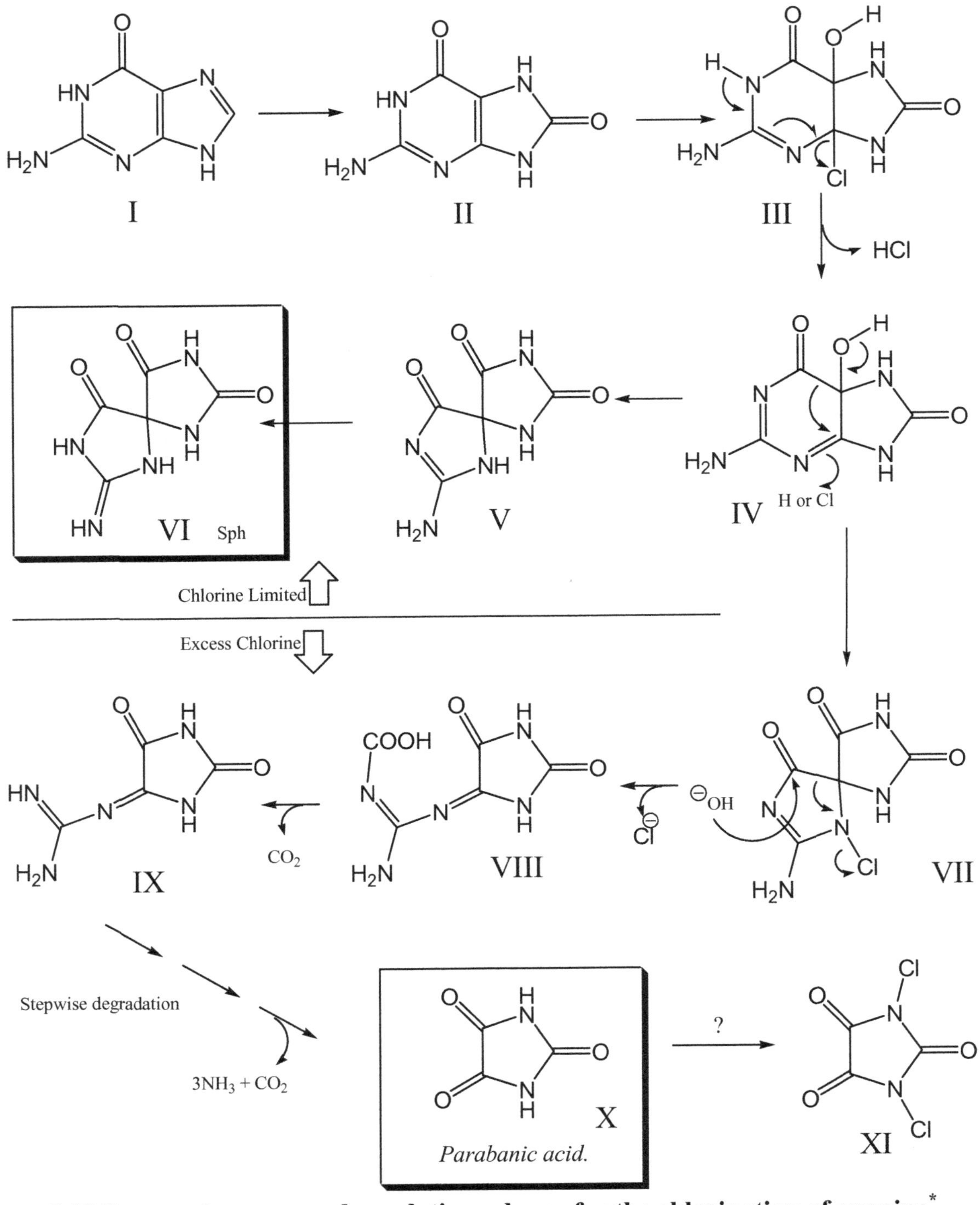

Figure 5.19 Proposed summary degradation scheme for the chlorination of guanine*

---

* A note about chemical schematics: We will follow conventions from the field of chemistry in these diagrams. Curved arrows within structures indicate movement of electrons. The focus is on the rearrangement of electrons and onds. Half arrowheads indicate movement of a single electron and full arrowheads an electron pair, either of which could result in the breaking of an existing bond and/or formation of a new one.

The formation of parabanic acid from guanine requires the consumption of three molecules of chlorine, as well as any additional chlorine consumed by the 3 released nitrogen atoms. If one assumes the classic breakpoint stoichiometry (1.65 M/M), we would calculate a total chlorine demand of 7.95 M/M. This is quite close to our experimental observations of 7.73 M/M (Table 5.4). In addition, we do not observe any persistent TOX. Unfortunately, guanine is the one compound tested that was poorly recovered in TOC/TN analysis, so the elemental balance data may not be reliable.

The above figure might also shed some light on possible reactions under various chloramination scenarios. Based on the kinetics of Dennis et al. (1979), we might expect the initial halogenation to occur according to a second order rate of 8 $M^{-1}s^{-1}$ at neutral pH. Therefore, a Ct of 30 mg/L-min (e.g., as might be used in a pre-chlorination, chloramination scenario) should result in about 20% degradation of the starting material. Suzuki's work suggests that once this first halogenation occurs, subsequent reactions will occur at rates that are nearly as fast. Therefore addition of ammonia would occur at a point where there is already a broad mix of intermediates present. Of course, degradation reactions not involving free chlorine would continue, but products like Sph are likely to remain.

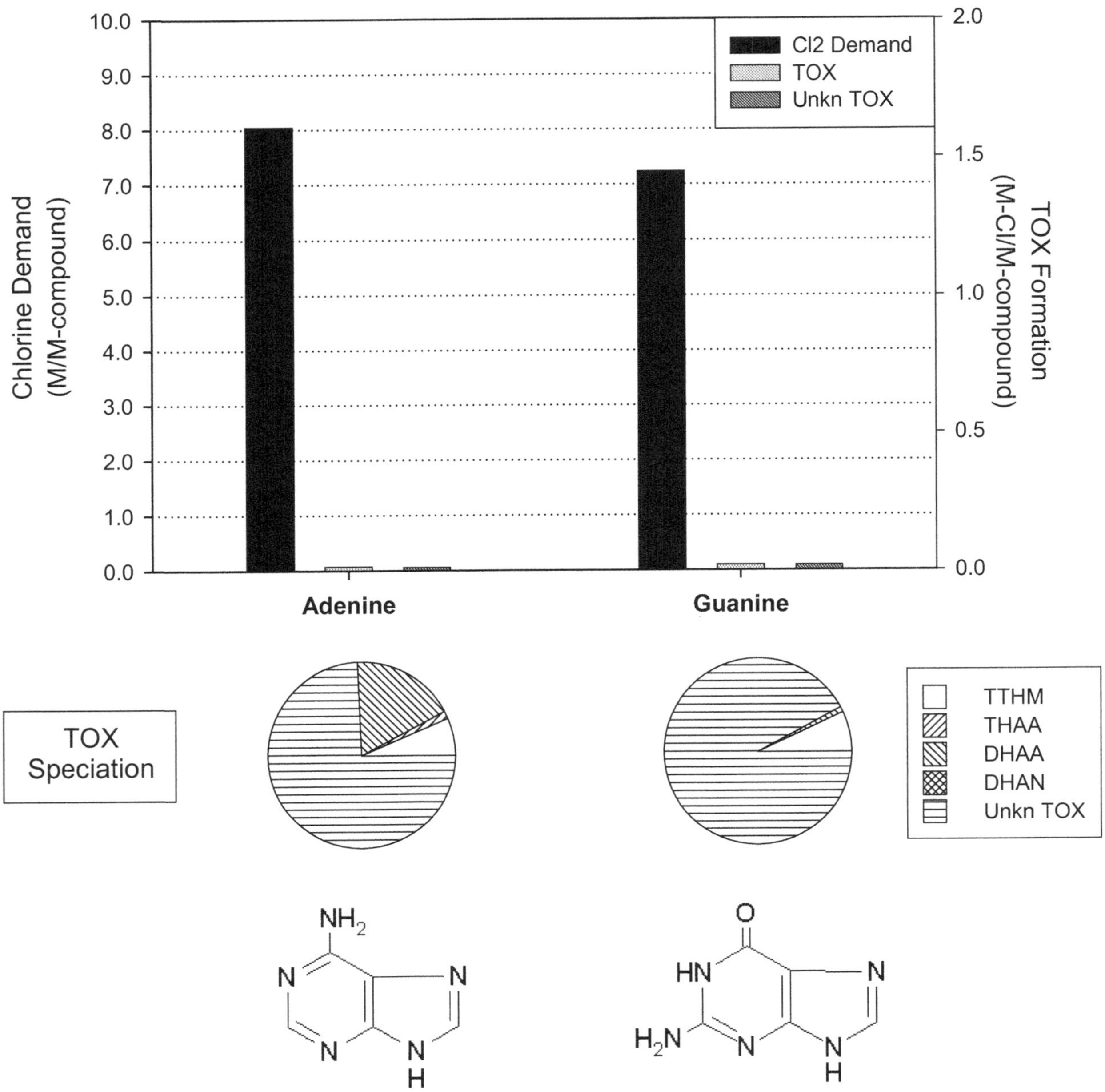

Figure 5.20 Reaction of free chlorine with purines

Adenine. Adenine has been little studied, but Whiteman and coworkers have shown that at least one chlorinated product, 8-chloro-adenine, can form under chlorine limited conditions (Whiteman et al. 1999). They calculated that this by-product accounted for about 4% of the total "measurable DNA modification products".

If parabanic acid, nitrogen gas, and carbon dioxide were the only products, we would expect to see a chlorine demand of about 8.95 M/M. Experimentally we see about 8 M/M with very little TOX formation (Figure 5.20). Also the yield of parabanic acid may be only 20% as reported by Hoyano et al. (1973). If it is accepted that adenine reacts in a fashion similar to guanine, one can propose a full degradation scheme, even though the experimental data are lacking (Figure 5.21).

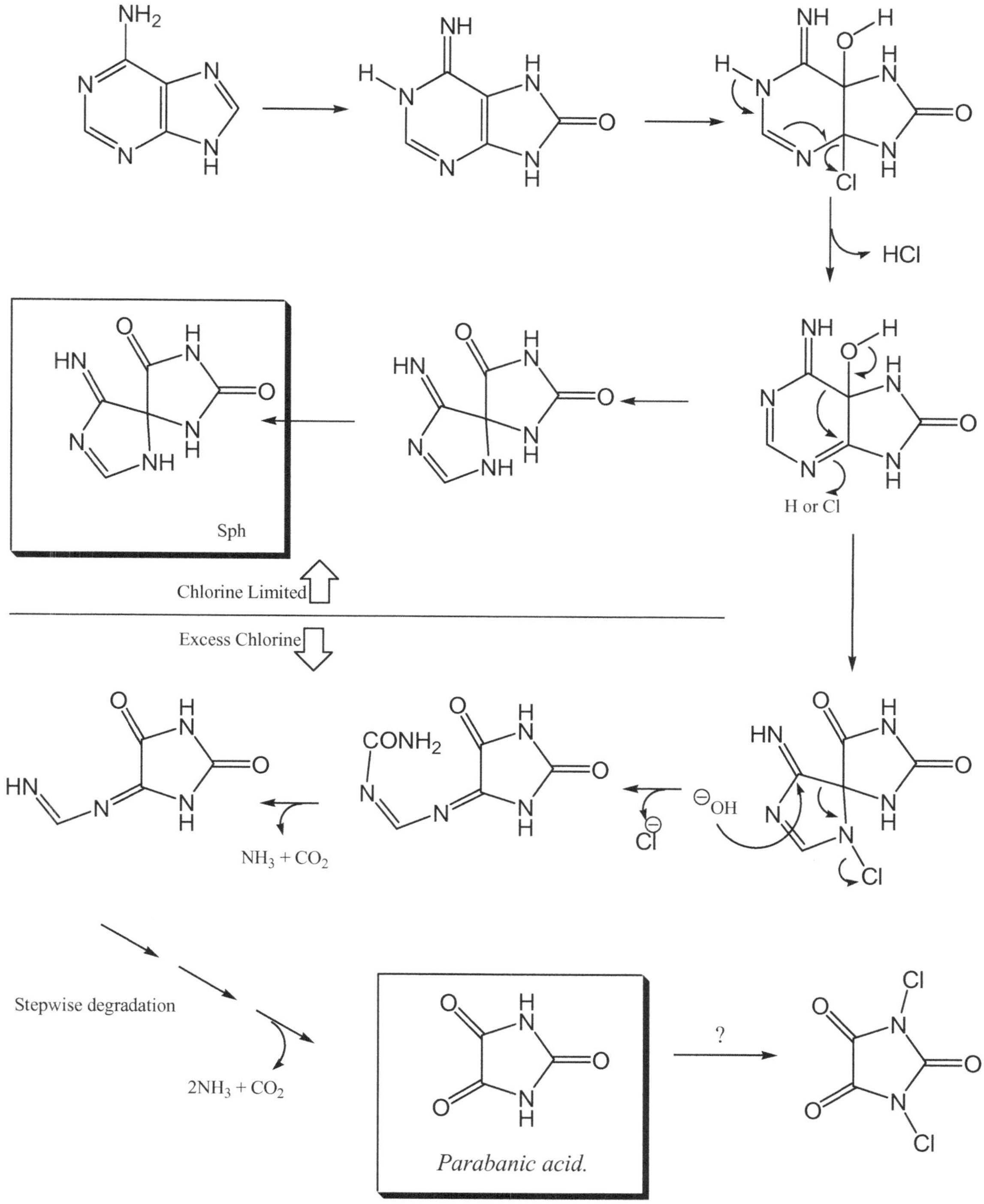

Figure 5.21 Proposed degradation scheme for adenine based on literature for guanine

Nucleic Acids. Available evidence suggests that purines and pyrimidines react within nucleic acids much as they do in the free form. Reaction of chlorine with DNA and RNA has demonstrated the formation of 8-chloroguanosine, 5-chlorocytidine and 8-chloroadenosine as predicted by studies with simple nucleosides (Masuda et al. 2001). Chlorinated cytosine (5-chloro) was observed in double stranded DNA and RNA under the same conditions (high chloride) that led to its formation in solutions of the nucleoside (Henderson et al. 1999). Using Calf Thymus DNA, Whiteman and colleagues (1997) showed the formation of chlorinated forms of uracil (5-Chloro) and oxo or hydroxyl forms of all five DNA bases. These authors also noted that iron (either Fe(II) or Fe(III)) and especially copper (either Cu(I) or Cu(II)) shifted the product distribution away from chlorinated and glycol products and toward oxo and hydroxyl compounds (hypoxanthine, 8-hydroxyguanine). This is presumably due to catalytic formation of hydroxyl radicals from the reaction of chlorine and the metal (Candeias et al. 1994).

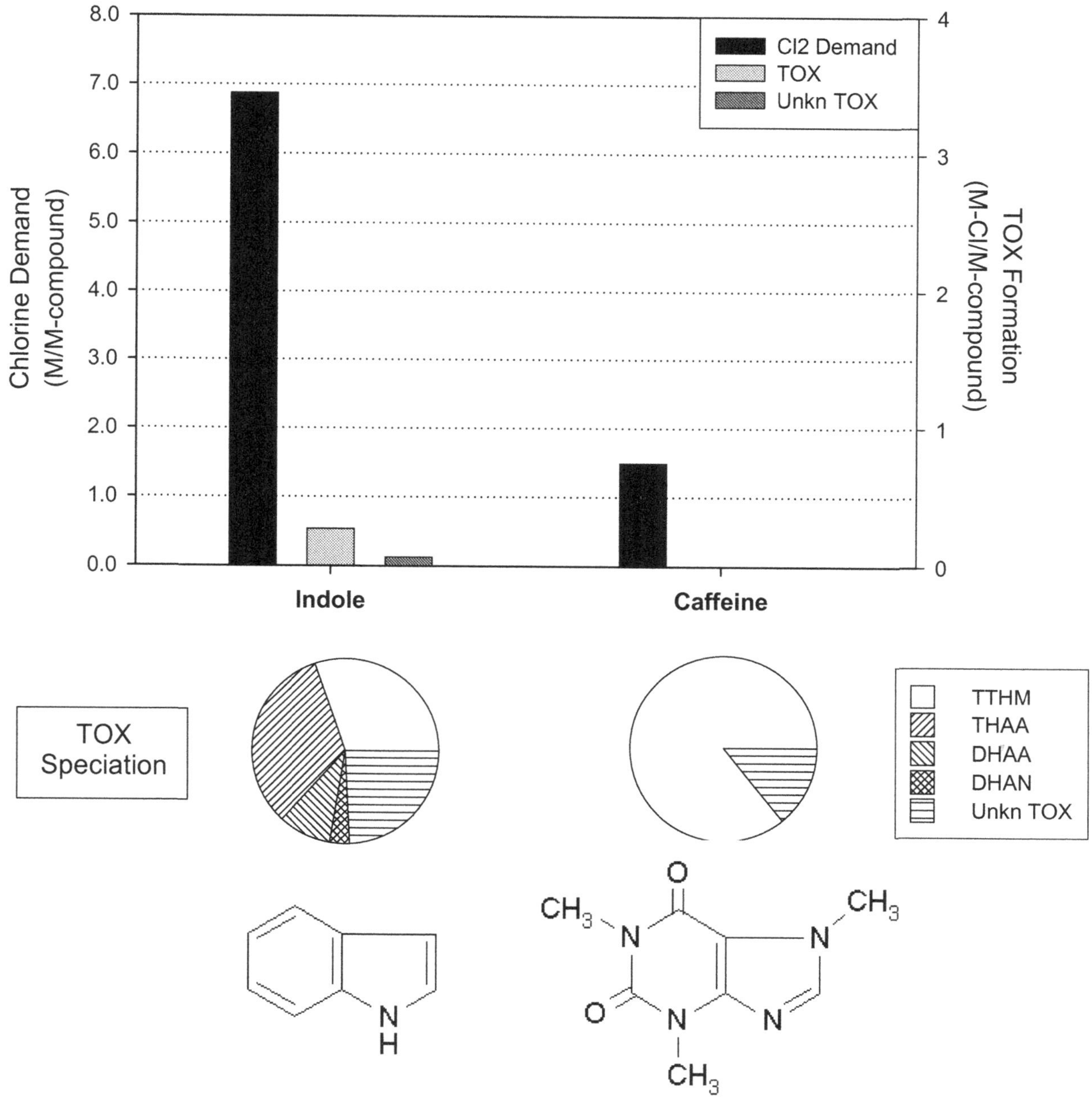

Figure 5.22 Reactivity of free chlorine with indole and caffeine

Indole. Despite the presence of only one heteroatom, indole is as reactive as any of the nucleic acid bases (Figure 5.22). A likely site of attack would be the pyrrole ring leading to 2-aminophenol or its N-substituted derivatives as a major product. It will be shown below that 2-aminophenol reacts with about 3.5 M/M of chlorine producing about 0.3 M/M of TOX, at least half of which is found in the form of THMs and THAAs and DHAAs. It's quite reasonable that destruction of the pyrrole ring would take about 3M/M of chlorine. Thus, the oxidative degradation of indole to 2-aminophenol is consistent with the chlorine degradation data.

Caffeine. The chlorination of caffeine has been studied by Struck and Elving (1965), Hoyano and coworkers (1973) and Gould and Hay (1982). Initial attack and chlorine substitution probably occurs at the 8 carbon, just as it does for guanine (Figure 5.23).

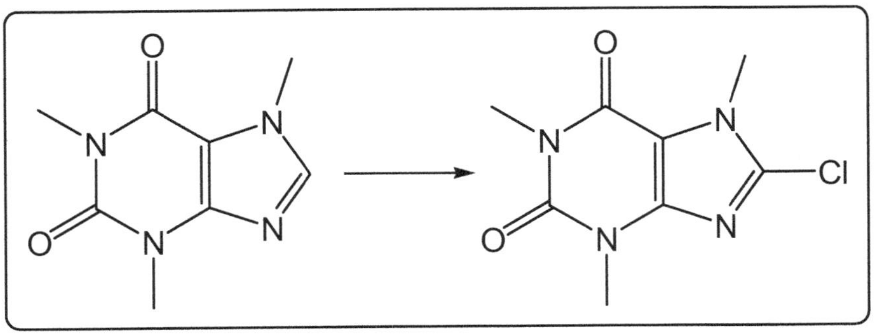

Figure 5.23 Initial attack of chlorine on guanine

Then, subsequent multi-chlorination and hydrolysis, followed by cleavage of the five-member ring, will lead to some intermediate products such as Alloxan and N, N-dimethyl parabanic acid (Figure 5.24).

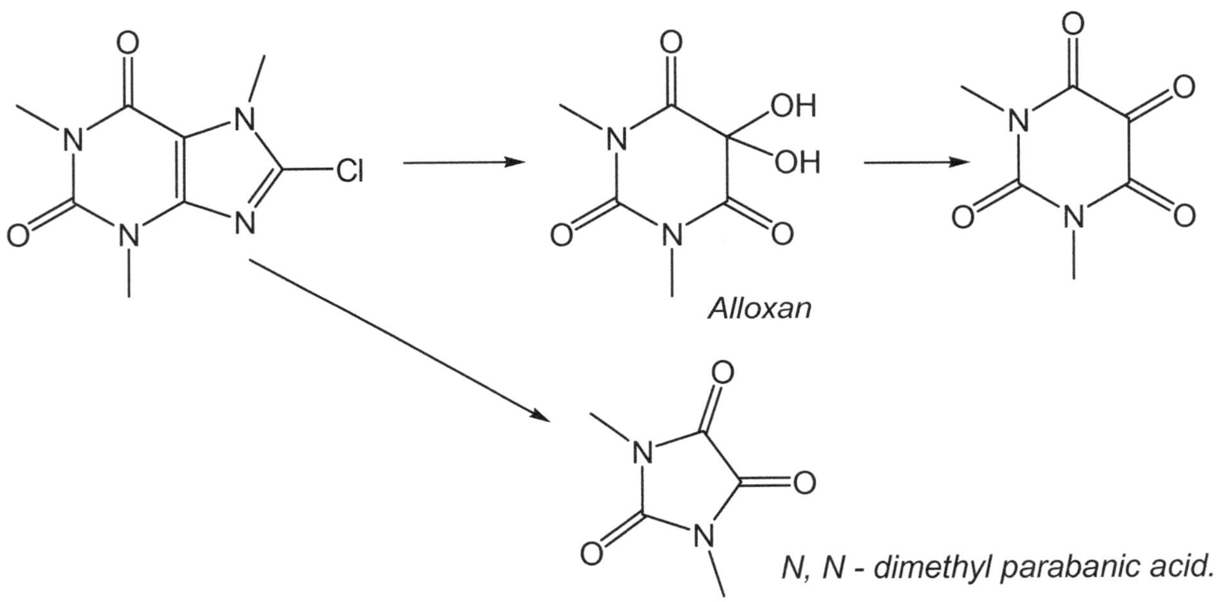

Figure 5.24 Ring rupture and fragmentation products from caffeine

With two tertiary ring nitrogens, the N, N-dimethyl parabanic acid is probably quite stable to further chlorination. Any further degradation is likely to involve the release of carbon dioxide (Figure 5.25).

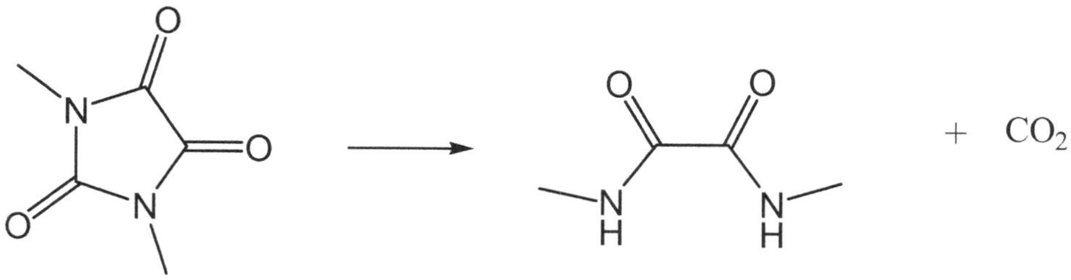

Figure 5.25 Possible further degradation of N,N-dimethyl parabanic acid with chlorination

There is a broad range of aromatic structures in NOM that are likely to be reactive with chlorine during drinking water treatment. Many of these are subunits within lignin, but the potential list includes many others. A few of these groups are listed below:

Coumarins
Lignins
Tannins
Simple phenols
Flavoniods, including Flavanones and Flavanonols
Other Phloroglucinol derivatives
Chromones
Quinones, especially anthraquinones
Catechins
Anthocyanins
Stilbenes

In this section some key activated aromatic structures will be examined, ending with a discussion of aromatic amines.

Monohydroxy Benzenes.

While simple monohydroxy benzenes (phenol and alkyl phenols) are found in many different natural plant products, they are quite often associated with degraded lignin structures of the type that derive from grasses and non-woody plants.

Aromatic substituents play a key role in chlorination reactions. Benzene is almost completely unreactive with chlorine under typical drinking water conditions. However, addition of a single electron donating group such as OH (i.e., phenol) results in a structure that is rapidly and extensively degraded by chlorine (Reckhow et al. 1990). The rate of reaction may be represented by a simple 2nd order rate constant, and the pH-dependent form of the reactants. Gallard and von Gunten (2002) concluded that only the reaction between HOCl and the phenolate ion forms are important under conditions typical of drinking water treatment. Using these data, along with data from Lee and Morris (1962), and similar dihydroxyphenol rate constants from Rebenne et al. (1996), they determined that these two groups of compounds show nearly the same Hammett slope of 3 (Figure 5.26; plotted versus the sum of the substituent constants, $\Sigma\sigma_{o,m,p}$).

The large negative slope in Figure 5.26 suggests a highly electrophilic reaction with aqueous chlorine. Initial attack is well established as a ring substitution at the 2, 4 and 6 positions. With simple phenol, 2,4,6-trichlorophenol is formed which undergoes further reaction and ring cleavage. Using Gallard and von Gunten's data, one can show that the decreasing rate constants going from the initial phenol, to the monochloro, dichloro and trichloro products are more than compensated by the increasing acidity of the halogenated products. This places more of the phenolic compound in the reactive phenolate form. Stated another way, as halogens and other electron withdrawing groups are added to the ring, electron density drops (for both phenolic and phenolate species), but the acidity increases shifting speciation toward the more reactive phenolate form. The end result is a series of products that are slightly more reactive than the starting compound. Half-lives based on a 1 mg/L chlorine residual and pH 7 range from about one half hour to less than 10 minutes. This suggests that systems using free chlorine as a secondary disinfectant will not contain halogenated phenolics, but rather their ring degradation products.

Table 5.10.
Rate constants for the initial chlorination of monohydroxybenzenes in aqueous media

#	Phenolic compound	pKa	k_5 (M^{-1} s^{-1})	$t_{1/2}$ (hr)*	Source
1	4-methyl (*p*-cresol)	10.26	2.71 (±0.49) × 10^4	0.9	Gallard & von Gunten, 2002
2	unsubstituted	9.99	2.19 (±0.08) × 10^4	0.6	Gallard & von Gunten, 2002
3	4-iodo	9.20	2.01 (±0.43) × 10^3	1.1	Gallard & von Gunten, 2002
4	4-chloro	9.43	2.17 (±0.33) × 10^3	1.7	Gallard & von Gunten, 2002
5	4-cyano	7.86	8.46 (±0.38) × 10^1	1.2	Gallard & von Gunten, 2002
6	unsubstituted	9.99	3.52 (±0.19) × 10^4	0.4	Lee & Morris, 1962
7	4-chloro	9.43	3.16 (±0.22) × 10^3	1.2	Lee & Morris, 1962
8	2-chloro	8.56	2.42 (±0.08) × 10^3	0.2	Lee & Morris, 1962
9	2,4-dichloro	7.85	3.03 (±0.09) × 10^2	0.3	Lee & Morris, 1962
10	2,6-dichloro	6.97	1.94 (±0.11) × 10^2	0.1	Lee & Morris, 1962
11	2,4,6-trichloro	6.15	12.84 (±0.69)	0.2	Lee & Morris, 1962

* Assuming about 1 mg/L chlorine residual at pH 7

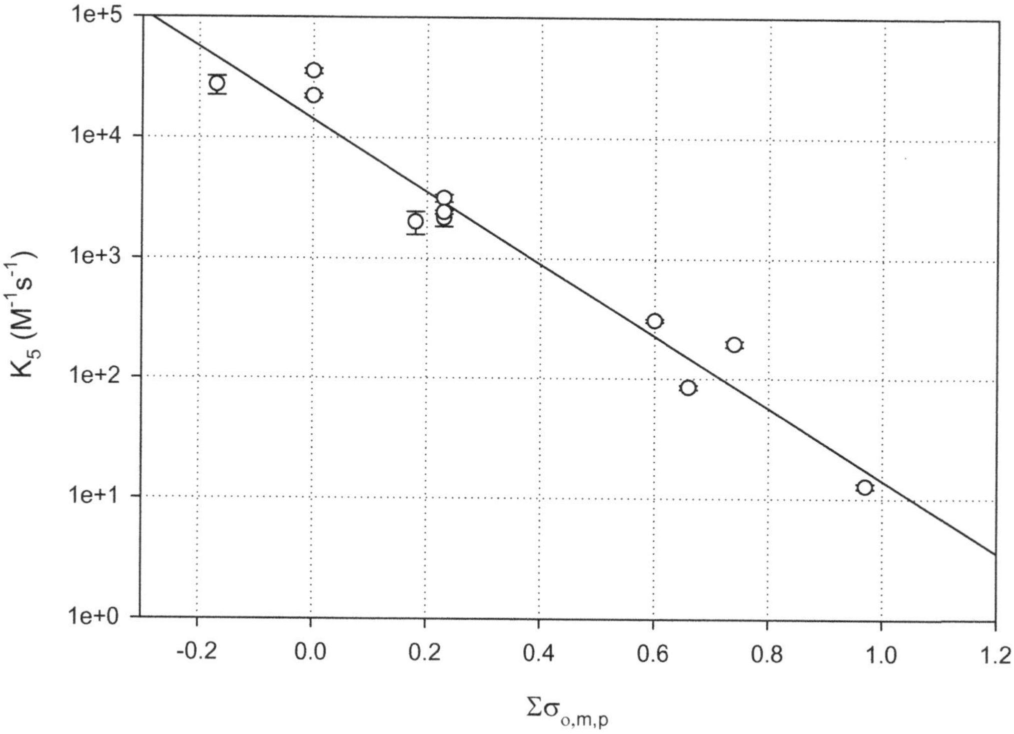

Source: Redrawn from Gaillard and von Gunten 2002

Figure 5.26 LFER plot for the reaction of HOCl with anionic forms of hydroxybenzene compounds. (2nd order rate constants; data from Table 5.10)

There is considerable uncertainty as to the products formed from 2,4,6-trichlorophenol. Studies using unsubsituted phenol have suggested attack at the 4 position, possibly forming a 2,4,4,6 tetrachlorophenol, 2,6 dichlorobenzoquinone (by hydrolysis) and then followed by ring cleavage (Smith et al. 1975). Subsequent products in the literature include 2-chloromaleic acid (Onodera et al. 1984)

Dihydroxy Benzenes.

Major reaction products from the various dihydroxy benzene isomers and derivatives are distinctly different, depending on whether the hydroxy groups are in the meta position (resorcinol and derivatives) or in the ortho or para positions. Despite their higher 2nd order rate constants (Figure 5.26), all undergo less complete degradation than their monohydroxy forms. Compounds with the meta configuration are unique in that they efficiently produce chloroform (Rook 1977, Larson and Rockwell 1979). The site of halogenation is, as expected, the α-carbon which persists in the haloform fragment (Boyce and Hornig 1983). This can occur even when the α-carbon is substituted with a carboxyl group, because this can easily be removed by oxidative decarboxylation (Larson and Rockwell 1979).

The importance of these types of structures in NOM has not been well studied. Many of the aromatic families in the bulleted list above contain members with meta-dihydroxy structures. It has been proposed that all meta-dihydroxy structures could be derived from 3 molecules of acetic acid (Birch 1967, 1968). The bioactive resorcinolic lipids have attracted recent attention

(Kozubek 1999). Laboratory studies have shown that aquatic NOM typically produces about 50 ug-TTHM/mg-C. Meta-dihydroxy structures produce THMs quantitatively, which makes them about 30 times as efficient a THM-generator as NOM on average. This means that only about 3% of the organic carbon in a natural water need be in the form of these structures to fully account for all THM formed.

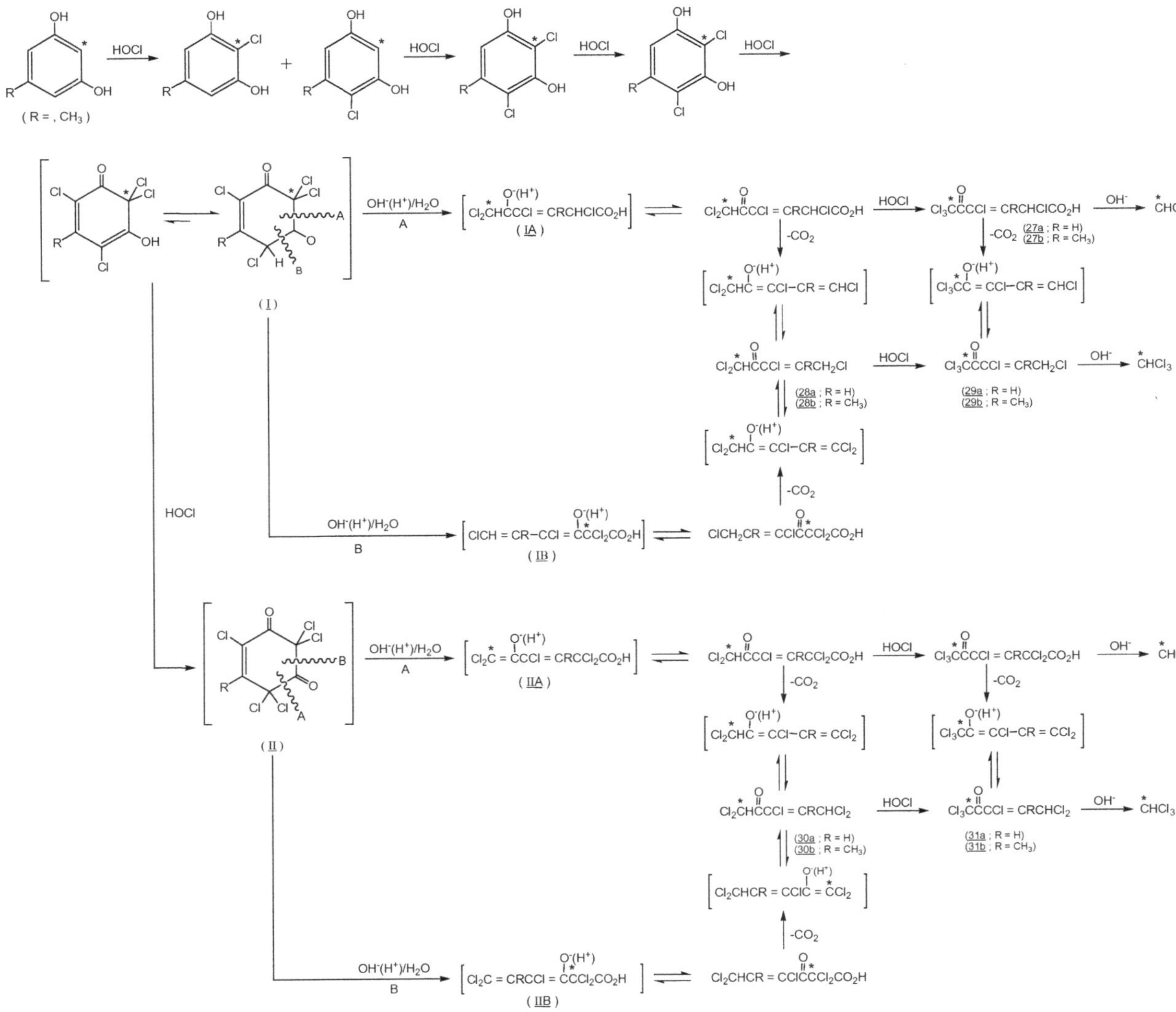

Source: Redrawn from Boyce and Hornig 1983

Figure 5.27 Degradation of resorcinol All structures were observed, except those in brackets.

Boyce and Hornig (1983) have proposed a set of parallel degradation pathways that account for the formation of several known products (Figure 5.27). However, laboratory studies have shown that fully 85% of the TOX is attributed to chloroform. As a result, the remaining organic fragments must be heavily oxidized and not populated with halogen atoms as indicated above. There is probably a substantial amount of hydrolysis and loss of organic halogen that is not considered in this mechanism. In addition, the observed formation of cyclopentene structures (Christman et al. 1985) as well as furanones and lactones (Tretyakova et al. 1994) is not considered.

Tretyakova and colleagues (1994) have tried to be more comprehensive in their proposed pathways (Figure 5.28). They were able to formulate plausible pathways to most of their major identified product at both low and high chlorine doses. However, based on empirical laboratory data, they still show too much substitution and too little oxidation for most meta-dihydroxy compounds.

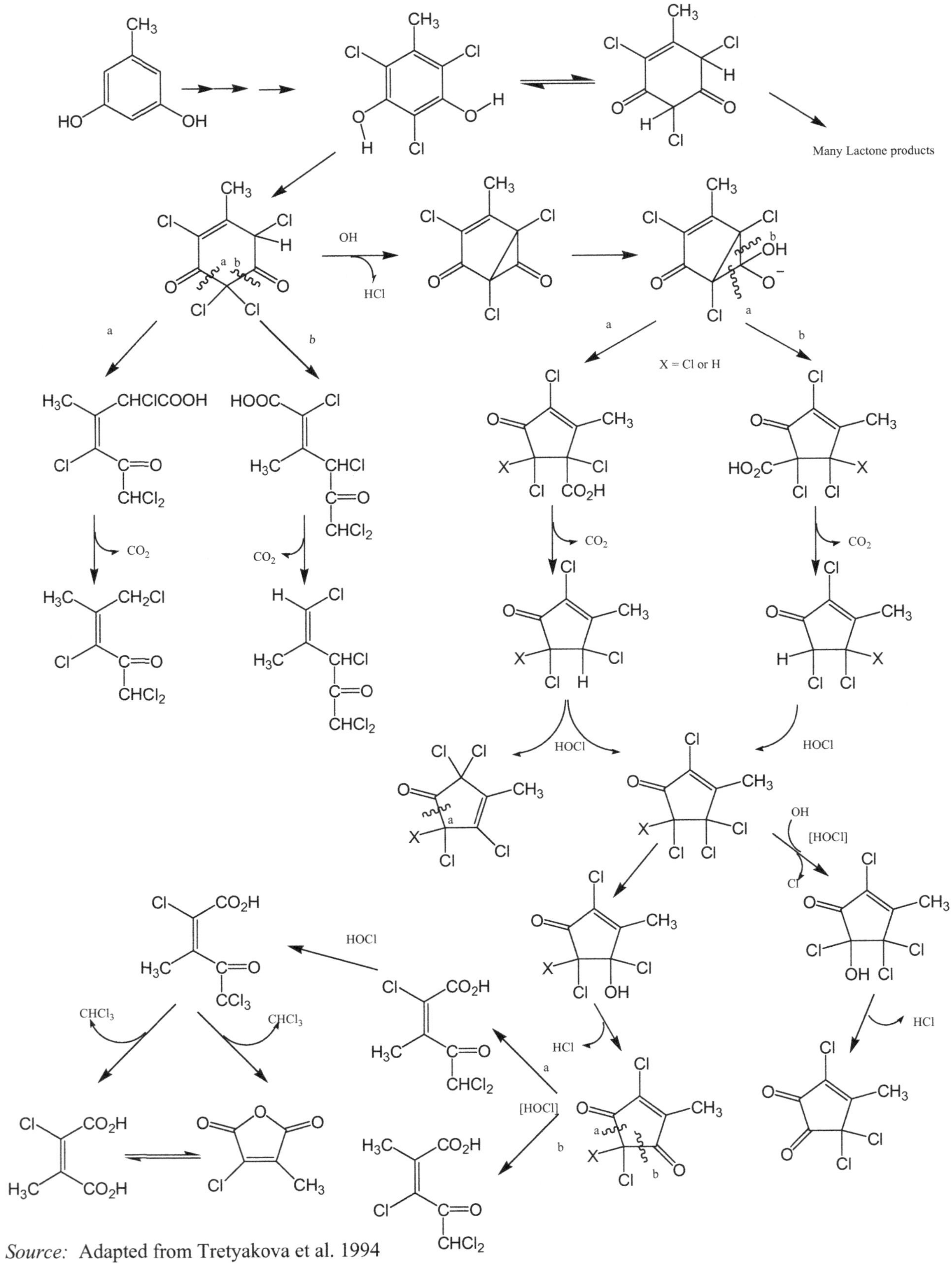

Source: Adapted from Tretyakova et al. 1994

Figure 5.28 Proposed degradation pathways from the chlorination of orcinol

The ortho and para dihydroxy-benzenes (catechol and hydroquinone, respectively) are also quite reactive but differ from resorcinol in the specifics of the pathways. Although they haven't been studied to the extent of resorcinol, there have been some laboratory tests on general reactivity. Both of these compounds will consume slightly less than 4 M/M of chlorine, ultimately forming small amounts of TOX. Thus the degradation results in largely oxidized products. One alternative reaction pathway for these compounds is through quinone formation followed by addition of HOCl across one or both double bonds. Figure 5.29 shows this for hydroquinone, but an analogous reaction could take place for catechol. If this occurs, there must be subsequent loss of HCl, giving largely oxygenated aliphatic products. It's also possible that more extensive ring chlorination could precede quinine formation as shown in Figure 5.30a for catechol. A logical subsequent degradation pathway is shown in Figure 5.30b. This follows roughly the scheme proposed by Zou et al. (2000). These authors were investigating a rather minor pathway leading to the potent mutagen, MX. While model compound data do not support this as the major pathway, it may explain some of the observed TOX formation.

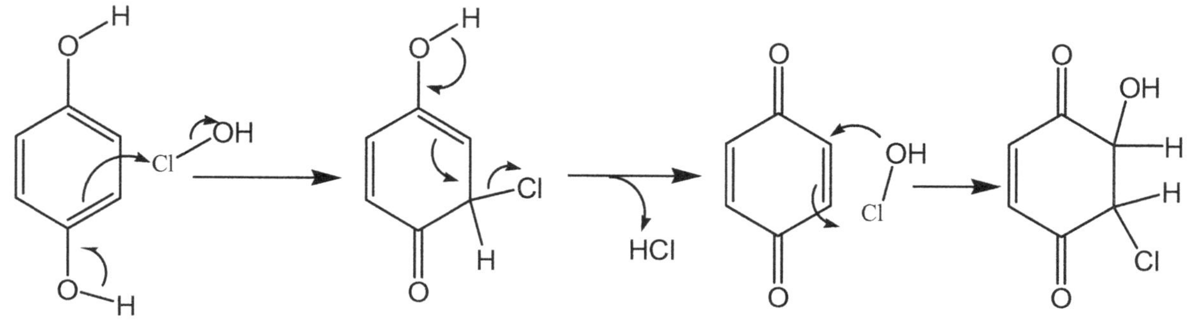

Figure 5.29 Possible early degradation of 1,4-Dihydroxybenzene

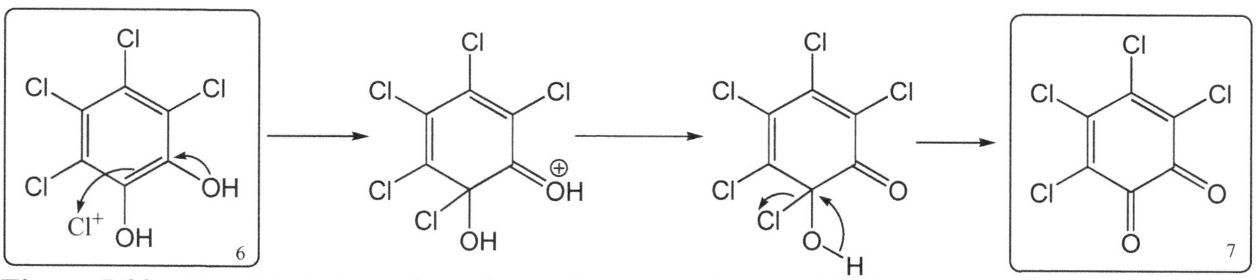

Figure 5.30a Possible halo-orthoquinone formation from 1,2-Dihydoxybenzene

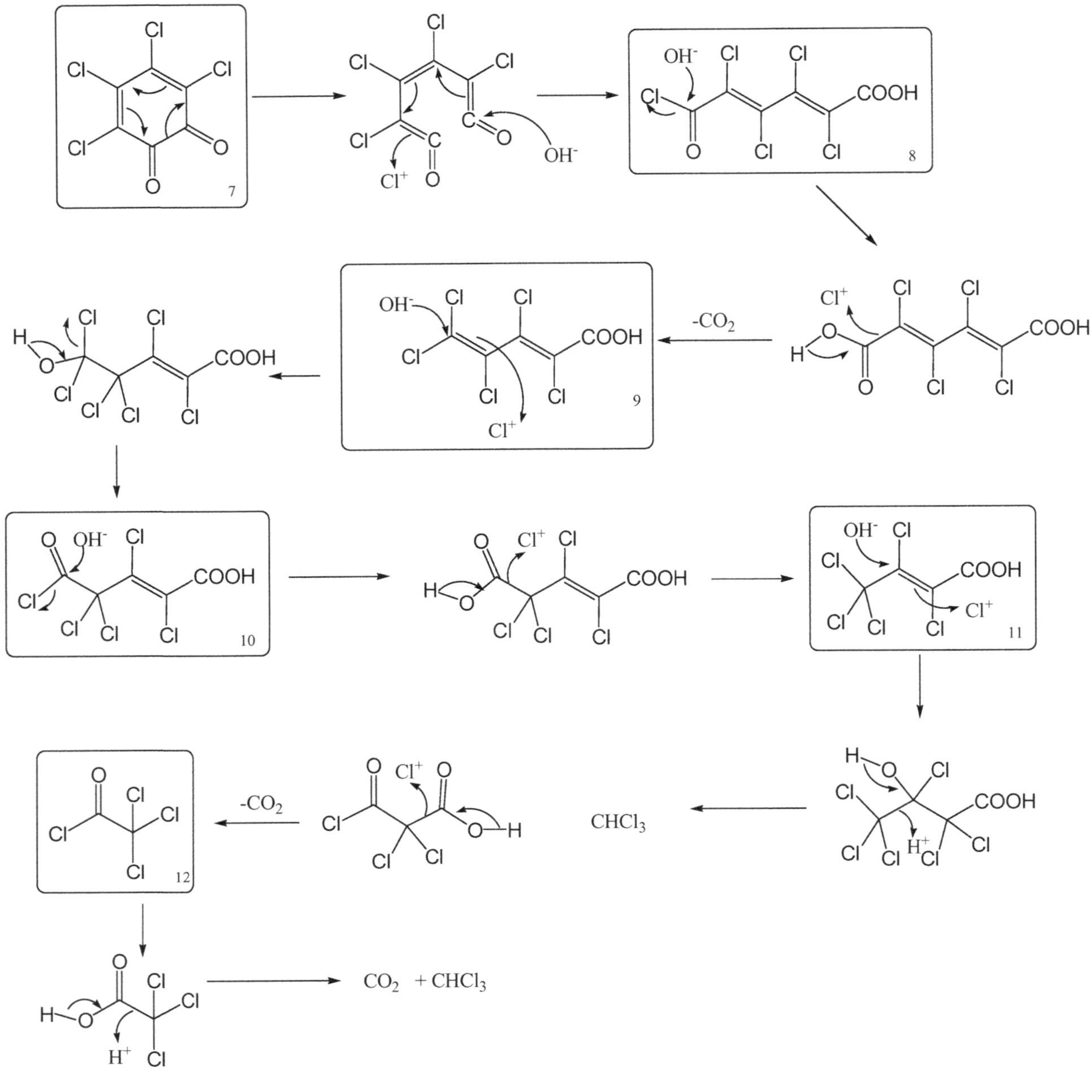

Figure 5.30b Subsequent degradation of tetrachloro-orthoquinone

p-Substituted Phenols

Like phenol itself, most substituted mono-hydroxy benzenes undergo extensive degradation as represented by chlorine reactivities of about 9 M/M. Complete oxidation of hydroxy benzoic acid to CO_2 and water would require 14 M/M of chlorine. These three compounds show a remarkably consistent TCAA yield of 0.26 to 0.27 M/M. Except for the hydroxyl acetophenone, about 10% of that (0.028) figure shows up as chloroform. In the case of the acetophenone an additional 0.075 M/M is produced, presumably from halogenation of the terminal activated methane. The remaining 0.24 to 0.43 M/M is present as unknown TOX.

Larson and Rockwell (1979) showed that p-hydroxybenzoic acid will undergo a facile decarboxylation during chlorination, ultimately giving the same reaction products as phenol itself. The initial reaction leads to the normal substitution products at the activated ring sites (Figure 5.31). Structures 1-5 in Figure 5.31 are all likely to persist in drinking waters, especially in underchlorinated waters (e.g., low chlorine dose, use of chloramines).

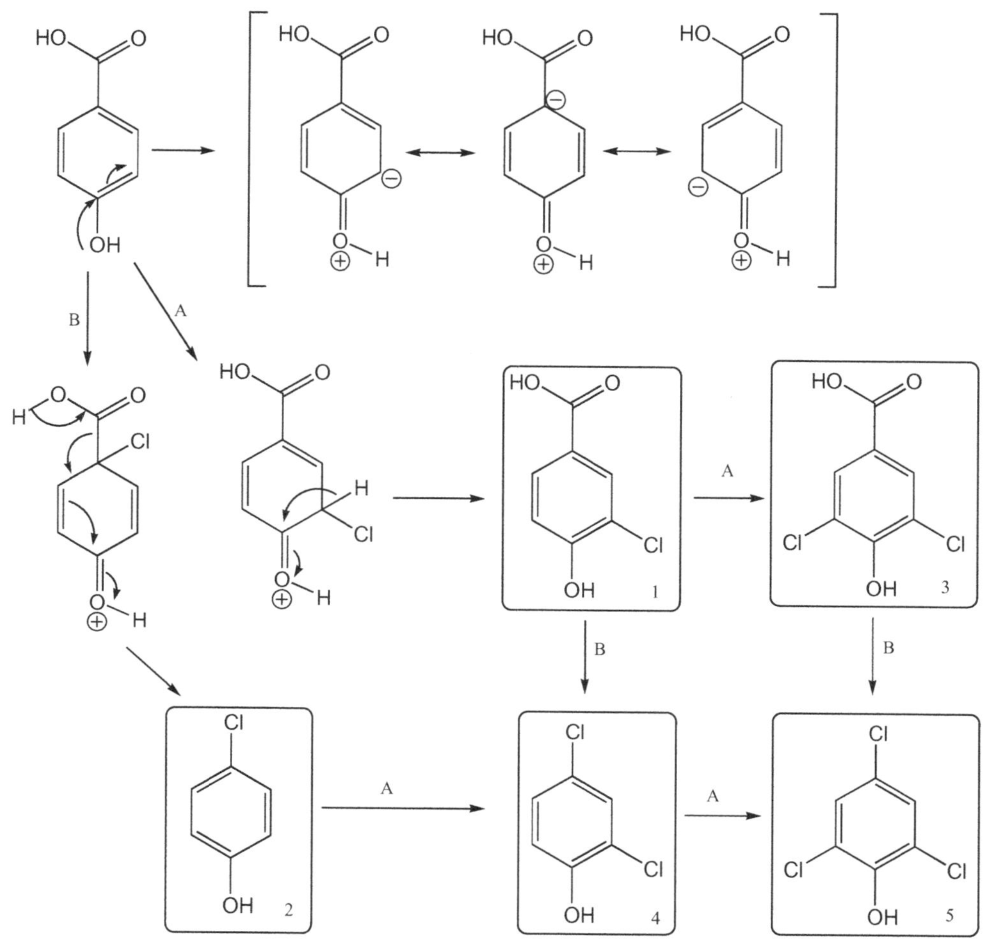

Source: Data from Larson and Rockwell 1979

Figure 5.31 Chlorination of p-hydroxybenzoic acid "A" represents electrophilic aromatic substitution, "B" is oxidative decarboxylation.

Then, addition of HOCl across the double bonds causes the loss of some aromaticity. Once the aromatic stabilization has been sufficiently weakened, the six-member ring will be broken. An oxidative rupture followed by HOCl addition, HCl elimination, and another HOCl addition produces a structure that is quite susceptible to oxidative fragmentation. The end result is TCAA and a chlorinated butenoic acid. A key feature of this mechanism is the facile oxidative fragmentation which should predominate over hydrolysis of the CCl_3 group. Aside from the trichlorophenol (#5), the two dienes (#6 & #7) could likely persist in drinking waters (Figure 5.32).

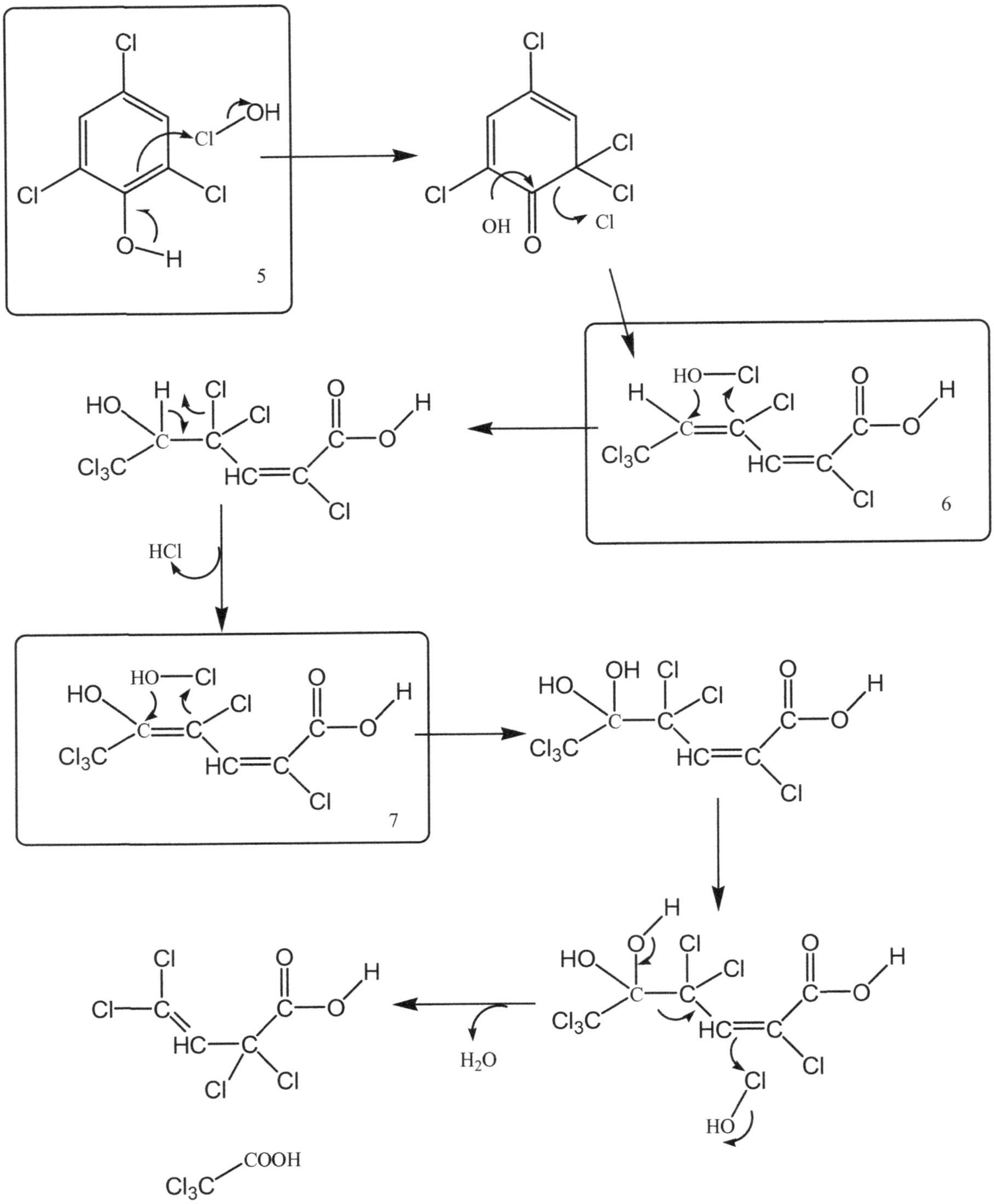

Figure 5.32 Formation of TCAA and other products from the trichlorophenol intermediate

Almost all reactive aromatics produce TCAA in substantial excess over chloroform. The well-known exceptions are the metadihydroxy benzenes (resorcinols). One possible reason for this difference in pathway can be linked to the site of ring cleavage. Inferring from their observed degradation products, Boyce and Hornig (1983) stated that the polychlorinated cyclohexene cleaves at locations a and b (Figure 5.33). Based on their pathways, this produces products that cannot undergo the same facile oxidative fragmentation proposed above.

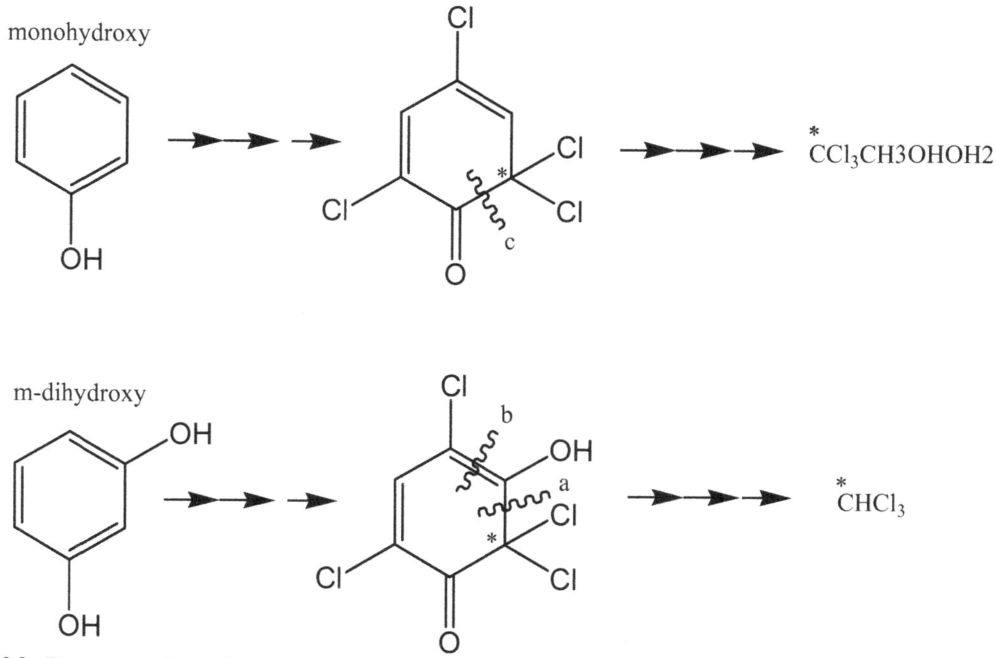

Figure 5.33 Proposed early departure between monohydroxy and dihydroxy benzene degradation pathways

Nevertheless, it should be recognized that a small amount of chloroform is produced from nearly all aromatic molecules and 4-hydroxybenzoic acid is no exception. One possible pathway might be through formation of a resorcinol-type structure (Figure 5.34).

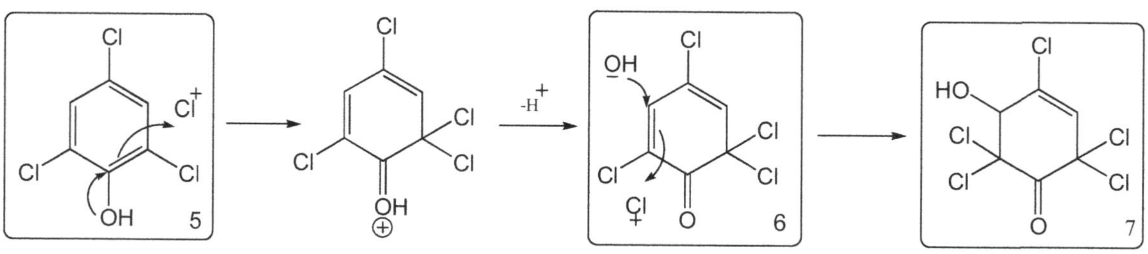

Figure 5.34 Formation of a resorcinol degradation intermediate from the monohydroxybenezene degradation pathway

Compound #7 continues to react by elimination of HCl (Figure 5.35). Aside from chloroform, a series of halogenated acids and diacids are formed through stepwise additions and decarboxylations.

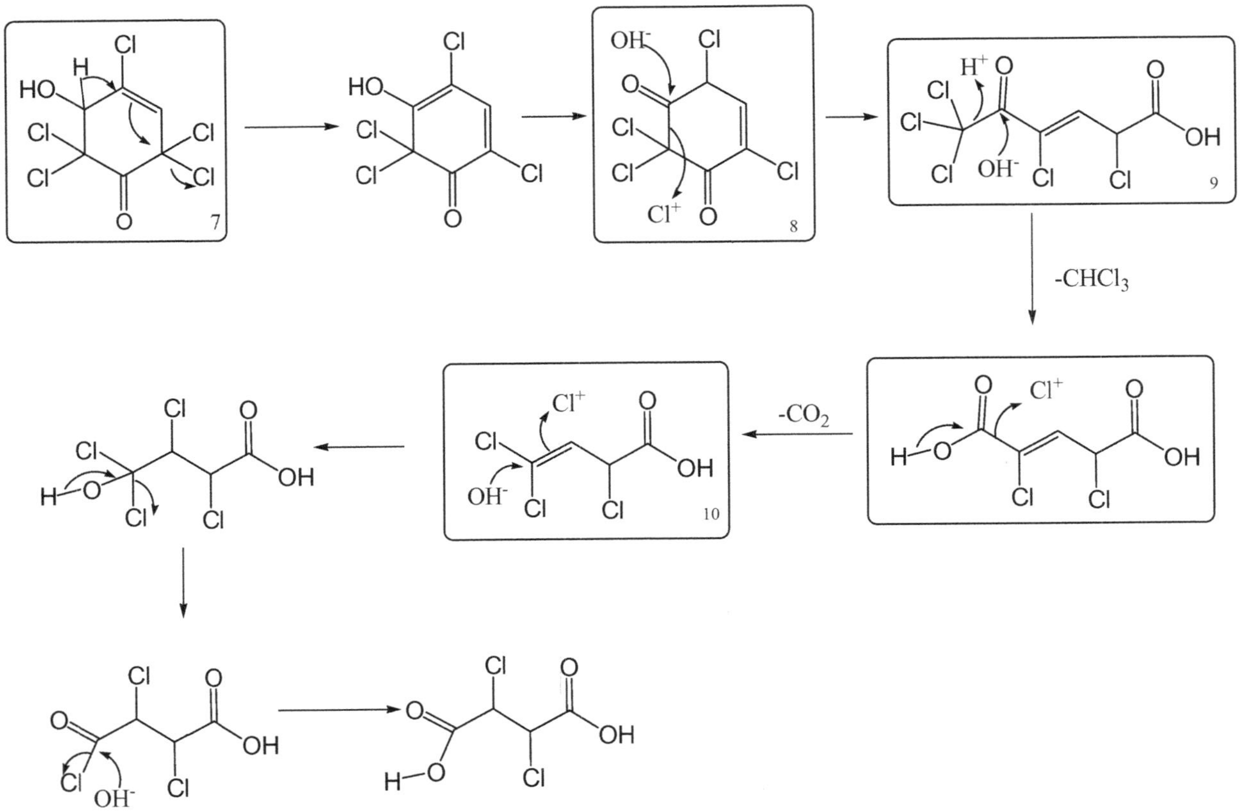

Figure 5.35 Resorcinol degradation pathway: scheme #1

Further chlorination followed by ring cleavage at another site could take place and the products would be slightly different (Figure 5.36).

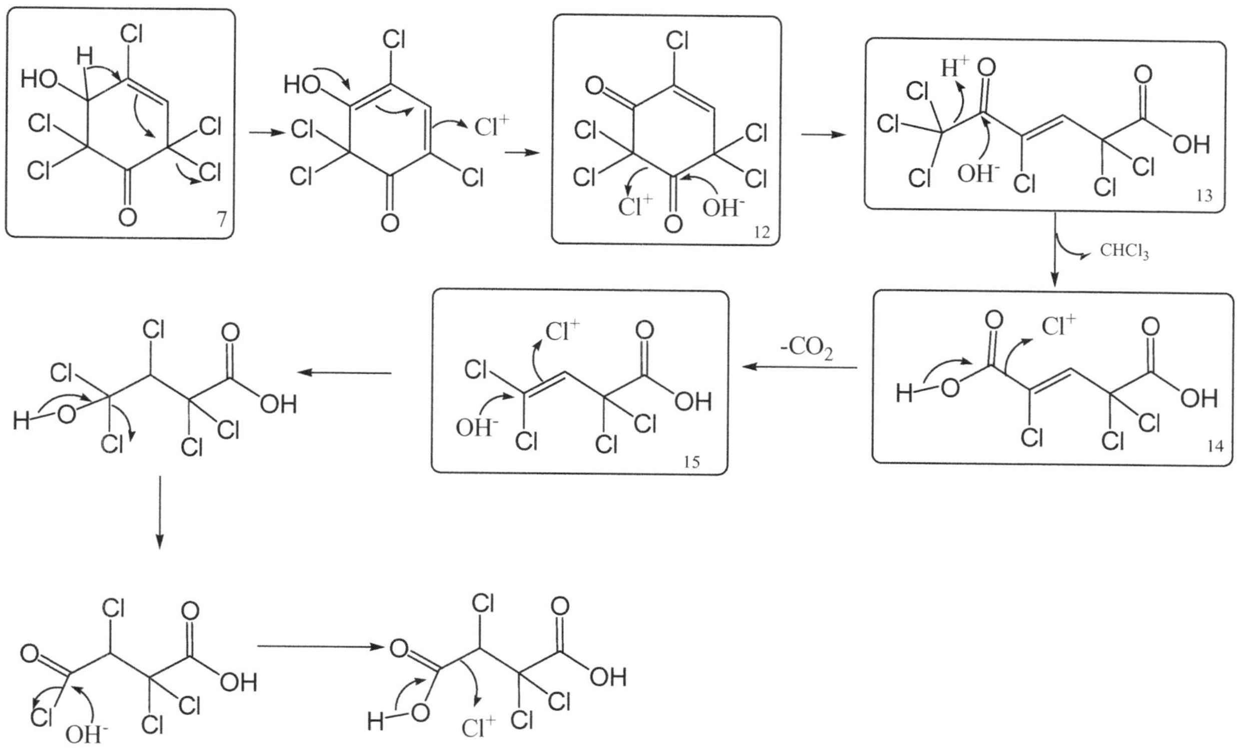

Figure 5.36 Resorcinol degradation pathway: scheme #2

As noted by many, it is also possible to form some five-member ring intermediates. These structures, including both cyclopentenes and furans, have been identified and noted in the literature (Christman et al. 1983, Tretyakova at al. 1994). Entry in to the pathway below (Figure 5.37) could be via compound #5 (starting with a monhydroxy benzene; probably a minor pathway) or compound #7 (starting with a resorcinol structure; possibly a major pathway). This pathway is likely in competition with the open ring pathways (above) that give rise to chloroform early on.

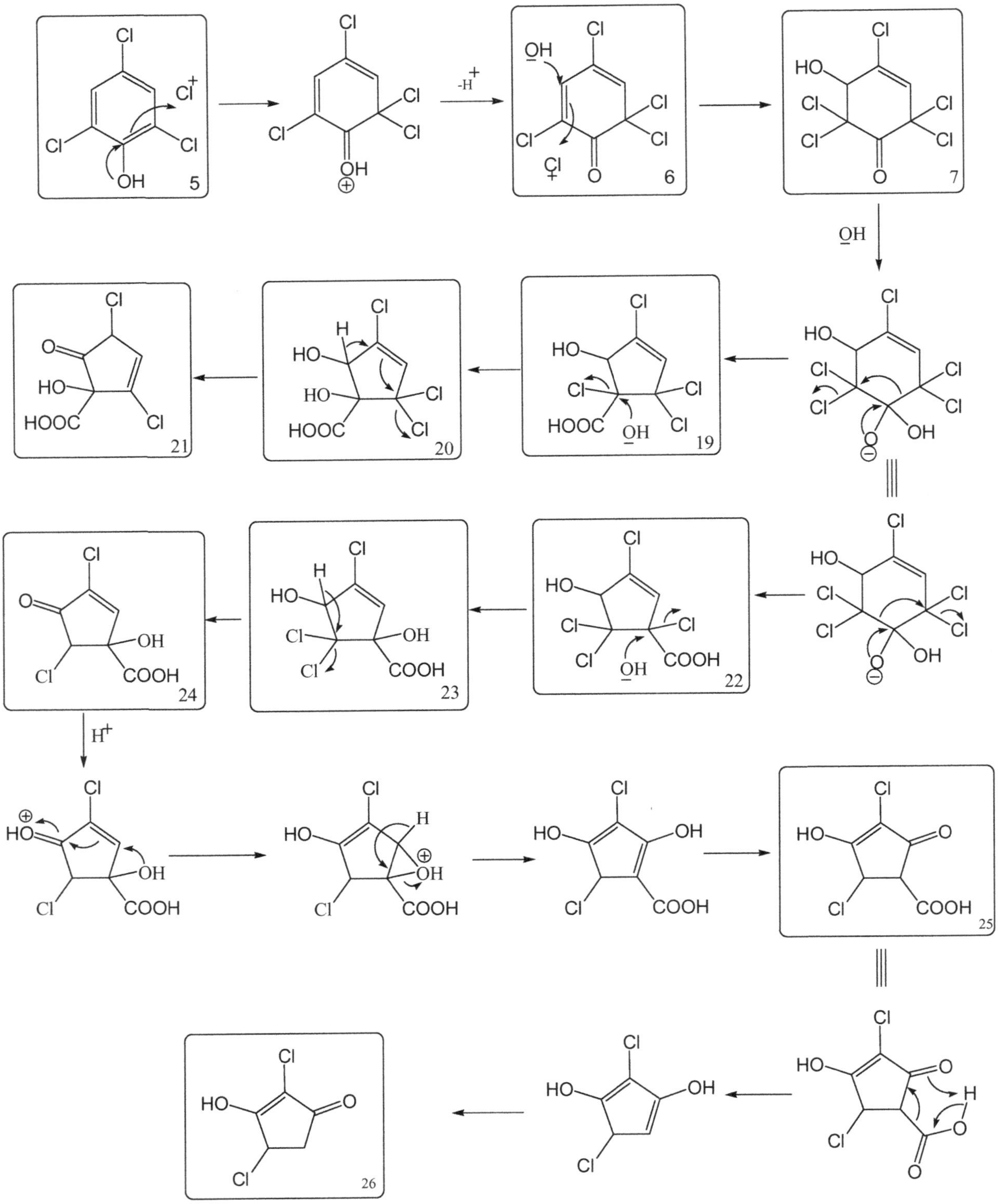

Source: based on Moye and Sternhel, 1966

Figure 5.37 Formation of cyclopentene products by chlorination of hydroxyl aromatics

Some quantitative data from laboratory studies using compounds of this type may be found in the literature (Larson and Rockwell 1979, De Laat 1981, Boyce and Hornig 1983, Reckhow and Singer 1985, Gallard and von Gunten 2002, Hartmann and Reckhow 2005) as well as from new studies conducted as part of this work (Figure 5.38). In general, these studies are in agreement and present a more comprehensive picture than was previously available. Both simple phenol and all three 4-hydroxy derivatives consume 9-10 M/M of chlorine, and produce a 27-36% molar yield of TCAA, with traces of DCAA. All produce about a 3% yield of chloroform, except the hydroxyacetophenone which produces a 10% yield. In addition, there is 0.24-0.43 M/M of unknown TOX formed, depending on the particular compound. This means that there is about a 15% incorporation and therefore 85% of the chlorine goes to oxidation reactions.

This information strongly supports a common set of pathways for these four related compounds. Larson and Rockwell (1979) showed that 4-hydroxbenzoic acid undergoes decarboxylation during chlorination to form the same trichlorophenol intermediate that is formed during chlorination of phenol. Thus, if both structures react quickly to form the same intermediate with few competing pathways, the distribution of products should be quite similar. Based on the model compound work, this appears to be the case. Furthermore, the other derivatives, 4-hydroxybenzaldehyde and 4-hydroxyacetophenone also appear to have the same product yields. In the case of the aldehyde, this is likely to occur as a result of oxidation to the acid, and decarboxylation. In the case of the acetophenone, one can imagine electrophilic attack by HOCl at the 4 position, leading to a chloro substituted product and acetic acid. Before this could occur, some formation of chloroform would occur (~7% yield) via a haloform type reaction on the terminal methyl group.

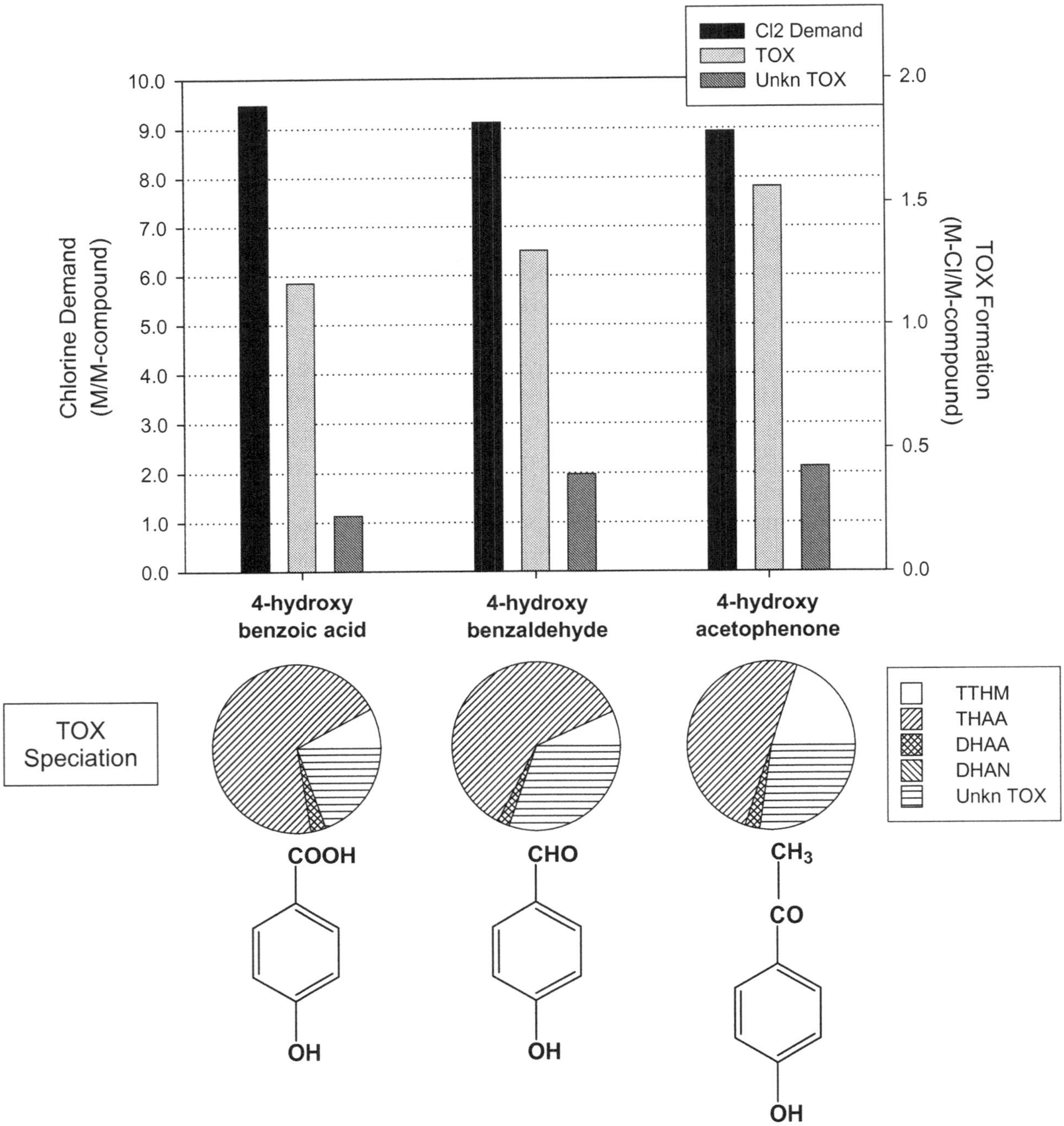

Figure 5.38 Reaction of free chlorine with 4-hydroxy phenols

Vanillyl Compounds

Vanillyl structures commonly found in nature include vanillic acid, vannillin and acetvanillone. These represent the o-methoxy substituted derivatives of the three p-substituted phenols. They are generally present in lignin from all types of vascular plants, including gymnosperms as well as angiosperms. These methoxy derivatives will not have the same extent of ring activation as the free phenols. For this reason, reaction rates will be much slower, and perhaps reduced to insignificance. The simplest illustration of the deactivating effects of methoxy group comes from comparison of the highly reactive phenol (9.5-9.8 M/M; De Laat 1981, Reckhow and Singer, 1985, Gallard and von Gunten, 2002) with the minimally reactive anisole (0.6-1.0 M/M; De Laat 1981, Reckhow and Singer, 1985). In the case of anisole, some TOX is formed (~0.35 M/M), but no THMs or HAAs, which suggests some ring halogenation, but very little ring rupture. The small amount of TCAA that is formed suggests ring cleavage products at a yield of about 2-3% under the conditions used (data from Reckhow and Singer, 1985).

An example mechanism for methoxy aromatics is illustrated below for the case of 4-methoxybenzoic acid. By analogy to 4-hydroxybenzoic acid, this compound would be expected to produce ring halogenation products, with some decarboxylation (Figure 5.39):

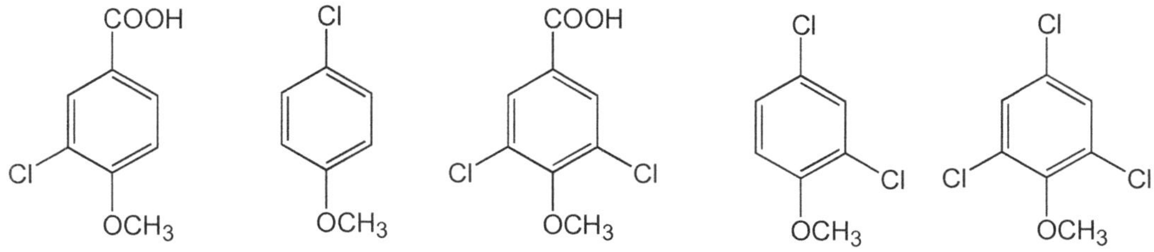

Figure 5.39 Expected products from chlorination of 4-methoxy benzoic acid

Then the methoxy group could be protonated and replaced by water, in what is probably a very slow reaction (Figure 5.40). To the extent this occurs, it would rapidly lead to ring cleavage products much as one would expect for the chlorination of phenol. The low reactivity of anisole suggests that this reaction is not significant under conditions typical of drinking water.

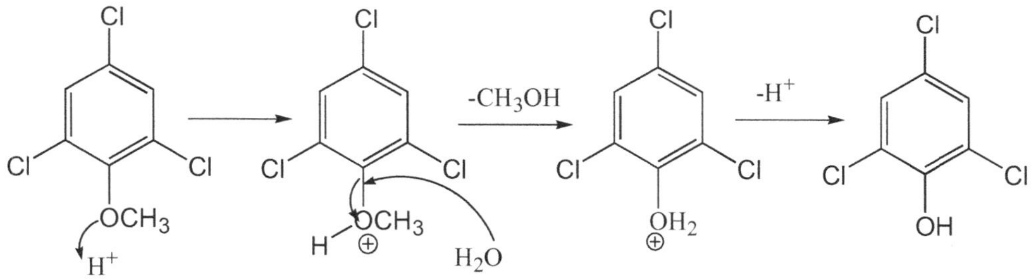

Figure 5.40 Possible hydrolysis of trichloranisole.

Vanillic acid. With vanillic acid, one would expect the same set of halogenation and decarboxylation reactions as noted for 4-hydroxybenzoic acid. It's also possible that the methoxy group could be hydrolyzed to another hydroxyl group, giving the molecule a catechol type structure. However, model compound data show major reactivity and yields for vanillic acid (Figure 5.41) resemble 4-hydroxybenzoic acid (Figure 5.38), and not catechol (Reckhow and Singer 1985). De Laat (1981) showed that while the methoxy phenols are extremely reactive with chlorine, their reactivity and products do not mirror the analogous dihydroxy benzenes. All this suggests that simple hydrolysis of the methoxy group is not likely to occur along a major pathway.

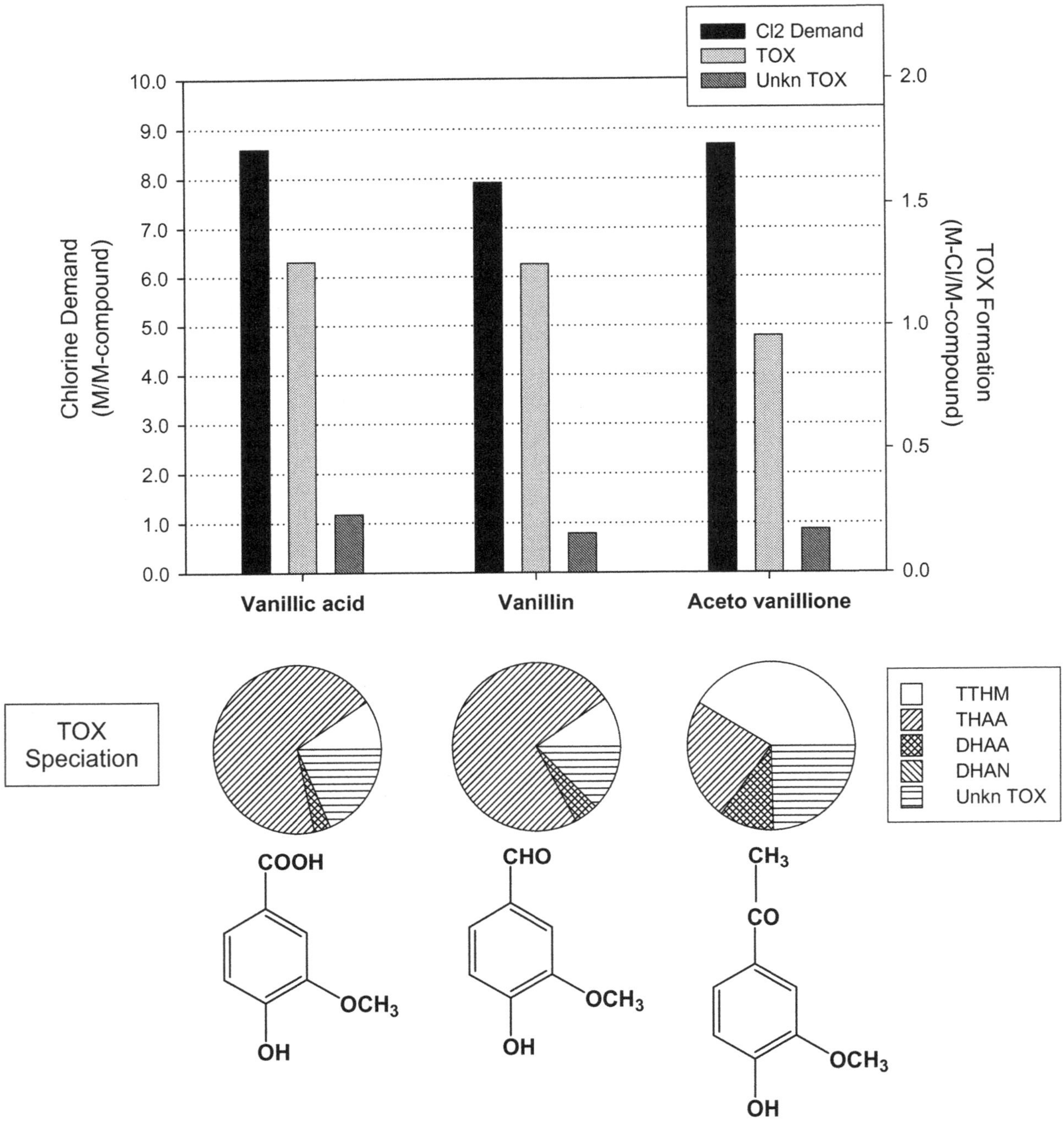

Figure 5.41 Reaction of free chlorine with vanillin derivatives

Chlorine reactions with vanillic acid are probably preceded by ring halogenation (Figure 5.42) leading to compounds 1-3, but little or no 4-6.

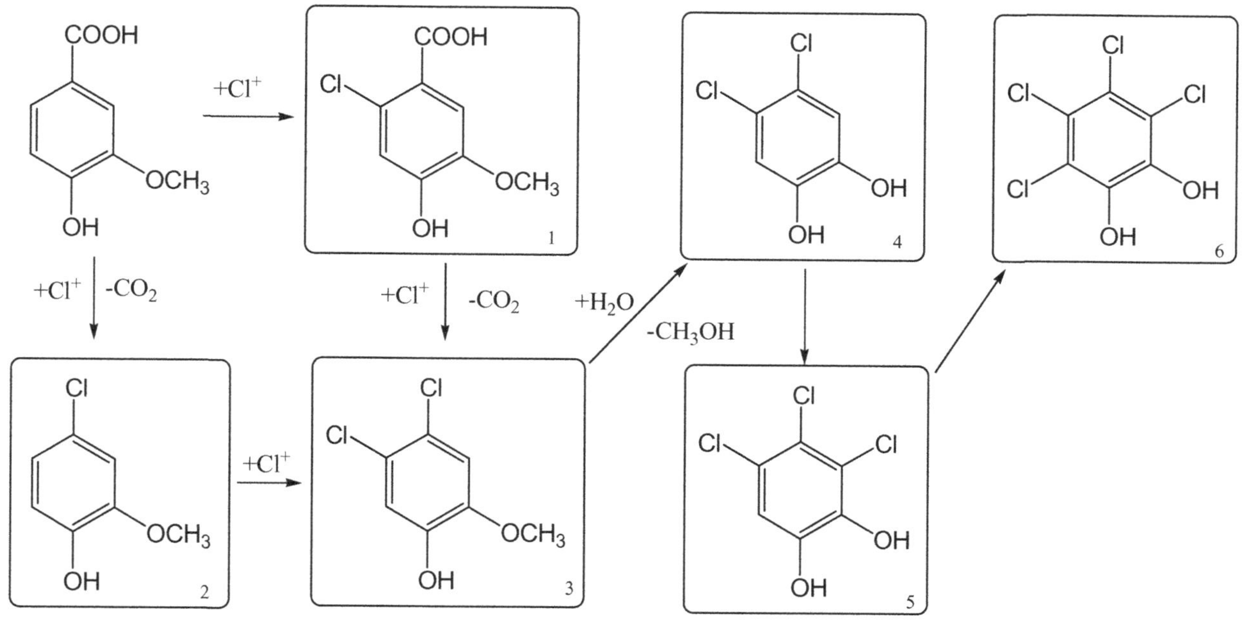

Source: Based on Sarkanen and Dence 1960

Figure 5.42 Initial degradation of vanillic acid

Subsequent degradation would be analogous to that proposed for p-hydroxybenzoic acid. Intermediate products would include a series of unsaturated methoxy acids (Figure 5.43).

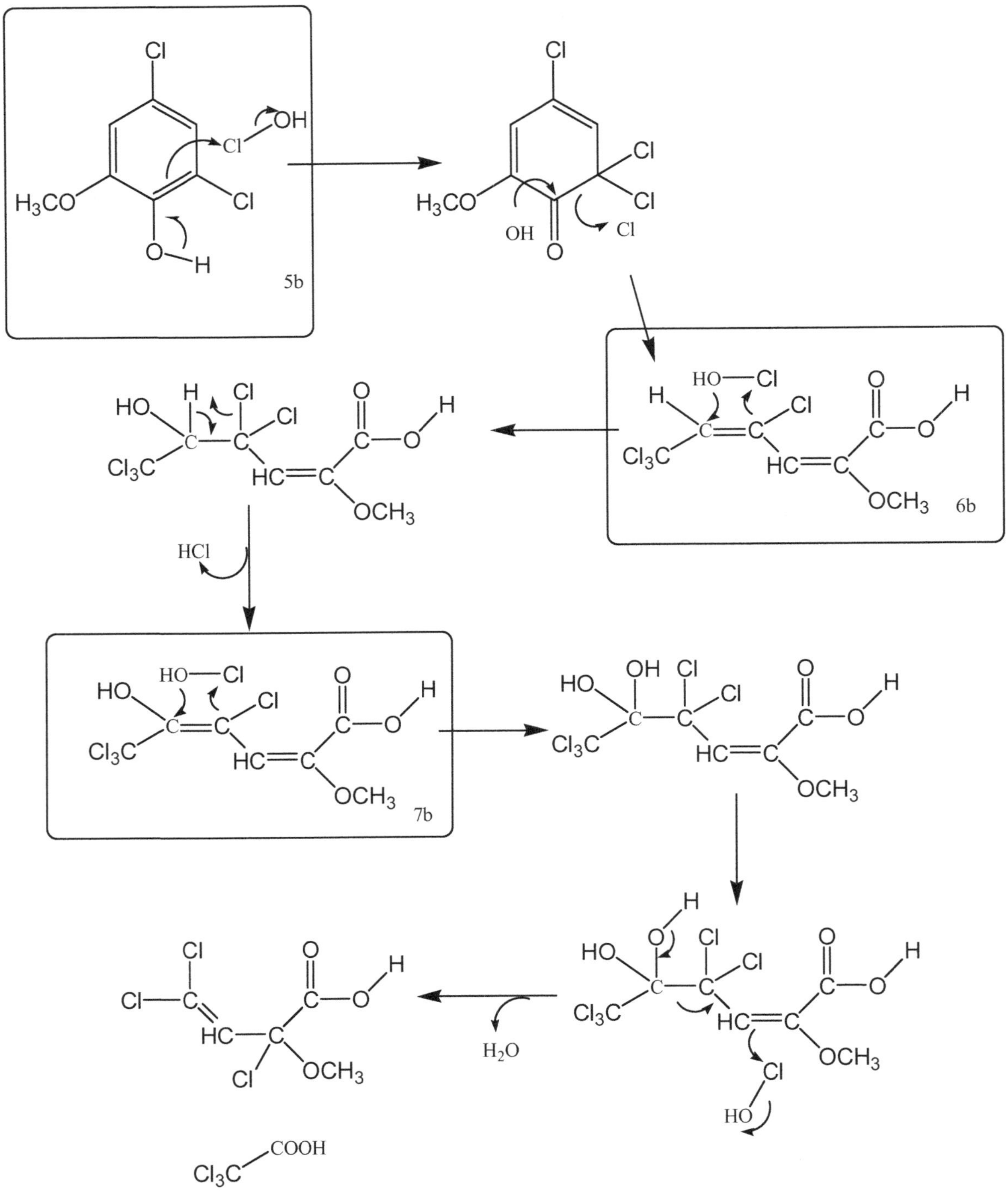

Figure 5.43 Possible ring rupture products of vanillic acid

Coniferyl Alcohol

Coniferyl alcohol is a well-known monomer of lignin. It has also been widely identified in aquatic humic substances by means of the CuO degradation method of Hedges and Ertel (1982). This compound contains an activated aromatic ring (strongly electron-donating OH group), as well as an unsaturated alkyl chain and an easily oxidized alcohol functionality. All three of these offer prime opportunities for reaction with chlorine, and the molecule is large and sufficiently multi-functional to have the potential for production of some previously unidentified DBPs.

Careful examination of this compound and the appropriate organic chemistry literature has led to the following postulated degradation pathway (Figure 5.44).

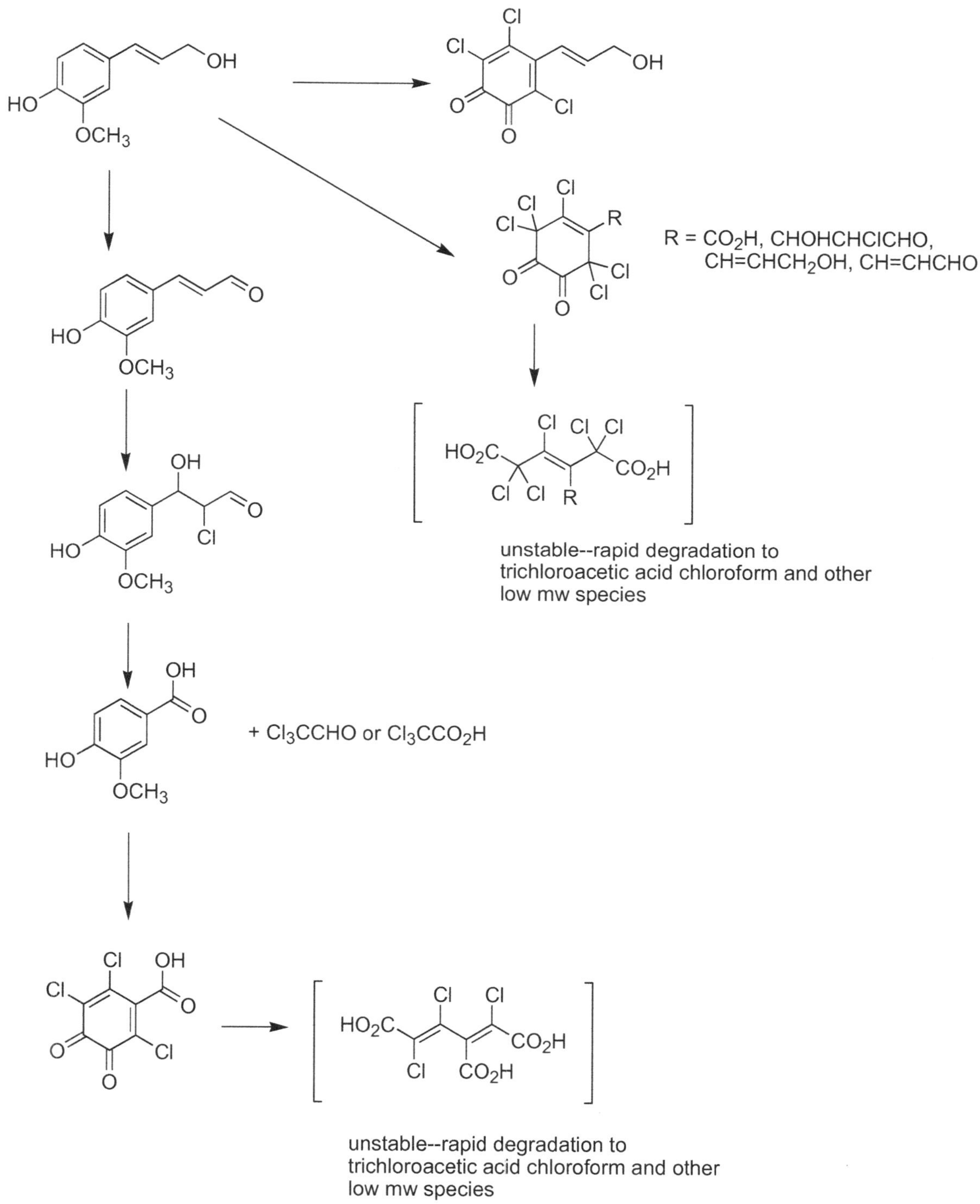

Figure 5.44 Proposed degradation pathway for coniferyl alcohol

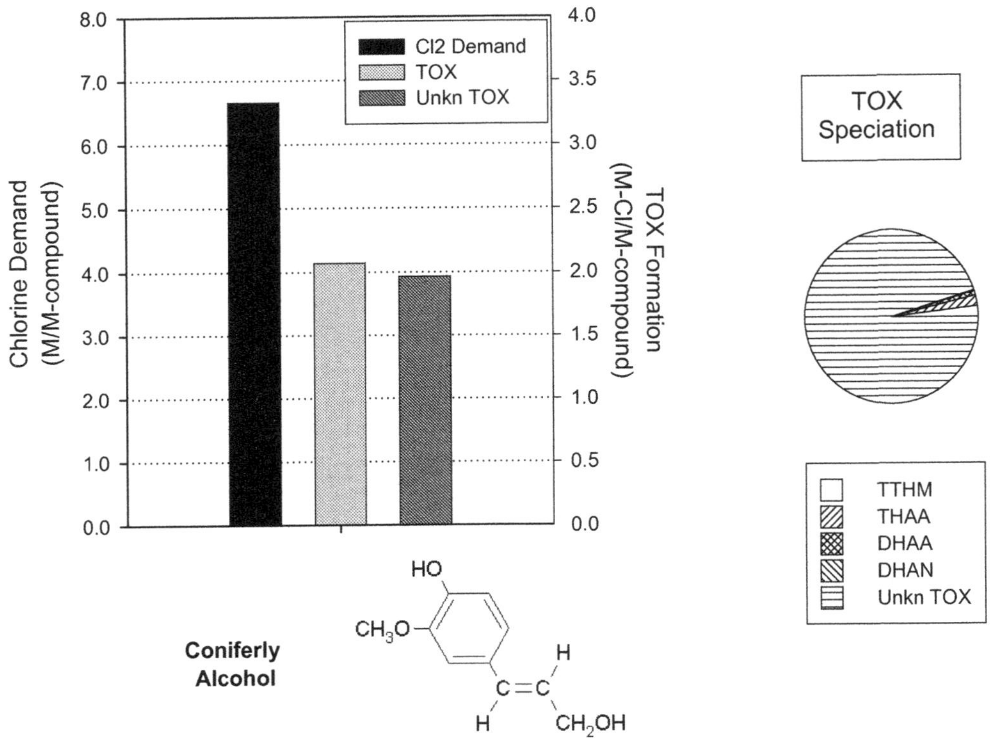

Figure 5.45 Reaction of free chlorine with coniferyl alcohol

Laboratory chlorination experiments showed that coniferyl alcohol is indeed quite reactive with chlorine under conditions appropriate to drinking water treatment. The average demand was 6.6 moles of chlorine per mole of compound (Figure 5.45). Of this, nearly one third resulted in organic chlorine bonds. The remaining reaction presumably resulted in chloride and oxidized organic matter. Despite the large amount of TOX (organically bound chlorine) there was very little of the regulated by-products. Combined, the THMs and the HAAs only amounted to 1.8% of the TOX. Therefore, 98.2% of the TOX was in the form of non-regulated, or "unknown" compounds.

Syringyl Compounds

Syringyl compounds are characteristic of angiosperm-derived lignins. These structures are very common in nature, and have been observed in aquatic NOM using CuO degradation methods. The three compounds (below) represent the same structures as the vanillyl group, except for the addition of a second methoxy substituent. This seems to have a measurable, but poorly understood effect on reactivity and by-product formation. The acid form (syringic acid) shows a lower level of reactivity (6.9 compared to 8.6 M/M) and substantially less TOX formation (0.70 compared to 1.26 M/M) as compared to the other two (Figure 5.46). All of that difference could be attributed to differences in TCAA formation (11% versus 30% yield). In contrast the aldehyde (syringaldehyde) exhibited higher TOX formation (2.07 vs 1.25 M/M) and

TCAA (48% vs 30% yield). The methyl ketones show greater reactivity in the vanillyl form, yet more TOX as the syringyl structure. This compound (syringone) also produces very large amounts of trichloropropanone (25% yield vs. 8%).

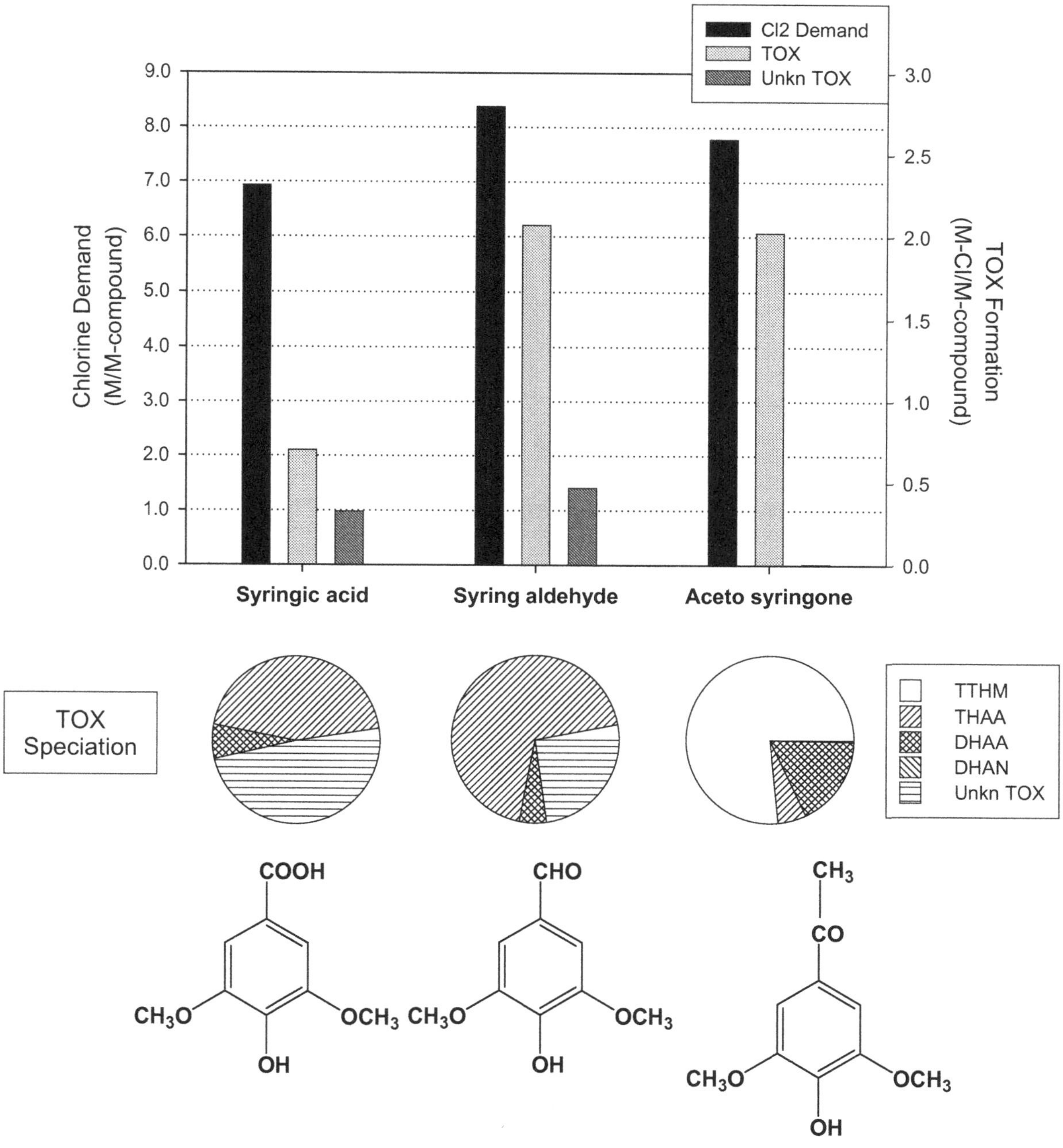

Figure 5.46 Reaction of free chlorine with syringyl derivatives.

Syringic acid. Chlorination of syringic acid likely proceeds via electrophilic aromatic substitution on the two available ring carbons, followed by decarboxylation (Figure 5.47 and Figure 5.48). This is analogous to the chlorination of 4-hydroxybenzoic acid.

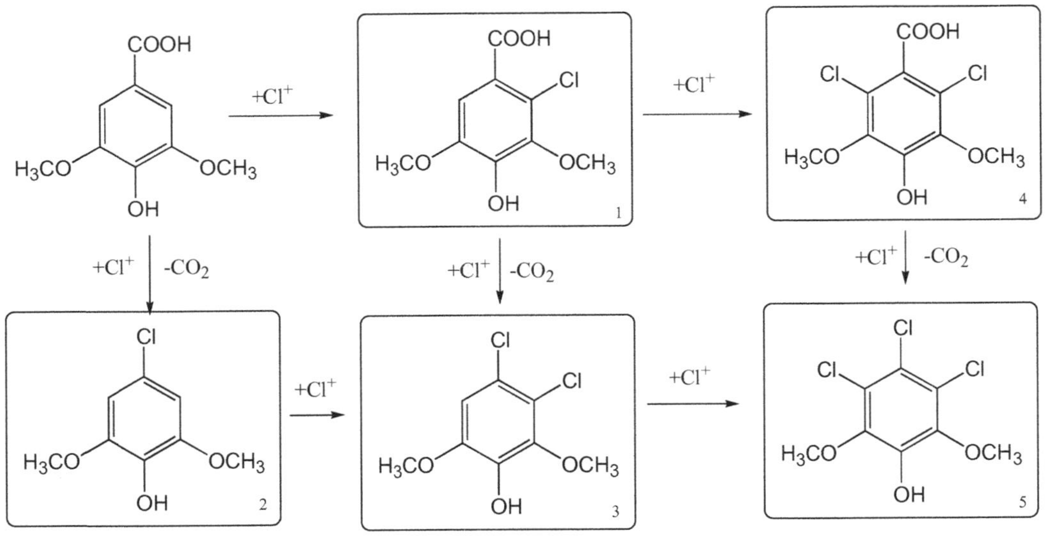

Figure 5.47 Proposed early pathway for syringic acid

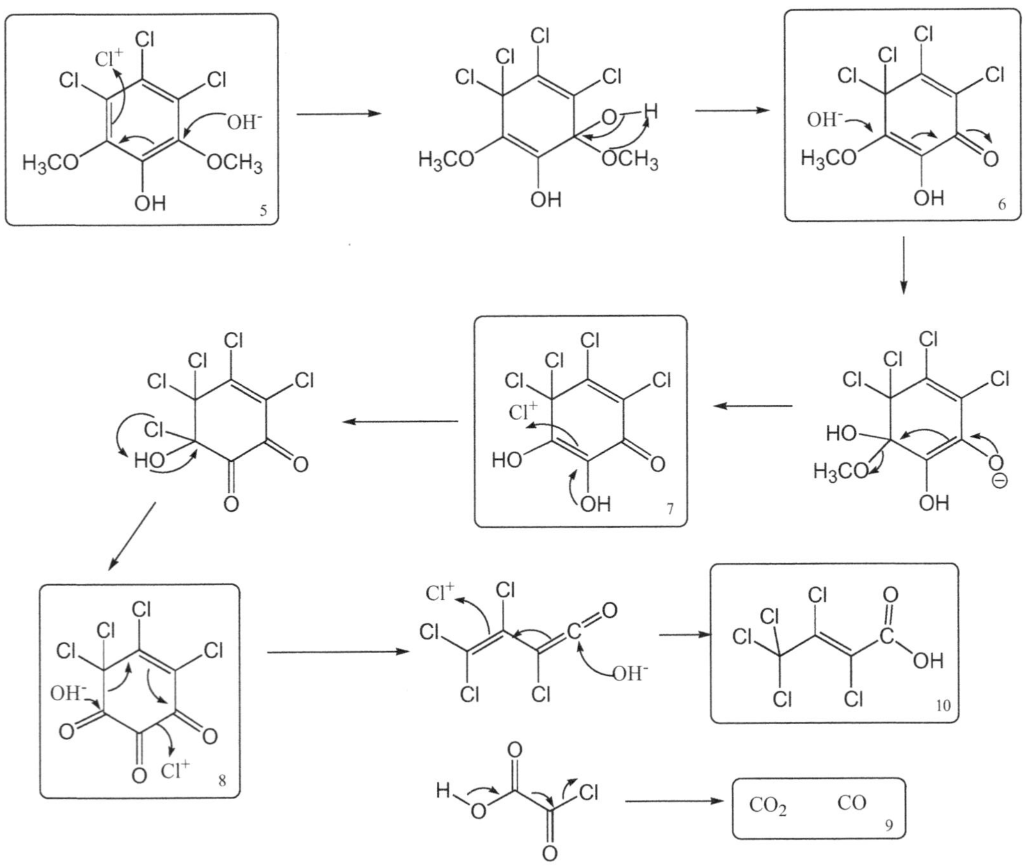

Figure 5.48 Proposed reaction pathway for syringic acid showing loss of aromaticity and fragmentation

Compound 5 undergoes addition of HOCl ultimately leading to loss of the methoxy group. This renders the other methoxy group susceptible to hydrolysis forming 7. This compound then undergoes chlorine oxidation to 8, and followed by fragmentation. Finally, compound 10 can be further cleaved as described previously.

Hydroxy Cinnamic Acids

The hydroxyl cinnamic acids represent another group of lignin monomers. These two have the p-hydroxy and vannilyl structures with a propenioc acid substituent para to the hydroxyl group. The hydroxy cinnamic acids are, of course, activated aromatics, with side chains that may also be reactive. Both compounds are quite reactive (9.3-10.3 M/M) with substantial TOX formation (1.3 and 2 M/M). Beyond this the two differ to a surprising extent (Figure 5.49). TCAA is a major product from ferulic acid (50% yield), but rather minor from p-coumaric (4%). For p-coumaric acid it is DCAA (~14% yield) and the unknown TOX (68% of total) that are most prominent, but nearly absent with ferulic acid.

There are interesting comparisons to make with the unsubstituted forms as well. For example p-coumaric acid has nearly the same reactivity and TOX formation as phenol and p-hydroxy benzoic acid, but the by-product distribution is different. Whereas the unsubstituted forms yield mostly TCAA (27-36%), the propenoic acid substituted compound tends to form DCAA (14%). Perhaps this points to the #4 carbon as the site for halogenation in HAA formation.

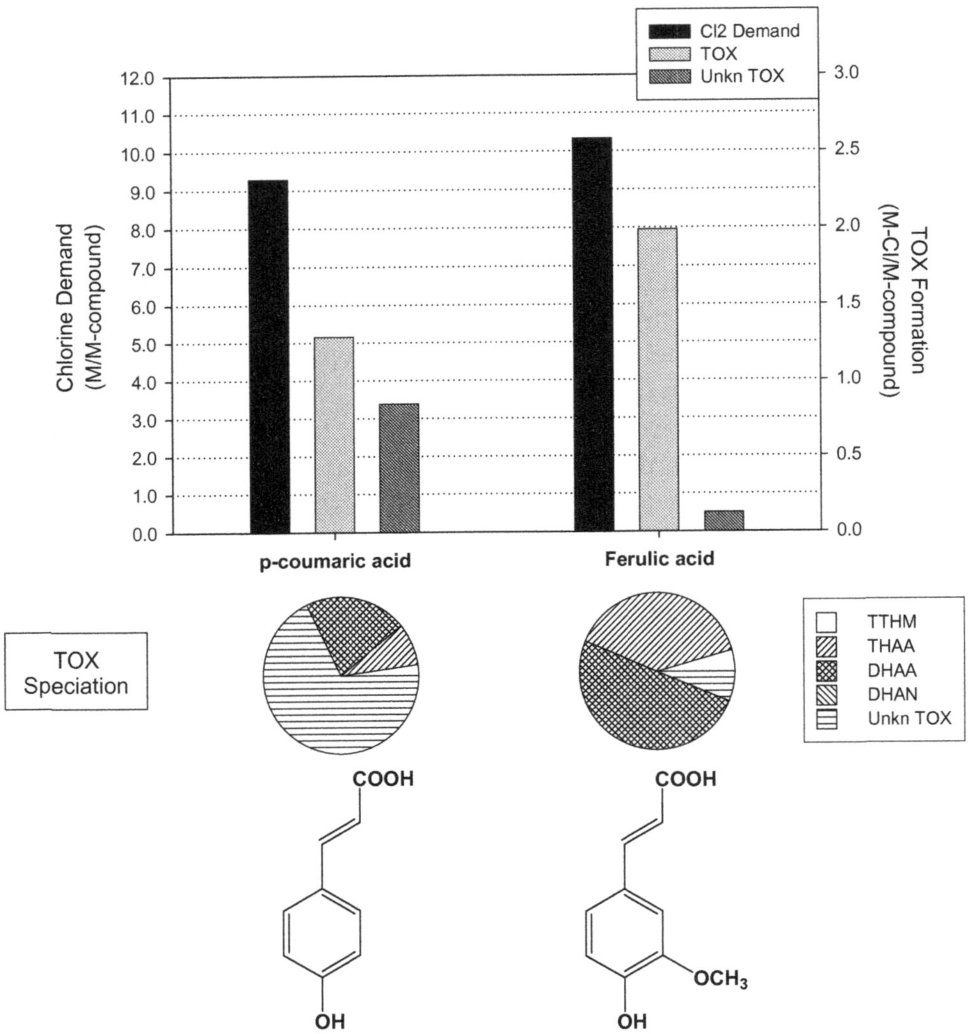

Figure 5.49 Reaction of free chlorine with hydroxycinnamic acid.

Chlorination of Amino Benzene Compounds

Aromatic amines were selected for study due to the recognized toxicology of the group, and the role of some aromatic amines in the causation of bladder cancer. The first compound to be discussed is aniline. The three aminobenzoic acid isomers will then be presented, along with an aminophenol. This section will conclude with a brief presentation of acetanilide. The role of pH will also be introduced.

Aniline

Chlorination tests with aniline showed similar high overall reactivity (Figure 5.50) indicating extensive oxidation, without substantial halogen incorporation at neutral pH. More halide incorporation was observed at elevated pHs, and a substantial portion of this could be attributed to THM formation. The reactivity of aniline bears some resemblance to that of phenol.

Both are activated aromatics with a single ring substituent. However, phenol gives rise to elevated levels of TCAA (~35% yield; Reckhow and Singer, 1985), whereas the yield with aniline is limited (~2.5% yield).

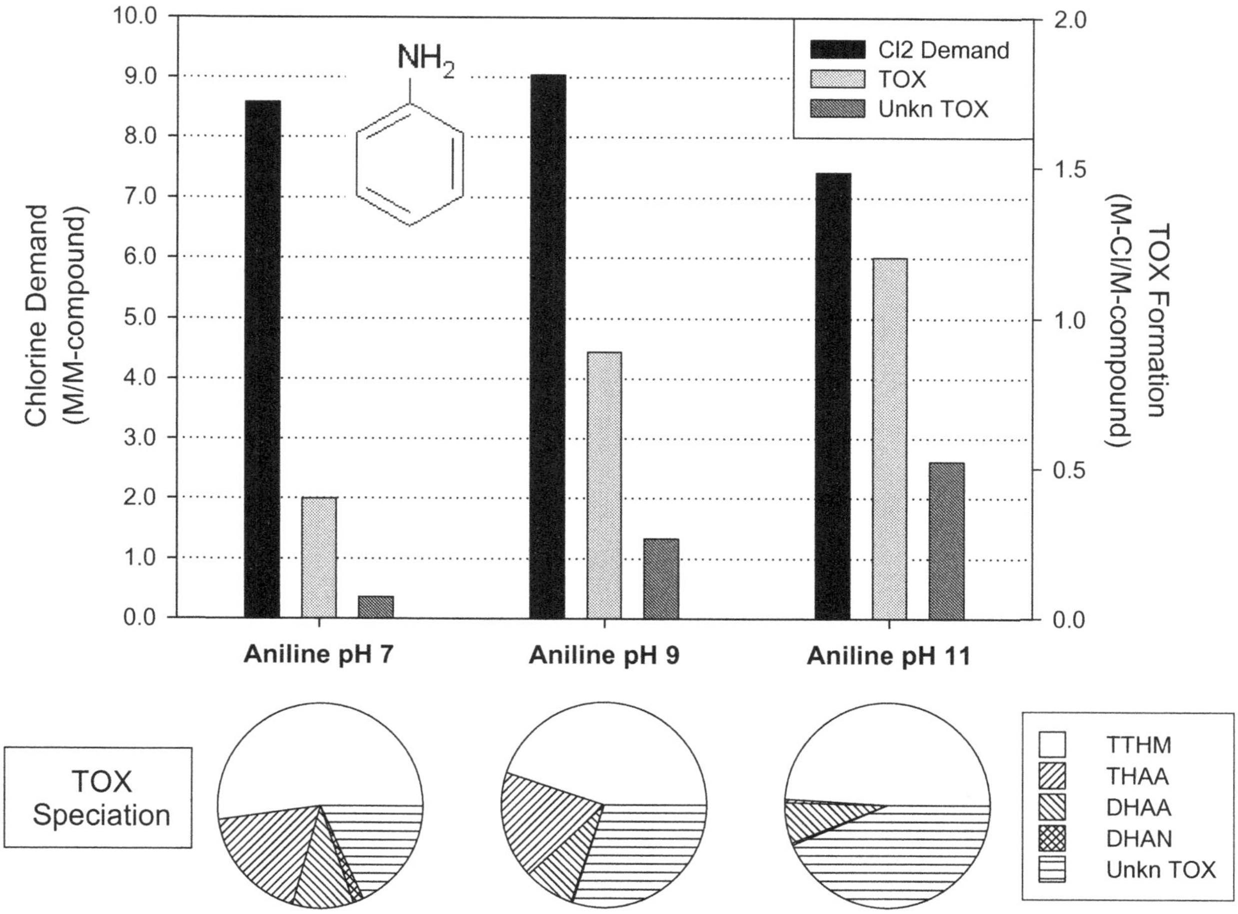

Figure 5.50 Reaction of free chlorine with aniline at 3 pHs

3-Amino Benzoic Acid

Classically, the presence of the amine group activates the ortho and para positions for electrophilic attack. (Figure 5.51). Eventually, the amine is converted to an imine, which then leads to hydrolysis and loss of the exocyclic nitrogen. At some point the ring is broken, and oxidation products remain. Decarboxylation as described by Larson and Rockwell (1979) is less likely here, due to lack of resonance stabilization.

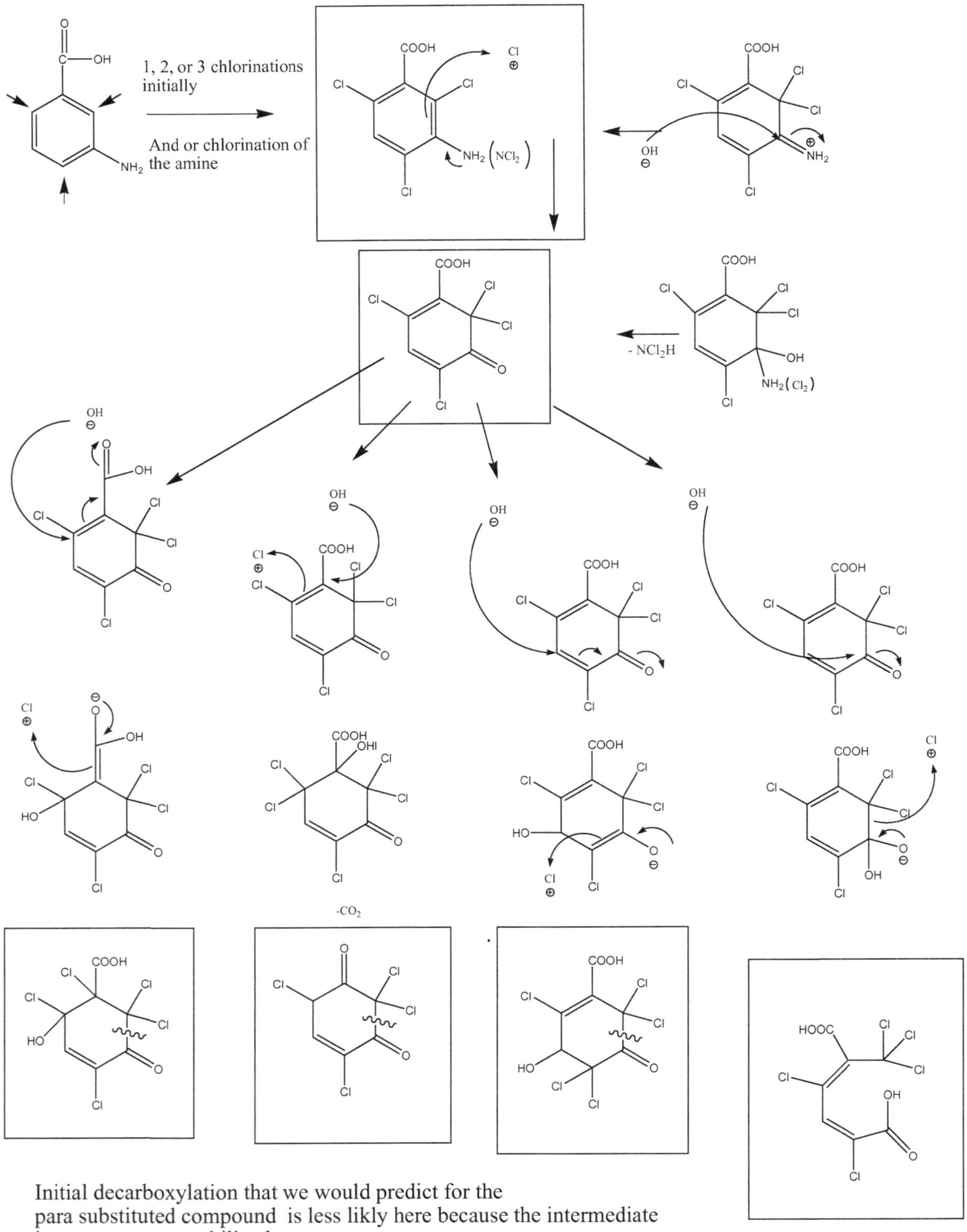

Initial decarboxylation that we would predict for the para substituted compound is less likly here because the intermediate is not resonance stabilized

Figure 5.51 Proposed degradation pathway for 3-amino benzoic acid

Laboratory experiments with 3-aminobenzoic acid showed extensive reactivity with relatively little TOX formation (Figure 5.52). This suggests that the end products of free chlorination are far more oxidized and contain less halogen than the proposed scheme shows.

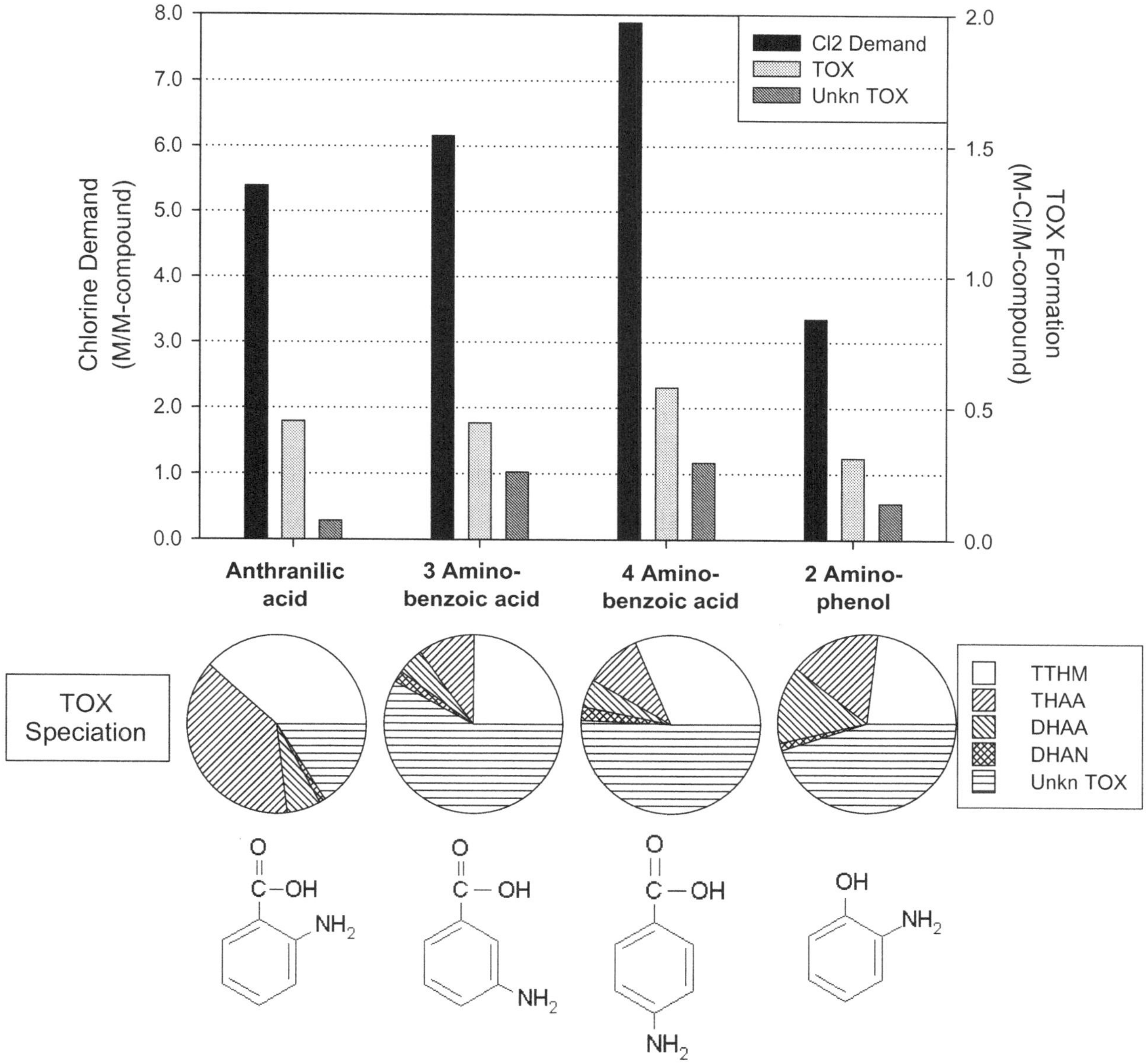

Figure 5.52 Reaction of free chlorine with substituted anilines

4-Amino Benzoic Acid

The para aminobenzoic acid experiment shows a higher level of reactivity, nearly matching that of aniline. This supports the presumption that the decarboxylation pathway is important for this particular isomer. Figure 5.53 shows how it might occur.

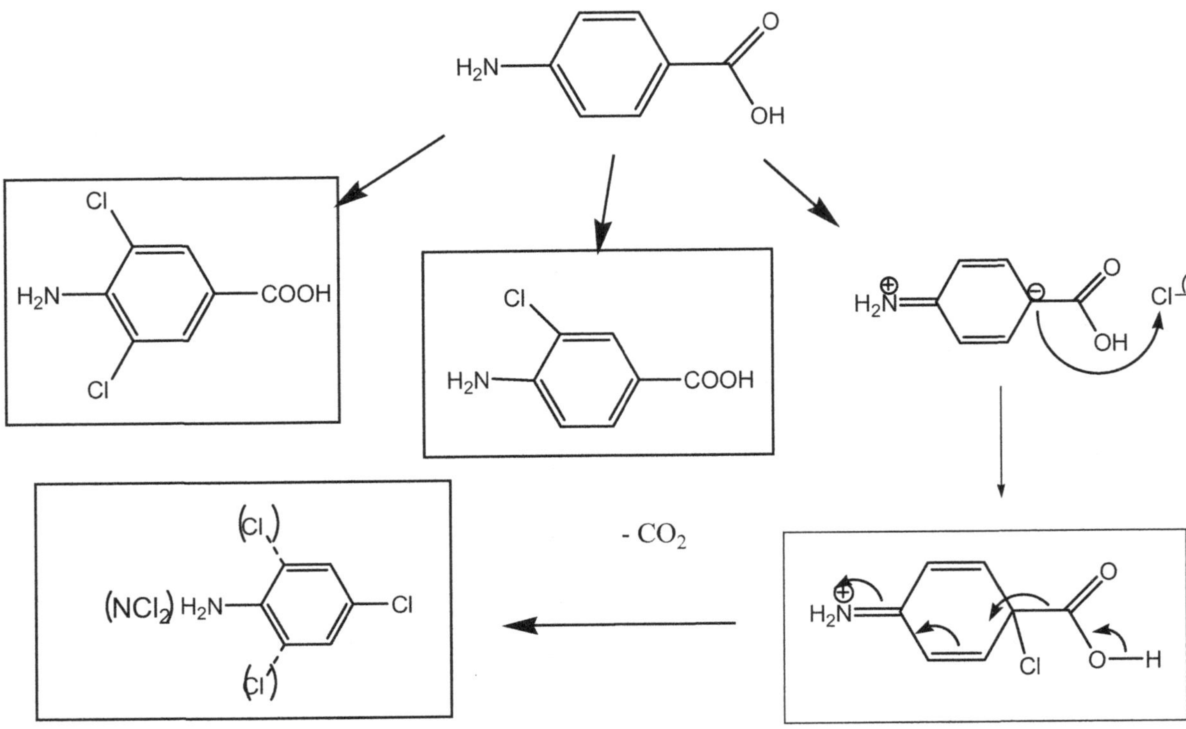

Figure 5.53 Proposed degradation pathway for 4-amino benzoic acid

2-Aminophenol

Ortho aminophenol resembles catechol in that it has two electron donating groups in the ortho position. The chlorine reactivity is also similar: 3.4 M/M (Figure 5.52) versus 4.0 M/M (Reckhow & Singer 1985). TOX production is small for both, although 2-aminophenol shows slightly higher levels of identifiable products.

Acetanilide

Chlorination tests with acetanilide showed the overall reactivity to be markedly decreased as compared to aniline (Figure 5.54). This was expected as a result of the electron withdrawing properties of the acetyl group. This renders the amine far less activating to the ring, and thereby greatly slows electrophilic aromatic substitution. Nevertheless, there is more than 1 M/M of chlorine demand at pH 7. It is significant that this chlorine demand is not accompanied by any measurable TOX formation. This suggests that the ring is not the site of attack, nor is it the terminal methyl group. The latter would rapidly form chloroform if it was to react. Instead this leaves the amine which could be the site of oxidation and formation of N-halo products. In organic solvent chlorine will react with acetanilide to form a stable N-chlorinated derivative. Our experiments in water under conditions of excess chlorine showed the formation of combined chlorine (suggesting the formation of the N-chloramine) as well as about 1.2 M/M of oxidative chlorine demand. This suggests that the molecule could have undergone additional hydrolysis and oxidation of the nitrogen; possibly forming the hydroxylamine, the nitroso, and the nitro form. This is analogous to the biodegradative pathway proposed by Kaufman et al. (1973). It is also quite possible that a second chlorine added to the amine, causing loss of acetate and formation of N,N-dichloroaniline. The presence of the two chlorines would render the aromatic ring far less reactive to electrophilic attack, and therefore explain why further degradation does not occur as it does with simple aniline.

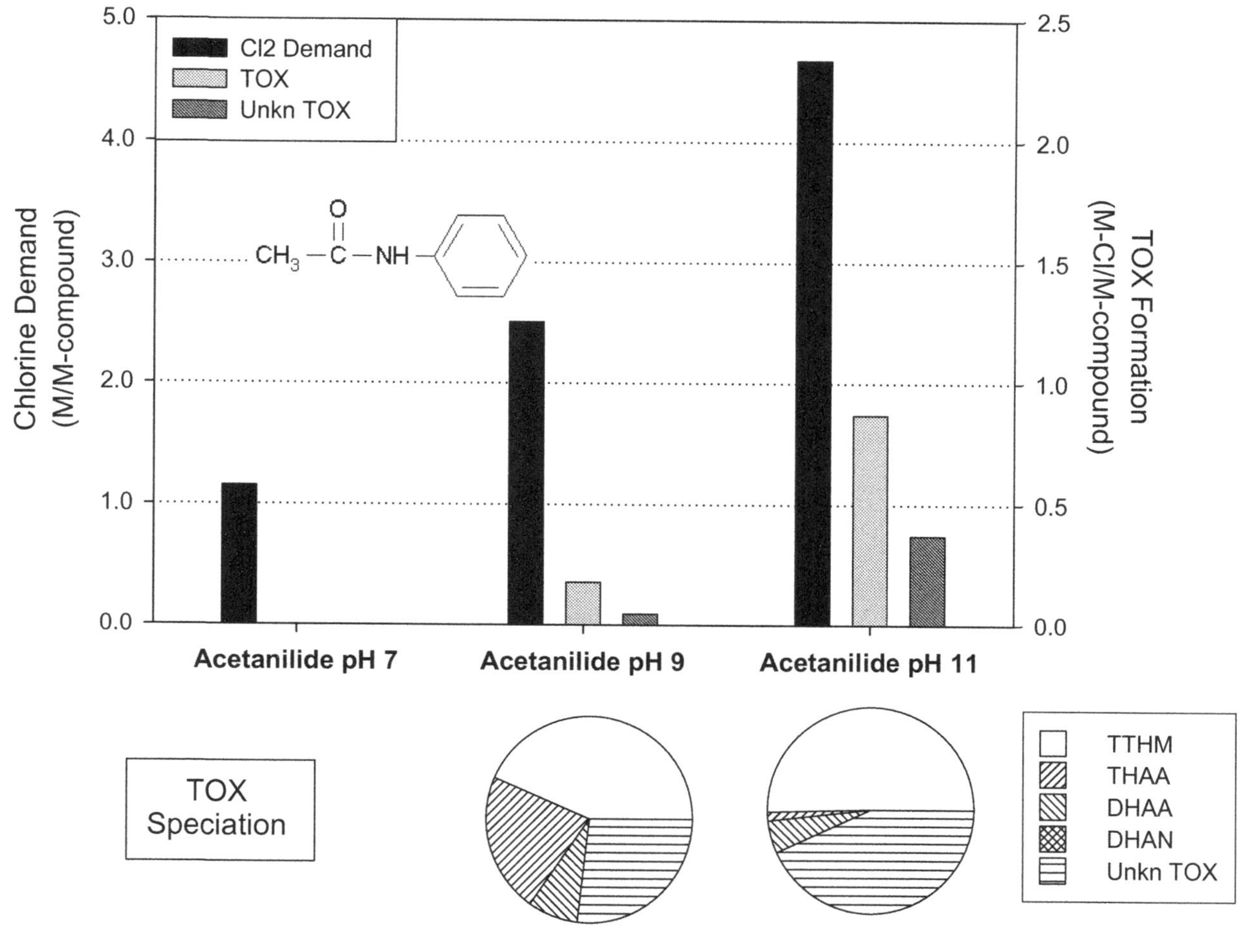

Figure 5.54 Reaction of free chlorine with acetanilide at 3 pHs

At higher pH, more chlorine demand is observed. Along with this is a higher level of TOX and THM formation. This would be expected as the terminal methyl group would become rapidly halogenated at high pH. Subsequent hydrolysis and decarboxylation would give rise to chloroform, CO_2 and aniline. The free aniline would then be subject to extensive chlorine consumption and degradation. It's also quite likely that some hydrolysis will occur even without halogenation of the acetyl group. Again this would lead to aniline and ring fragmentation products.

ILLUSTRATION OF USE OF PRIORITIZATION TOOL

Feed-forward Process

The feed-forward approach is partly illustrated with the work on selected nitrogenous compounds. For example, consider the following analysis of the reactions of chlorine with purines and pyrimidines. At the beginning of this chapter, we proposed a range of anticipated concentrations of the five nucleic acid bases. Using these figures along with the measured DBP yields, one can estimate the DBP formation that might be attributed to these precursor structures. Table 5.11 shows this type of analysis applied to the Unknown TOX formation. This suggests that uracil, and to a lesser extent, thymine, produce substantial amounts of potentially toxic halogenated organic by-products.

Table 5.11
Unknown TOX expected from nucleic acid bases (µg/L)

Base	50%ile	90%ile	99%ile
Adenine	0.04	0.11	0.41
Thymine	1.40	4.48	15.99
Guanine	0.05	0.17	0.61
Cytosine	0.29	0.93	3.31
Uracil	4.78	15.30	54.63

Once a precursor structure or group of structures are so identified as problematic, the next step is to examine the known or postulated chemistry of degradation of these compounds. As noted earlier in this chapter, Young and Uden (1994) found uracil to produce chloral hydrate at a 19% molar yield with lesser amounts of TCAA and almost no THMs. In laboratory studies during this research, we found that about 3.4 M/M of TOX was formed; with only a 6% yield of DCAA and traces of TCAA and THMs (chloral hydrate was not measured). Presuming that both DCAA and chloral hydrate are formed in the reported yields, we still have about 75% of the TOX unaccounted for. Further, the carbon and nitrogen analysis suggests that extensive loss of both occurred during acidification and gas purge. This points to either extensive formation of CO_2 & N_2 or some other highly volatile forms (probably low charge, low polarity products). If our reasoning is correct, there are substantial amounts of some very small, highly halogenated by-products that have yet to be recognized. Some of this could be NCl_3 as measured by Gould and coworkers (1984) using the non-specific DPD-FAS method. However, there remain some inconsistencies among the various studies that do not permit any simple explanation such as quantitative NCl_3 formation.

By examining the case of uracil, it's clear that the feed-forward approach can point to some critical gaps in our understanding of chlorine by-products. In this way, it gives us guidance in the form of well-defined precursors that require further investigation.

Feed-back Process

A partial test of the feed-back approach is exemplified by the work on aniline derivatives. Here, certain structures were postulated as capable of forming reactive nitrogenous species in vivo and ultimately leading to DNA adducts. One of these was the class of N-hydroxy aromatic amines (e.g., N-hydroxy aniline). N-hydroxylation is a likely outcome of reaction of amine with free chlorine. However, careful consideration of the reaction of aromatic amines suggested that N-hydroxylation would be kinetically defavored as compared to ring oxidation and cleavage. However, deactivation of the ring by diverting the amine's electron density to an appropriate N-substituted group could make hydroxylation favorable. This postulate led to a series of model compound tests which were summarized in the preceding text. These data indicate that the N-acetylated form is sufficiently unreactive at neutral pH to preserve the ring and possibly form an N-hydroxy derivative (Figure 5.55).

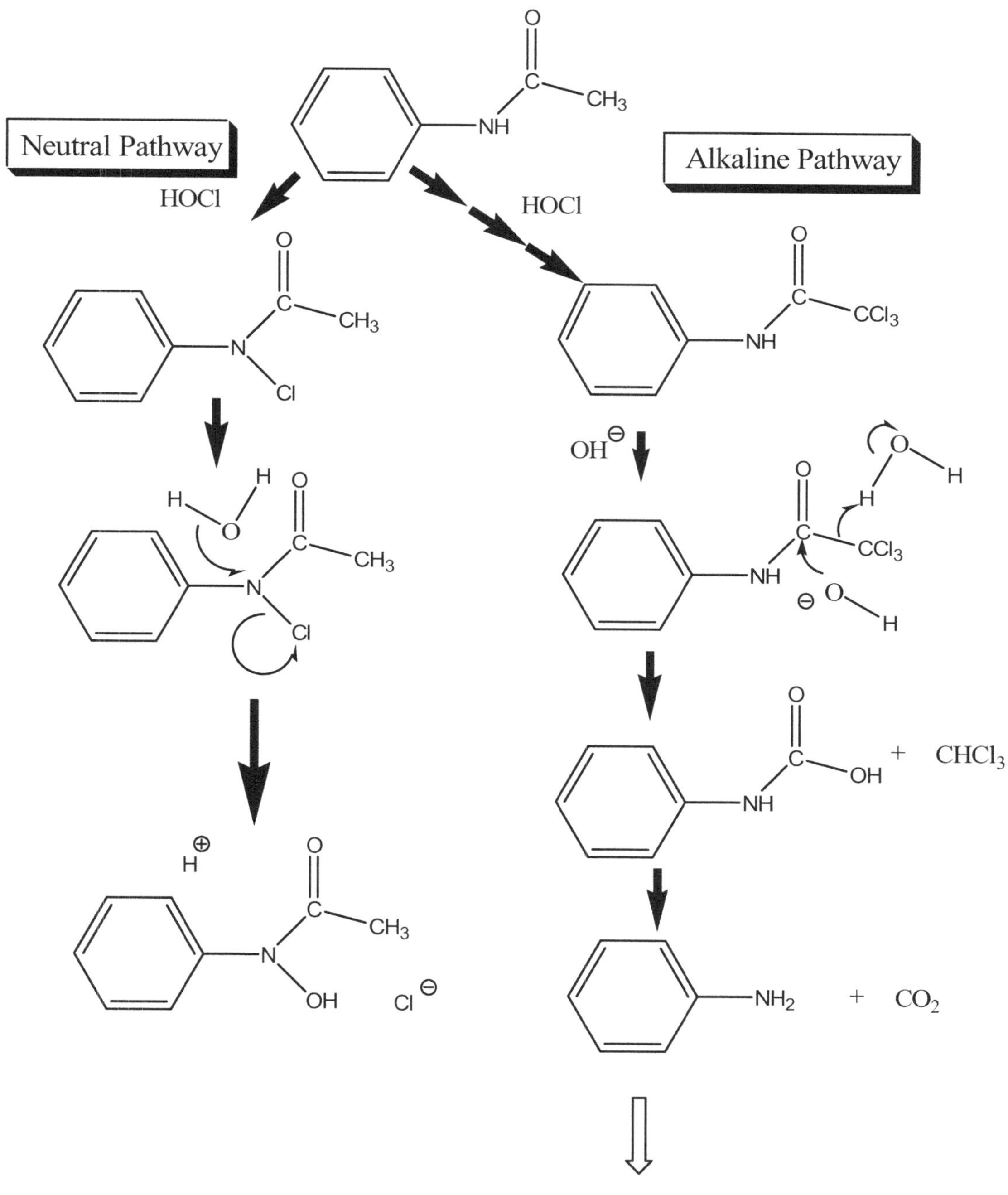

Figure 5.55 Reaction scheme for example of feed-back approach with aromatic amines

CHAPTER 6. OVERALL RESULTS – TOXICOLOGY

DESCRIPTION OF THE DATABASE

Appendix A summarizes data obtained from the literature indicating the general toxicity of those disinfection by-products (DBPs) that have received substantive study. Studies that have specifically addressed reproductive and/or developmental toxicities of DBPs have been identified. No observed adverse effect levels (NOAELs) and lowest observed adverse effect levels (LOAELs) for established DBPs in the key studies have been provided. The LOAEL will form the basis of comparisons with other data that have been generated in the project through quantitative structure-toxicity relationships (QSTR).

In Appendix A, those DBPs that have been studied sufficiently to detect carcinogenic activity in experimental animals are identified and the TD_{50} in each species listed. The TD_{50}s listed were largely drawn from the carcinogen potency database (CPDB), a website maintained at the University of California (Gold 2005). Also identified are the target organs. For simplicity, the target organs were not identified by species. As to be expected, the target organs vary considerably among species and this information is readily accessible online. In several instances, the CPDB database was not up to date, so it was necessary to estimate the TD_{50} from the available data adjusting for the portion of the lifetime of the animal on which the study was conducted. The general formula for calculating the TD_{50} is provided on the CPDB website.

Appendix E, Table E.1 is a listing of the chemicals subjected to QSTR analysis. A database was developed that includes the probability that a DBP is carcinogenic for each of eight cancer models, an Ames' mutagenicity model, a chronic LOAEL model, and a developmental toxicity potential model in TOPKAT®. Table E.1 should be consulted to determine whether a given putative DBP was subjected to analysis or not. The actual database will be available through RJ Bull on CD upon request.

Ambiguities arose in the literature making it was necessary to assign specific structures to a compound before a chemical could be tested in TOPKAT®. Generally, this was because isomers were frequently not identified in the original studies. Some variants in structural isomers were specifically explored in TOPKAT® to broaden the depth of the analysis, but not exhaustively so. TOPKAT® does not distinguish between stereoisomers. Although some drawings may suggest a certain configuration, the program strips this information from the structure before the model is run. Therefore, it should not be assumed that a prediction applies to the specific stereoisomer that may be implied by a structure drawn for this report.

VARIATION IN POTENCY AMONG HALOGENATED CARCINOGENS

It is important to appreciate some of the limitations of the basic data upon which QSTR predictions are based. Uncertainties arise for a variety of reasons. Some of these reasons are illustrated in this section.

First, there are substantive differences among species in their sensitivities to carcinogens. Figure 6.1 compares the distribution of the potencies of all compounds that have been found to produce cancer in rats or mice that are included in the CPDB. The observed distribution suggests that rats are somewhat more sensitive than mice as the median TD_{50} for rats is 11.5-mg/kg day^{-1}, whereas it is 53.8-mg/kg day^{-1} in mice. In part, this relates to the fact that the sets of chemicals that have been tested in different species differ significantly.

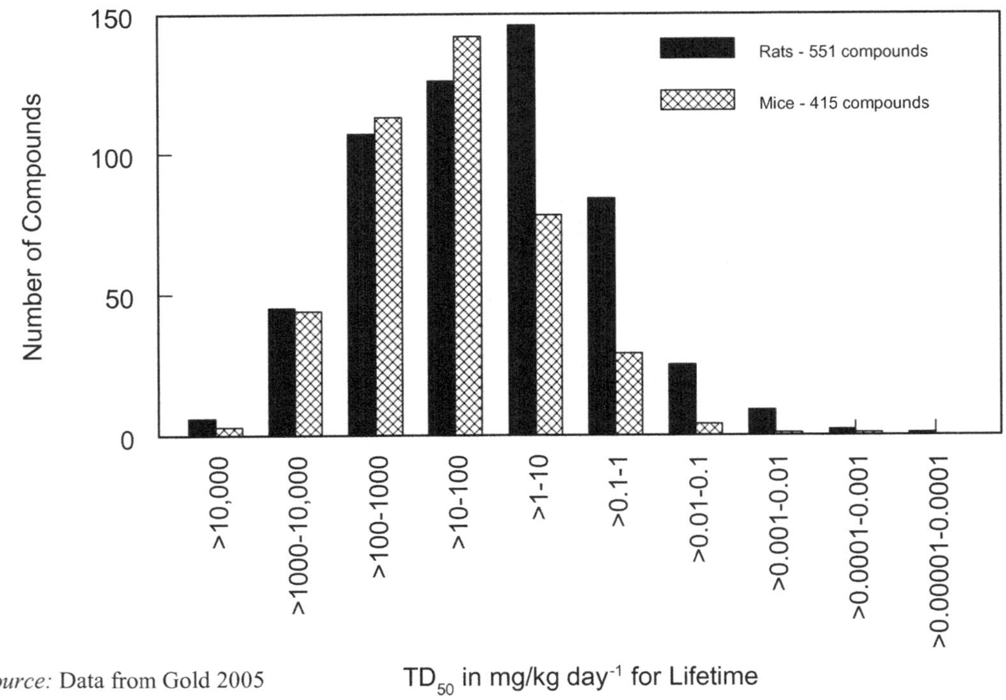

Source: Data from Gold 2005

Figure 6.1 Distribution of carcinogenic potencies of carcinogens that tested positive in rats or mice This figure neglects those chemicals for which results were discordant as well as those compounds that did not test positive.

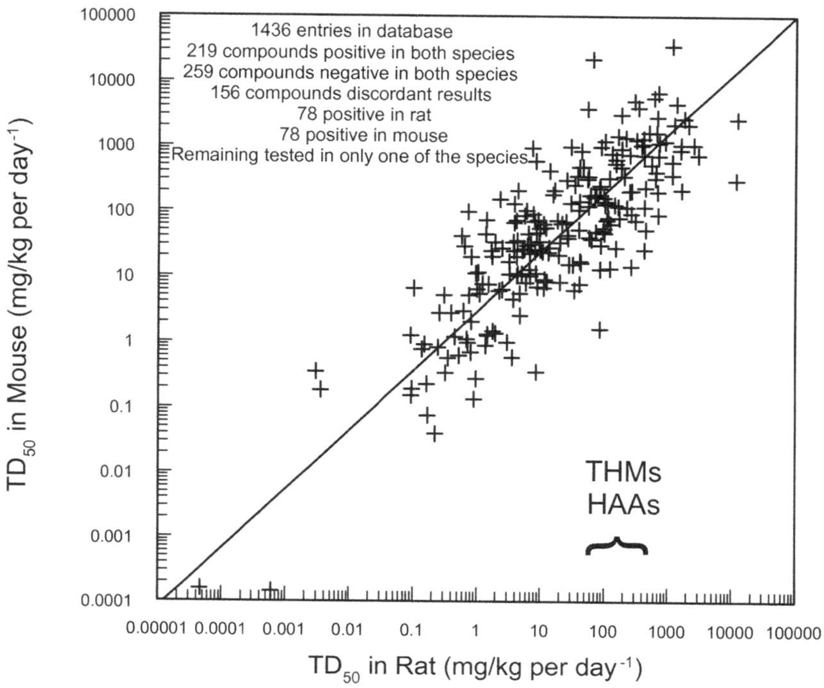

Source: Data from Gold 2005

Figure 6.2 Correlation between the potencies of chemicals that test positive in mice vs. rats. The figure only includes positive compounds in both species. The numbers of discordant findings between species, negative compounds, and compounds not tested in both species are described in the inset.

However, it is useful to determine how the sensitivity among species varies for chemicals that have been tested in both species.

In Figure 6.2 the potencies of 219 compounds that have been tested in both rats and mice are included in a scatter plot. As might be expected, there is a pretty good correlation in potency between the two species with an $r^2 = 0.74$. Examination of the graph, however, indicates that the difference in sensitivity between mice and rats for a particular compound differs by as much as 100-fold and excursions of 10-fold are not unusual. These differences are further accentuated by the fact there were 156 compounds in which the data were discordant – i.e. the chemical was found carcinogenic in one species and not the other. Of interest is the fact that the apparent false negative rates among these chemicals were exactly the same, 78 compounds were positive in the rat, but negative in mice and 78 were positive in mice, but negative in rats. As the responsiveness to a given carcinogen can vary among species and by sex, QSTR predictions of carcinogenicity must vary in the same way. These data indicate why it is difficult to be precise in estimating carcinogenic risk to humans from particular chemicals when using data in experimental animals. Put another way, there is a reasonably robust relationship to build upon in general, but the responses to a particular chemical can vary substantially among species and sex. In the human population, variation among individuals further complicates predictions.

As a point of reference, the range of TD_{50}s for regulated DBPs is indicated in Figure 6.2. This illustrates the fact that regulated compounds are clearly not unusually potent as carcinogens and are found to be on the weak side of the median potency of all carcinogens in both species.

The issue of interspecies variation in sensitivity to carcinogens is explored further in Figure 6.3 where the potency of carcinogens that tested positive in hamsters is compared to the potency of the same compounds that have tested positive in rats. There were only 25 compounds that fit these criteria in the CPDB and the overall correlation was poor ($r^2 = 0.48$). The dashed line indicates the curve where the TD_{50}s in the rat would have been equivalent to the hamster. All but four of the compounds fall below this line and only one is substantially above it. This indicates that the rat was more sensitive to most of the compounds tested. Thus, based on this limited dataset, the hamster appears to be less sensitive to carcinogens than the rat.

A limited number of compounds have been tested in primates. Comparisons of the potencies of compounds that test positive in either Rhesus or Cynomulgus monkeys are compared to their potency in the rat (Figure 6.4). Overall, the correlation is good with an $r^2 = 0.71$, very close to the correlation observed between rats and mice. Within this limited data set (total of 12 chemicals) the relative sensitivity of the primate and rat appear similar. N-Nitroso-N-dimethylamine (NDMA) is reported to be negative in a study in Rhesus monkeys (Gold 2005), but this study apparently only employed 2 animals and a single low dose rate (1 mg/kg day). More extensive studies of N-nitroso-N-diethylamine (NDEA) produced tumors in Rhesus or Cynomulgus monkeys and bush babies with TD_{50}s of 0.0536, 0.00725, and 0.0122, respectively. Therefore, it is likely that primates are as sensitive to NDMA as rats (Table 1.1).

A practical decision has been made by regulatory agencies to use cancer potency estimates from the most sensitive species for estimating cancer risk in humans in the absence of data to the contrary and the rat is generally the most sensitive species. Therefore, the comparisons of the potencies of carcinogens in this study utilized rat data; this is also the largest database. Taking this approach formally eliminates a number of DBPs that have only been found carcinogenic in mice. The practical impact of this convention is small, as a chemical is not treated as a carcinogen when it produces tumors in only a single species when it has been adequately tested in both.

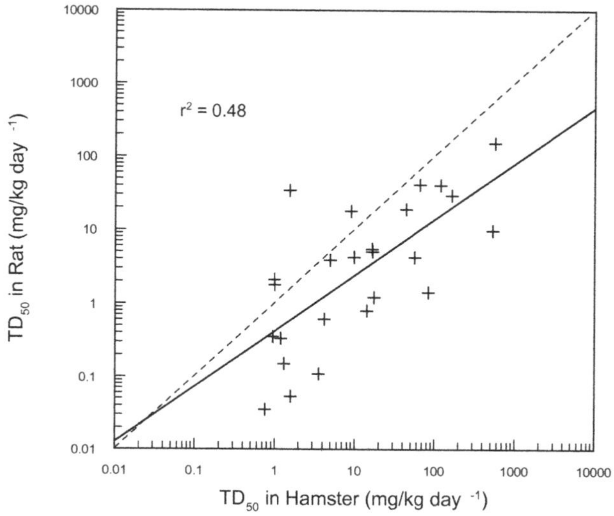

Source: Data from Gold 2005

Figure 6.3 Correlation between the potencies of chemicals that test positive in hamsters vs. rats (solid line). Dashed line displays equivalent sensitivity. The figure only includes positive compounds in both species.

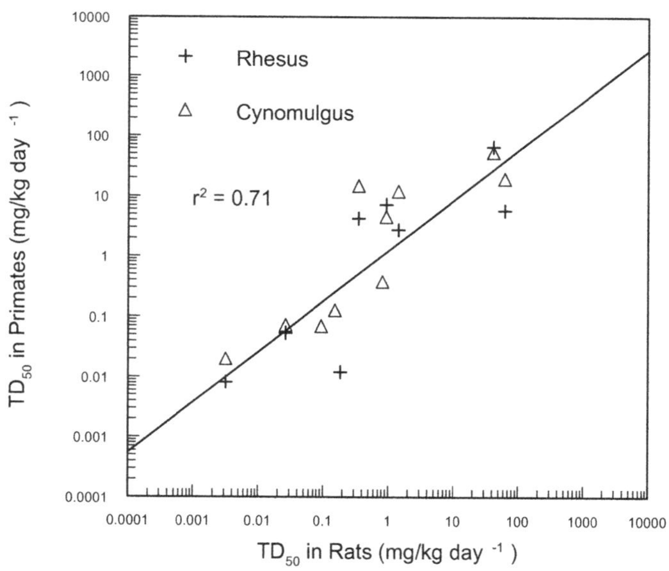

Source: Data from Gold 2005

Figure 6.4 Correlation between the potencies of chemicals that test positive in primates vs. rats. Two species were used in these studies, Rhesus and Cynomulgus monkeys. The figure only includes positive compounds in both species.

Surrogate parameters are often employed to assess the performance of a water treatment process. The USEPA indicated its belief that the occurrence of chemicals in selected members of two DBP classes used to develop a regulation was predictive of the risk associated with chlorination. To test whether or not this hypothesis is reasonable, the distributions of potencies of different classes of halogenated DBPs have been compared to the potencies of all chemicals that are in the CPDB in the following pages.

The potencies of halogenated compounds in the CPDB were compared to the distribution of potencies of all carcinogens that tested positive in rats. The first step was to determine the extent to which carcinogenic halogenated chemicals might be differentiated from carcinogens in other classes. Figure 6.5 compares the distribution of potencies of all chemicals that tested positive as carcinogens in rats as tabulated in the CPDB with the median potency of all halogenated compounds in the CPDB that tested positive as carcinogens. The median potency of all rat carcinogens was 11.2-mg/kg day^{-1}. The median potency the halogenated compounds was 20-mg/kg day^{-1}. However, the distribution of potencies of the halogenated compounds is quite different having an extended tail that includes some very potent carcinogens.

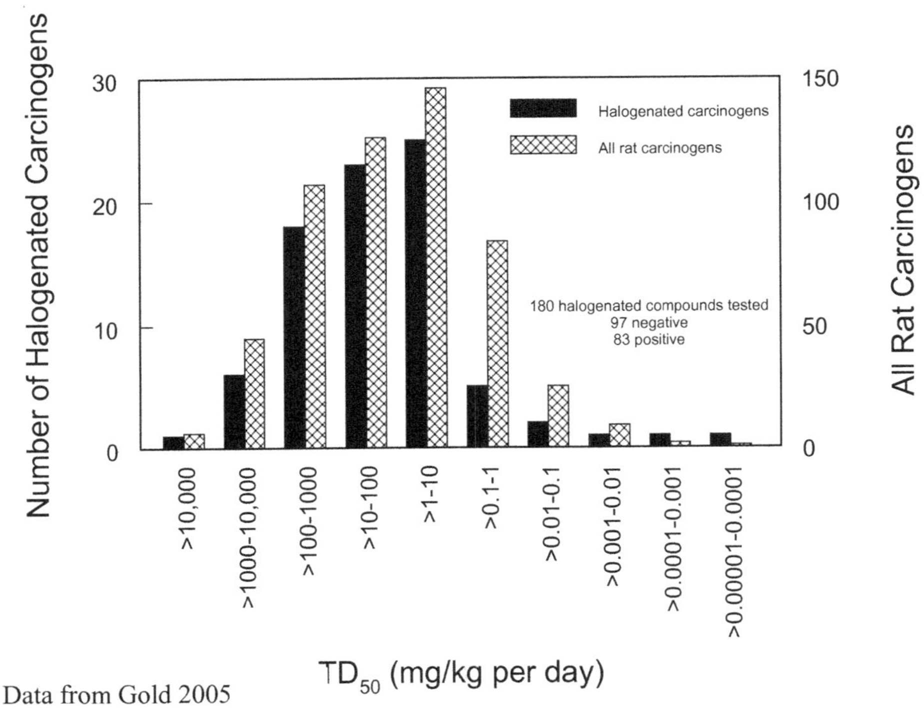

Source: Data from Gold 2005

Figure 6.5 The distribution of potencies of all halogenated chemicals found in the carcinogen potency database that tested positive in rats compared to the distributions of potencies of all positive compounds tested in the rats The numbers of tested halogenated compounds, the number that tested positive and negative in the rat are provided in the legend.

Figure 6.6 displays the distribution of the predicted chronic LOAELs of halogenated DBPs plotted against the TD_{50}s of all the halogenated rat carcinogens found in the CPDB. The mean of predicted LOAELs for halogenated DBPs is somewhat greater than the mean of the TD_{50}s for halogenated carcinogens, as expected. However, there is an obvious tail to the distribution of the TD_{50}s for halogenated carcinogens that is not apparent in the predicted chronic LOAELs for DBPs. Within the group with TD_{50}s < 1-mg/kg day^{-1} are found the seven chemotherapeutic agents:

- nitrogen mustard and its oxide
- chlorambucil
- mephalan
- isophosphoramide
- cyclophosphamide
- chlorozotocin

A polyhalgenated aromatic hydrocarbons and a mixture of such compounds:

- polybrominated biphenyl mixture HCDD mixture, and
- 2,3,7,8-tetraclorodibenzo-*p*-dioxin

One halogenated hydrocarbon:

- 1,2-dibromo-3-chloropropane (DBCP)

One halogenated ether:

- Bis-(chloromethyl)ether

Five nitroso compounds:

- 2-fluoroethyl-nitrosourea
- N-nitroso-N-methyl-4-fluoroaniline
- 1-nitroso-1-(2-hydroxypropyl)-3-chloroethylurea
- 1-nitroso-1-hydroxyethyl-3-chloroethylurea
- 1-chloroethylnitroso-3-(2-hydroxypropyl)urea.

Two questions arise with many of these chemicals: 1) is it likely that the halogen plays a role in the carcinogenic activity [e.g. the chlorine on 1-chloroethylnitroso-3-(2-hydroxypropyl)urea has only a minimal effect on the TD_{50} when compared to its non-chlorinated analog] and 2) are any of these chemicals similar to those that have been observed as DBPs? It is unlikely that the halogens play a role in the carcinogenicity of the five nitroso compounds. There has been no clear demonstration of the formation of polyhalogenated aromatic hydrocarbons in the chlorination of drinking water. The formation of dioxins has been demonstrated in the bleaching of pulp, but this occurs when heat is applied to dry the pulp (Cleverly et al. 1996). The formation of dioxins with disinfection has not been found, despite considerable effort. Some potent halogenated compounds included were chemotherapeutic agents that have structures related to nitrogen mustards. Such structures have not been reported as the result of drinking water disinfection to the knowledge of the authors.

There are a limited number of halogenated compounds (Appendix E, Figure E.1) that have moderately high carcinogenic potencies that were more analogous to DBPs and were frequently prominent in the similarity analyses that were performed for positive predictions of the carcinogenicity of some DBPs. In addition to DBCP, these include:

- 1,2,3-trichloropropane
- 1,2-dibromoethane
- 1,2-dichloroethane
- tris(2,3-dibromopropyl)phosphate
- dichloroethylene
- dibromodulcitol
- dichlorvos

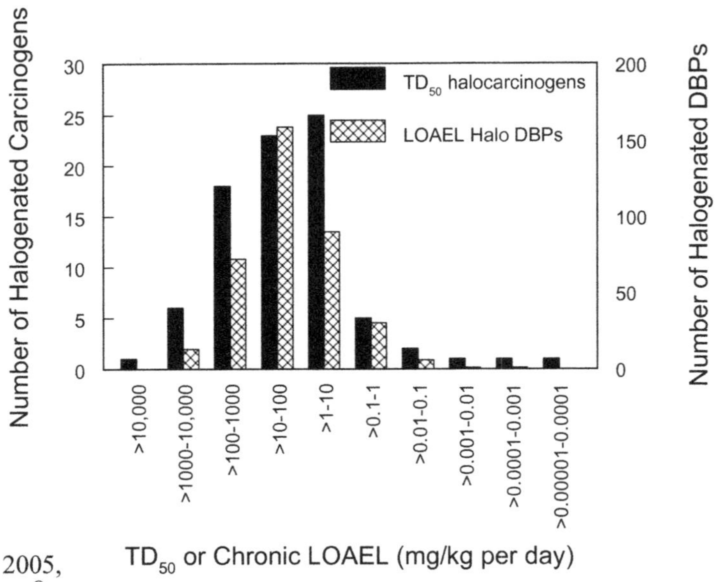

Source: TD_{50} data from Gold 2005, LOAELs predicted by TOPKAT®

Figure 6.6 A comparison of the distribution of the TD_{50}'s of halogenated rat carcinogens to the estimated chronic LOAELs predicted by TOPKAT® software

It is apparent that risks associated with chlorinated drinking water could well be caused by DBPs that are very potent and occur at very low concentrations. Such micro-contaminants may vary widely among water supplies and, as a result, not be reflected by surrogate measures of by-product formation. Therefore, the use of parameters such as the THMs, HAA5 or even TOX as a surrogate measure for risk associated with DBPs is naïve at best.

PREDICTIONS OF TOXIC PROPERTIES USING QSTR

Beyond the DBPs for which more or less completed in vivo data sets exist, there are literally hundreds of chemicals that have been identified as DBPs for which there are no empirical data in humans or experimental animals. The potential carcinogenicity and developmental toxicities of these compounds were evaluated utilizing QSTR models in TOPKAT®. However, no model exists to predict the TD_{50}. TOPKAT® includes a model for predicting chronic LOAELs. The empirical reasons that chronic LOAELs may be considered as an approximation of the TD_{50} were addressed in the methods section of this report. Returning to Figure 6.6, the TD_{50}s for all halogenated compounds identified in the CPDB were compared with the predicted LOAELs for the toxicologically uncharacterized halogenated DBPs. The median TD_{50} and LOAEL were both found to be in the 10-100-mg/kg range (both were found to be 19.3 mg/kg day^{-1}). On theoretical grounds LOAELs should be consistently lower than TD_{50}s. The distribution of predicted LOAELs for halogenated DBPs appears to be a normal distribution, whereas that of TD_{50}s of the halogenated carcinogens has the appearance of a Poisson distribution. This may reflect some bias in the selection of halogenated chemicals for testing in animals and that fact that halogenated chemicals are only a subset of the carcinogens in the CPDB.

An empirical test of the reliability of the relationship between a predicted LOAEL from TOPKAT® and the TD_{50}'s from the CPDB was conducted. One hundred eleven compounds were selected in a random fashion, drawing broadly on different classes and ranges of values (listed in Table E.2, Appendix E). Seventeen of these chemicals were found to be outside the optimal predictive space (OPS) of TOPKAT® or had SMILES codes not recognized by the program and were eliminated, leaving 94 chemicals for which the predicted LOAELs could be compared to the TD_{50}s. The correlation coefficient was found to be significant, but low (r^2 = 0.14). The coefficient was found to be heavily influenced by three chemicals, two that had very low TD_{50}s (<0.1) and one with a very high TD_{50}. Eliminating these three compounds led to the relationship depicted in Figure 6.7, where the correlation coefficient of 0.21 was found. Since many of the compounds being evaluated in this project are halogenated, we narrowed the comparison to 46 chemicals in the selected group that were halogenated (identified in Table E.3, Appendix E) and found essentially the same correlation coefficient (0.20).

The above analysis demonstrates a tenuous relationship between the chronic LOAEL predicted by TOPKAT® and the TD_{50}. Examination of Figure 6.8 provides a partial explanation for this finding. The graph illustrates two points. The first is that the LOAEL and the TD_{50} occur in the same general range, although the variability spans roughly two orders of magnitude. Second, there is a marked departure of the dashed line representing the linear regression line for the data from the line that identifies equivalency (i.e. predicted LOAEL = TD_{50}) in the low dose range. This indicates that for a large number of the more potent carcinogens, the LOAEL is substantially higher than the TD_{50}. This is consistent with the finding that many potent carcinogens do not produce overt toxicity at dose rates that are strongly carcinogenic, whereas the two endpoints tend to be seen at similar doses with weaker carcinogens. However, it is also necessary to understand how much of this departure may arise from the structure of the model for predicting chronic LOAELs.

Among the 94 chemicals compared in Figure 6.7, 34 were in the training set and the actual chronic LOAELs were provided. A comparison between the actual chronic LOAELs and the predicted LOAELs for these compounds is provided in Figure 6.8. Since these compounds were part of the development of the model, it was not too surprising to find a better relationship, but it was far from perfect (r^2 = 0.53). The next step was to compare the actual LOAELs and the TD_{50}s that were reported in the CPDB. The result is depicted in Figure 6.9. Although the linear regression line still displays a departure of about one order of magnitude from the idealized chronic LOAEL = TD_{50}, the correlation coefficient is essentially the same as the relationship between predicted and actual LOAELs (r^2 = 0.52). Thus, the actual LOAELs have a better relationship to the TD_{50}s than the predicted LOAELs. In part, this can be attributed to the fact that these data are all drawn from the same carcinogenesis studies, whereas the construction of the TOPKAT program drew on a broader literature source in which toxicity may have not been as consistently reported. Modern studies, particularly those conducted by the National Toxicology Program (NTP), make much more consistent observations of non-tumor pathology than is and has been the case in the open literature as a whole. The comparison provided above includes data from many compounds in that database. Therefore, the better relationship most likely reflects a more consistent ability to identify LOAELs in modern carcinogenesis studies.

The relationship between the LOAEL and TD_{50} is still far from perfect. This raises questions about its use as a decision criterion when attempting to rank the potency of compounds predicted to be carcinogenic as well as those that are potentially developmental toxicants. To compensate for this difference, we utilized a cutoff for chemicals predicted to be carcinogenic or

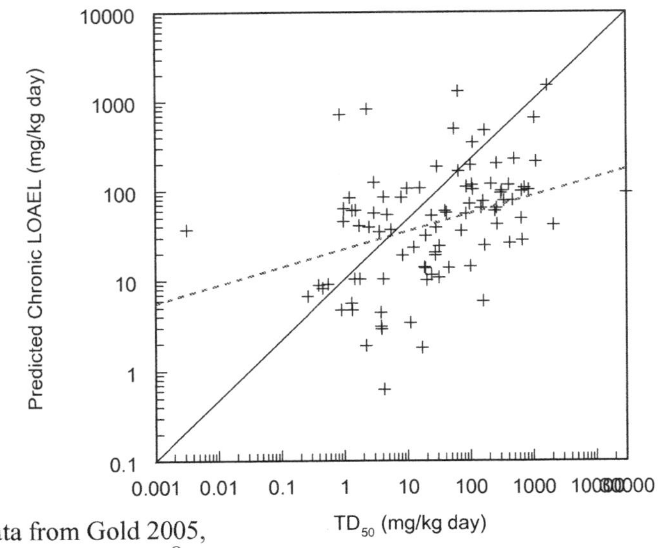

Source: TD_{50} data from Gold 2005, LOAELs predicted by TOPKAT®

Figure 6.7 Relationship between the chronic LOAEL predicted by TOPKAT® and the TD_{50} calculated for rat studies that have been placed in the CPDB. Correlation coefficient $r^2 = 0.21$, represented by dashed line. Solid line reflects chronic LOAEL = rat TD_{50}.

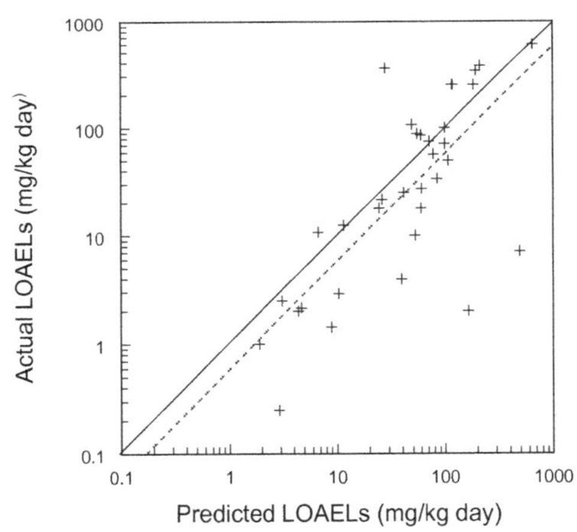

Source: Data from TOPKAT® predictions and training set

Figure 6.8 Relationship between chronic LOAELs predicted by TOPKAT® and the actual LOAELs for chemicals that were in the training set drawn from the list used to construct Figure 6.7. Correlation coefficient $r^2 = 0.53$, represented by dashed line. Solid line reflects actual LOAEL = predicted LOAEL.

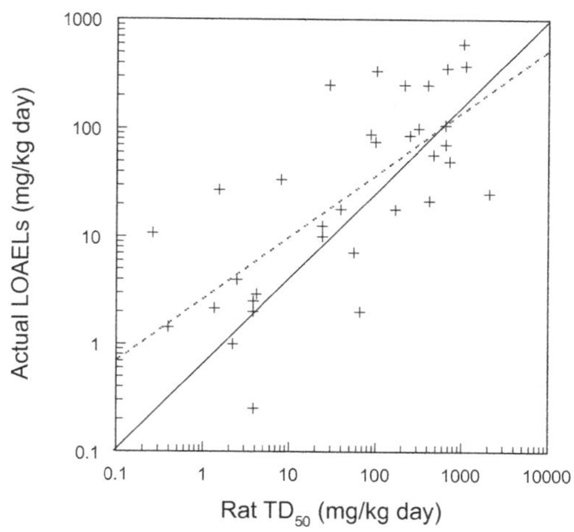

Source: TD50s from Gold 2005,
LOAELs from TOPKAT® training set

Figure 6.9 The relationship between the actual chronic LOAELs and the rat TD_{50} for compounds in the CPDB. Correlation coefficient $r^2 = 0.52$, with the fit indicated by the dashed line. The solid line reflects linear relationship between the actual LOAEL and the rat TD_{50}.

develop mental toxicants to those with a predicted LOAEL < 10 mg/kg, rather than using the 1 mg/kg cutoff used in the absence of such predictions.

Target Organs

The organs targeted by halogenated compounds vary widely (Gold 2005). Among the DBPs that have been examined experimentally, liver, kidney cancer and/or toxicity predominate (see Table 1.1). In some cases thyroid tumors are observed. Thyroid cancer is a common finding with halogenated compounds as they frequently affect the rate at which thyroid hormones are metabolized when given at high doses (Thomas and Williams 1999). Colon cancer has been produced in rats treated with bromodichloromethane and bromoform when administered by gavage in a corn oil vehicle. However, the bromodichloromethane result was not confirmed when it was administered in drinking water at concentrations that resulted in similar daily doses (NTP 2005b). Colorectal cancer has been associated with chlorinated water in some studies (Mills et al. 1998). A single study has associated chlorinated drinking water with brain cancer (Cantor et al. 1999). The only DBP found to produce nervous system tumors in animals has been NDMA.

De novo prediction of target organs of potential carcinogens was not possible with TOPKAT®. Since the models focus on the electrotopological characteristics of the molecule rather than the specific elements, this could not be justified. Many times the structure with the lowest similarity index to a halogenated query structure did not even contain halogen. As a consequence, the potential for possible congruity of target organs could only be based on empirical comparisons of the query structure with the target organs of carcinogens with similar structures identified in the open scientific literature.

The USEPA does not require or make use of target organ specificity when evaluating risks for carcinogens (USEPA 1994 & 2002). Lack of congruence in target organs among species is not uncommon. The differing genetic makeup among strains of the same species also frequently results in differing organs being affected by exposure to a carcinogen.

The bladder is a well established target organ for carcinogens in humans. The metabolic bases for responses to established bladder carcinogens have been substantially worked out, especially within the aromatic and heterocyclic amine classes, provides a basis for identifying chemicals as potential causes. This published work was used to evaluate which DBPs were candidates as potential bladder carcinogens.

Mechanisms

The mechanisms by which halogenated compounds produce cancer vary widely. The nature of the mechanism can be very important in estimating the extent to which a particular compound contributes to cancer at the levels found in drinking water. Halogens are frequently referred to as good leaving groups. In simple terms, this means that in the presence of other reactants, usually nucleophiles, the halogen is lost as a result of the reaction. Some halogenated compounds are sufficiently unstable that they react readily with water because of water's high concentration in biological systems. Other halogenated compounds are less reactive and preferentially react with strong nucleophiles found within cells. Some chemicals are not themselves electrophiles, but an electrophilic form of the chemical is produced enzymatically. A typical bulk nucleophile present in mammalian systems is the sulfhydryl group present in the tri-peptide, glutathione, or in the amino acid cysteine and amino groups within amino acids, proteins and nucleic acids. Halogenated compounds that are electrophiles also react with DNA bases if the active form reaches this target. Halogenated compounds that fall into this category are frequently mutagenic.

In some compounds, the halogen is not particularly reactive on its own or even through enzymatic action. When these compounds are carcinogenic it is because of some relatively specific interactions with biological molecules that modify physiological function. This may happen when substitution with halogens is dense, particularly if the halogens are on the same carbon or on an aromatic ring. In such cases the halogen confers non-polar characteristics on the molecule. Non-polar interactions are frequently important in receptor interactions. Thus, many chemicals with these characteristics produce cancer by interfering with normal homeostasis rather than producing a somatic mutation.

Compounds that produce cancer by interacting with DNA and that produce mutations are thought not to have thresholds. In other words, it is thought that there is a dose-dependent probability of inducing cancer at any non-zero dose. Chemicals that produce cancer by altering the physiology of the organ in which cancer develops are generally thought to have non-linear dose-response curves, which can be thought of in terms of a threshold dose below which no cancer would result.

A third group of halogenated compounds can also be imagined. These would be compounds that are sufficiently reactive to be cytotoxic and kill cells. For a variety of reasons these compounds do not reach DNA or the products they form with DNA do not have characteristics that will result in a mutation when DNA is replicated. In sensitive species, repeated cytotoxic damage to organs can result in a carcinogenic response as a result of repeated bouts of reparative hyperplasia.

A fourth group of chemicals can be identified that have both mutagenic and cytotoxic properties. If cytotoxicity or other modes of action (e.g. specific inhibition of DNA repair) can be shown to occur at much lower doses than chemically-mediated mutagenic effects, a non-linear approach to risk assessment might be justified. As indicated in Chapter 1, several of the regulated DBPs may fit into this category (e.g. THMs, HAAs, and bromate).

In reality, it is easy to identify halogenated chemicals that clearly fit into all of these categories. However, it is not possible to classify all halogenated compounds into one group or the other. Rather it is necessary to experimentally demonstrate the nature of the compound's mode of action. This can be an expensive and sometimes difficult process. Within the group of DBPs, MX is an example of a mutagenic carcinogen, chloroform as one that acts as a cytotoxic carcinogen, and dichloroacetic acid as a carcinogen that acts largely through alteration of the physiological of an organ. Theoretically, the model used to extrapolate risks from high doses in animals to low levels that humans are exposed to in drinking water should differ with these three groups of compounds.

Role of Metabolism and Kinetics

The diversity of halogenated chemicals makes it very difficult to generalize about metabolism. Halogenated compounds may be activated or inactivated by metabolism. This is largely dependant on the nature of intermediates that are formed. Nevertheless, consideration of the metabolism or lack of metabolism that a halogenated compound undergoes is essential for identifying mechanisms. Local metabolism can also be an important determinant of which organs are affected. The number of compounds addressed in this effort has precluded a review of potential routes of metabolism for most DBP groups. Metabolism has been shown to play important roles within those classes that have been studied extensively (trihalomethanes, haloacetic acids, and haloacetonitriles). Metabolism is not explicitly addressed in TOPKAT® models. However, it is implicitly embedded in that the electrotopological properties upon which equations are built within TOPKAT® are related to toxicity of whatever form. Nevertheless, this is a shortcoming of the predictions and becomes problematic when a chemical has fragments that are not represented in the training set of chemicals for each model.

Carcinogenic Potency

Carcinogenic halogenated compounds vary widely in their potency. The distribution of potencies for halogenated carcinogens displayed in Figure 6.5 illustrates the reason why it is necessary to focus at a minimum on the individual classes of halogenated by-products, if not individual by-products, to resolve the issue of whether or not the cancer risk from chlorination of water is as high as epidemiological studies suggest.

Structure-Activity Considerations

Placement of halogenated hydrocarbons into smaller groups facilitates assessment of structure-toxicity relationships. Secondary functional groups on halogenated compounds can have large effects on toxicity. Such differences are mediated in a variety of ways. In general, polar groups can result in large differences in uptake, distribution, metabolism, and elimination of halogenated compounds. Multiple substitutions of halogen on a single carbon substantially affect the rate of metabolism and the probable reactivity of its metabolic intermediates that may be responsible for toxicity and carcinogenicity. Is the halogen on a carbon atom or another atom (e.g. oxygen or nitrogen)? The nature of the carbon backbone of a halogenated compound can be very important. Does it contain ring structures or is it a simple hydrocarbon? Are these structures saturated or unsaturated? What is the spatial relationship between the halogen and other special features of the molecule in question? These differences in structure are responsible for the very wide variation in carcinogenic potency that was displayed in Figure 6.5.

Halogenated DBPs were placed in the initial 30 categories and groups K49 and K50 that are identified in Table 6.1. Some of these groups reflect differences in functional groups and others reflect differing degrees of halogen substitution within a class that is based on its functional group. THMs remained separate from other haloalkanes simply because these are regulated compounds and have received reasonably complete study of their basic toxicological properties. For reference to the basic QSTR database, we have included the index number that was arbitrarily assigned to each class of DBPs. The number of chemicals in each of these classes varied widely; this is also indicated in the table. The small numbers in some classes prevent a characterization of the group compared to other classes. Some regrouping of chemicals was done in the final analysis to simplify the presentation of data.

Table 6.1
Classification of disinfection by-products for which TOPKAT® was used to predict their toxicological properties

Index number	Chemical class	Number of chemicals examined (not evaluated)
K01	Saturated haloacids	51
K02	Unsaturated haloacids	39
K03	Haloalcohols	20
K04	Dihaloacetaldehdyes	4
K05	Unsaturated haloketones	6
K06	Haloalkenes	5
K07	Haloesters	6
K08	Halophenols	17
K09	Other haloalkanes	13
K10	Saturated haloaldehydes	14
K11	Saturated halodicarboxylic acids	25
K12	Saturated haloketones	29
K13	Trihaloacetaldehydes	4
K14	Trihaloacetates	4
K15	Unsaturated haloaldehydes	10
K16	Unsaturated halodicarboxylic acids	17
K17	Haloacetonitriles	7
K18	Halotricarboxylic acids	2
K19	Monohaloacetates	3
K20	Dihaloacetates	5

(continued)

Table 6.1 (Continued)

Index number	Chemical class	Number of chemicals examined (not evaluated)
K21	MX and analogs	12
K22	Aromatic haloacetaldehydes	5
K23	Haloamides	24
K24	Haloaromatic acids	31
K25	Halogenated aromatic amines	7
K26	Halonitromethanes	6
K27	Halopyrroles	3
K28	Halothiophenes	7
K29	Other haloaromatics	13
K30	Other halonitriles	20
K31	Alcohols	2
K32	Aliphatic aldehydes	7
K33	Ketones	4
K34	Saturated carboxylic acids	1 (2)
K35	Unsaturated carboxylic acids	1
K36	Saturated dicarboxylic acids	0 (13)
K37	Unsaturated dicarboxylic acids	0 (2)
K38	Aromatic aldehydes	0 (19)
K39	Aromatic acids	0 (2)
K40	Heterocyclic acids	0 (2)
K41	Other heterocyclic compounds	0 (3)
K42	Cyanocarboxylic acids	0 (2)
K46	Other aromatic compounds	0 (3)
K47	Amides	2
K48	Predicted nitrosamines	7
K49	Organic N-chloramines	16
K50	By-products predicted from substructures of NOM	42

HALOGENATED BY-PRODUCTS WITH PREDICTED LOAELS < 1 MG/KG DAY^{-1}.

The first decision criterion utilized was to identify those halogenated DBPs that have chronic LOAELs that were predicted by the TOPKAT® model to be lower than 1 mg/kg. As these compounds should be of concern regardless of their more specific toxicological properties, they are identified in Table 6.2. The groups from which these are drawn are included to facilitate discussion.

The 54 compounds in Table 6.2 represent approximately 11% of the total number of putative halogenated DBPs that were evaluated (one compound, K49C15, is not halogenated, but it is a product of N-chloramine degradation). These chemicals are identified here because many do not survive the subsequent decision criteria because predictions are outside of the OPS of subsequent models or there was a fragment/s in the query structure not represented in the training set. Some of these will be identified as research priorities. Most chemicals with similarity indices >0.3 were dropped if they did not survive application of criteria used to identify the highest priority probable carcinogens or developmental toxicants. An exception to this rule was made for several chemicals identified group K50. These chemicals were haloquinones and cyclopentenoic acid derivatives that were predicted to arise from chlorine and chloramine reactions with substructures in natural organic matter (NOM). They are retained because of the exceptionally low chronic LOAELs that were predicted and because review of literature on the toxicology of quinones suggested that compounds in this group could be bladder carcinogens. A more thorough analysis and discussion of the haloquinones is presented and discussed later (Table 6.14).

Table 6.2.
Disinfection by-products with TOPKAT® predicted chronic LOAELs < 1 mg/kg day^{-1}

Index Number	Chemical	LOAEL mg/kg	Simil. Index	LOAEL Lower Limit mg/kg	LOAEL Upper Limit mg/kg
K01	**Saturated haloacids**				
K01C14	2,4-Dichloro-3-hydroxybutanoic acid	0.950	0.134	0.16	5.7
K02	**Unsaturated haloacid**				
K02C38	2,3,4-Trichlorocyclohexadien-5-one carboxylic acid	0.220	0.319	0.05	0.90
K02C05	2,3-Dichloro-4-oxopentenoic acid	0.61†	0.181	0.09	4.1
K02C18	3,5,5-Trichloro-4-oxo-2-pentenoic acid	0.64	0.181	0.10	4.0
K05	**Unsaturated haloketones**				
K05C06	Trichlorocyclopentenedione	0.50	0.233	0.15	1.7
K06	**Haloalkenes**				
K06C03	Hexachlorocyclopentadiene	0.010		0.00	0.04
K10	**Saturated haloaldehydes**				
K10C07	2,3,4-Trichlorobutanal	0.960	0.137	0.16	5.9
K11	**Saturated halodicarboxylic acids**				
K11C23	Chlorobutanedioic acid	0.850	0.131	0.14	5.1
K11C13	2,3-Dichloropentanedioc acid	0.870	0.142	0.14	5.4
K15	**Unsaturated haloaldehydes**				
K15C03	2,3-Dichloropropenal	0.970	0.164	0.16	5.9
K16	**Unsaturated halodicarboxylic acids**				
K16C04	2-Chloro-3-pentenedioc acid	0.220	0.198	0.03	1.6
K17	**Haloacetonitriles**				
K17C01	Bromochloroacetonitrile (BCAN)	0.093†	0.194	0.01	0.69
K17C05	Dichloroacetonitrile (DCAN)	0.29†	0.170	0.04	2.0
K17C03	Dibromoacetonitrile (DBAN)	0.94*	0.149	0.15	5.9
K21	**MX and analogs**				
K21C05	3-Chloro-4-(chloromethyl)-5-hydroxy-2[5H]furanone (CMCF)	0.029†	0.298	0.00	0.20
K21C11	3-Chloro-4-(dichloromethyl)-2-furanone (Red-MX)	0.074†	0.304	0.01	0.43
K21C08	3-Chloro-4-(dichloromethyl)-5-hydroxy-2[5H]furanone (MX)	0.10†	0.233	0.02	0.44
K21C01	3-Chloro-4-(bromochloromethyl)-5-hdroxy-2[5H]furanone	0.16†	0.232	0.04	0.70
K21C03	5-Hydroxy-4-trichloromethyl-2-furanone	0.21†	0.271	0.04	1.2
K21C02	3-Chloro-4-(dibromomethyl)-5-hydroxy-2[5H]furanone	0.29†	0.233	0.06	1.1
K21C12	3-Chloro-4-(tribromomethyl)-5-hydroxy-2-[5H]furanone	0.41†	0.239	0.09	1.8
K21C06	E-2-Chloro-3-(dichloromethyl)-4-oxobutenoic acid (E-MX)	0.67	0.179	0.11	4.1

(continued)

Table 6.2 (Continued)

Index Number	Chemical	LOAEL mg/kg	Simil. Index	LOAEL Lower Limit mg/kg	LOAEL Upper Limit mg/kg
K21C07	3,4-(Dichloro)-5-hydroxy-2[5H]furanone (MCA)	0.79*	0.191	0.22	2.8
K23	**Haloamides**				
K23C01	3-Bromo-3-Chloropropamide	0.62	0.339	0.10	3.7
K23C20	2,3-Dichloropropamide	0.51	0.183	0.08	9.2
K30	**Other halonitriles**				
K30C03	2,3-Dichloropropionitrile	0.043†	0.205	0.01	0.36
K30C09	3,4-Dichlorobutanenitrile	0.074†	0.202	0.01	0.61
K30C04	2,3-Dichloro-3-bromopropionitrile	0.09†	0.200	0.01	0.70
K30C07	3,3-Dichloropropionitrile	0.11†	0.198	0.01	0.93
K30C02	2,3-Dibromopropionitrile	0.15	0.183	0.02	1.1
K30C13	Cyanogen chloride	0.23†	0.157	0.04	1.5
K30C12	Cyanogen bromide	0.25†	0.157	0.04	1.6
K30C06	3,3-Dibromopropionitrile	0.35†	0.430	0.05	2.5
K30C11	2,3,4-Trichloro-2-butenenitrile	0.68†	0.156	0.10	4.5
K30C08	3,3-Dichloropropenenitrile	0.77†	0.127	0.12	5.0
K30C15	3-Bromopropionitrile	0.89†	0.205	0.14	5.7
K30C17	3-Chloropropionitrile	0.93†	0.205	0.14	6.0
K49	**Organic N-Chloramines**				
K49C11	N-Chloroglycylaldimine	0.42‡	0.165	0.071	2.5
K49C14	N-Chloro-2-methylbutyrylaldimine	0.20‡	0.213	0.019	2.2
K49C15	2-Methylbutanenitrile	0.10†	0.274	0.0074	1.4
K50	**By-products predicted from NOM substructures**				
K50C04	3,4,5,6-Tetrachloro-o-quinone	0.0047†	0.413	0.00095	0.023
K50C05	1,2,3-Triketo-4,4,5,6-tetrachlorocyclohex-5-ene	0.050	0.335	0.012	0.21
K50C06	1,2,2,4-Tetrachloro-3-hydroxycyclopent-4-enoic acid	0.19†	0.411	0.023	1.6
K50C07	3,5-Dichloro-1-hydroxy-4-ketocyclopent-2-enoic acid	0.00025†	0.635	0.000002	0.037
K50C17	2-Hydroxy-2-methyl-3-chloro-4,6-diketopyrimidine	0.60†	0.217	0.107	3.4
K50C20	Alloxan	0.30†	0.262	0.05	1.9
K50C26	1,3,3,5,5-Pentachloro-4-keto-5-hydroxycyclohexene	0.21*	0.232	0.054	0.82
K50C35	2,6-Dichloro-1,4-benzoquinone	0.049†	0.313	0.008	0.30
K50C36	2,3,6-Trichloro-1,4-benzoquinone	0.033	0.342	0.007	0.15
K50C37	2,3,5,6-Tetrachloro-1,4-benzoquinone	0.005†	0.417	0.001	0.026
K50C38	2,6-Dichloro-3-methyl-1,4-benzoquinone	0.079	0.341	0.017	0.36
K50C40	2,3,6-Trichloro-1,4-benzoquinone-4-(N-chloro)imine	0.073‡	0.328	0.019	0.29
K50C41	2,3-Dichloro-4,5-diketobenzoprop-2-enol	0.064	0.280	0.15	0.28
K50C42	2,3,5-Trichloro-4,5-diketobenzoic acid	0.073	0.362	0.16	0.34

* Outside optimal prediction space, but within permissible limits.

† Outside optimal prediction space and exceeds permissible limits.

‡ Query structure contains fragment/s not represented in any of the training sets.

Two other groups of by-products are clearly of importance based upon this initial cut, analogs of 3-Chloro-4-(dichloromethyl)-5-hydroxy-2[5H]furanone (MX) and halonitriles. The analogs of MX are listed under K50. Compounds in this group include cyclopentenoic acids and one pyrimidine derivative rather than halofuranones, but the similarities in the structures are obvious. Only one of the analogs of MX survived the decision criteria directed at this group to identify probable carcinogens. MX was not among those that were identified as probable carcinogens based upon TOPKAT® predictions. The obvious differences between these predictions and the reality of MX's demonstrated carcinogenicity will be discussed in more detail in the section devoted to the potential carcinogenicity of members of this group of compounds. The chemicals in group K50 are predicted to arise from chlorination of substructures of natural organic matter in water. As with the halofuranones, the some of predictions were outside of the OPS of the model. However, the very low chronic LOAELs predicted were noted and were of concern. These predictions were given more credibility because an experimental LOAEL for MX in the literature was within the confidence interval of the TOPKAT® prediction for MX. The following paragraph explores the accuracy of the predicted chronic LOAELs for these two classes of halogenated DBPs that are outside the OPS of the model more fully.

MX was tested in a chronic cancer bioassay in Wistar rats at doses of 0.4, 1.3 and 5 mg/kg per day (Komulainen et al. 1997). Non-tumor pathology was not discussed in detail and statistics were not provided, but it would appear that thyroid enlargement was observed at the mid-dose. There was a significant increase in thyroid follicular cell adenomas in male rats treated at 0.4 mg/kg. The actual LOAEL for MX is within the confidence interval of that predicted by TOPKAT®. Therefore, MX and its analogs were retained in the high priority group. This conclusion is extended to the two cyclopentenoic acid by-products in Table 6.3 (K50C06 and K50C07) predicted to arise from the chlorination of substructures in NOM.

Within the halonitriles group most of the predicted chronic LOAELs exceeded the OPS of the automatically selected acyclic submodel. Ethylene cyanohydrin was the closest analog. This was the only model that included nitriles. The similarity was within the acceptable range (<0.2) for most of the halonitriles. Nevertheless, only in the cases of 2,3-dibromopropionitrile and bromochloroacetonitrile (BCAN) were the predictions within permissible limits of the OPS. The chronic LOAEL listed for ethylene cyanohydrin was 5 mg/kg and referred to an EPA document that was not readily retrievable [HEED ECAO CIN G039,8808,(1)]. A broad search of literature failed to provide data to support the development of that LOAEL. TOPKAT® predictions outside the OPS are not necessarily incorrect, but one can have little confidence in them. A cursory examination of the literature for haloacetonitriles that have received some study suggests that the estimates made with TOPKAT® values may be a little too low. Specifically, dibromoacetonitrile (DBAN) was found to have a NOAEL of 23-mg/kg day^{-1} and dichloroacetonitrile (DCAN) of 8-mg/kg day^{-1} in an older subchronic study (Hayes et al. 1986). A newer study indicated that the NOAEL for DBAN was 1.1-mg/kg day^{-1} (Poon et al. 2003). The observed subchronic LOAEL in the latter study would be 11-mg/kg day^{-1} as doses were spaced by an order of magnitude. TOPKAT® predicted the LOAELs of 0.29 and 0.94 mg/kg day^{-1} for DBAN and DCAN, respectively. A NOAEL of 1 mg/kg day^{-1} was also calculated for trichloroacetonitrile (TCAN) in a teratogenicity study (Smith et al. 1988). Significant increases in fetal resorptions were observed at 7.5-mg/kg day^{-1} in this study. Comparable values for NOAEL and LOAEL of DCAN in a developmental toxicity study were 15 and 45-mg/kg day^{-1}, respectively (Smith et al. 1989a). Data from teratogenicity studies are not strictly chronic LOAELs as the exposure periods were a matter of days to weeks rather than a lifetime. Nevertheless, these data suggest that the TOPKAT® predictions of chronic LOAELs may represent levels appropriate to the endpoints of concern, developmental toxicity and cancer.

Evaluation of the chronic LOAELs predicted for MX and its analogs is somewhat more complex largely because both ring and non-ring forms of the analogs were included. Table 6.3

displays the predicted chronic LOAEL and similarity indices for these two compounds in the relevant cyclic and alicyclic models. TOPKAT® automatically selected the alicyclic model for the closed ring structures whereas the acyclic model was selected for the open structures. Those with the basic furanone ring structure were most closely associated with captan with one exception, 5-hydroxy-4-trichloromethyl-2-furanone. In this case, dimethipin was identified as the closest analog (similarity index = 0.271), with captan as the second most similar compound in the database (similarity index = 0.293). The similarity indices for the remaining ring structures ranged from 0.191 with 3,4-(Dichloro)-5-hydroxy-2[5H]furanone (MCA) to 0.304 with 3-Chloro-4-(dichloromethyl)-2-furanone (Red-MX). Therefore, those compounds that that exceeded the OPS for the model had marginal similarity indices.

The structures in the TOPKAT® acyclic model training set identified as most similar to the open ring structure of E-MX were bromodichloromethane with a similarity index of 0.179, dibromochloromethane (0.186), and monochloroacetic acid (0.188). Therefore, the acyclic model prediction for E-MX appears to be well supported in the training set. As the similarity indices with members of the training set of the alicyclic model were marginal for most of the ring structures related to MX, it was necessary to validate these predictions in the open literature as described above.

The predicted chronic LOAELs for the haloquinone derivatives are discussed in the haloquinone section (see Table 6.14). A broader evaluation of the chronic LOAEL submodels was necessary for this class, so it was incorporated into discussion of model predictions for carcinogenicity and developmental toxicity.

Table 6.3
Comparison of alicyclic and acyclic submodel predictions of the chronic LOAELs for MX and selected related compounds

Index No.	Compound	Alicyclic	Simil index	Acyclic	Simil. index
K21C05	3-Chloro-4-(chloromethyl-5-hydroxy-2[5H]furanone (CMCF)	0.029†	0.298	14.9	0.028
K21C06	E-2-Chloro-3-(dichloromethyl)-4-oxobutenoic acid (E-MX)	30.3*	0.266	0.66	0.179
K21C08	3-Chloro-4-(dichloromethyl)-5-hydroxy-2[5H]furanone (MX)	0.10†	0.233	4.5	0.088
K21C11	3-Chloro-4-(dichloro)-2-furanone	0.074†	0.304	11.1	0.065
K50C06	1,2,2,4-Tetrachloro-3-hydroxy-cyclopent-4-enoic acid	0.19†	0.411	17.9*	0.038
K50C07	3,5-Dichloro-1-hydroxy-4-keto-cyclopent-4-enoic acid	0.25†	0.635	1.9†	0.128

* Outside the OPS of the indicated model, but within permissible limits
† Outside the OPS of the indicated model and exceeds permissible limits

HALOGENATED BY-PRODUCTS PREDICTED TO BE CARCINOGENIC OR DEVELOPMENTAL TOXICANTS

Saturated Haloacids

Saturated Haloacids Produced With Disinfection

Saturated haloacids are prominently listed as by-products in Richardson et al. (1998). The TOPKAT® evaluation included the trihaloacetates (K14), monohaloacetates (K19), and the dihaloacetates (K20). Some limited occurrence data have been developed for iodinated haloacids (Plewa et al. 2004a) and were included in the analysis. There were some actual data on all of the brominated dihaloacetates that are discussed in the section dealing with characterized DBPs, but the QSTR predictions for these chemicals were useful in model calibration. It should be noted that the only monohaloacetic acid tested in chronic studies was monochloroacetic acid (NTP 1990). With the inclusion of members of this class with longer carbon chains for QSTR, it was necessary to specify isomers in instances where the data were either not explicit or where several isomers were detected, but not specifically identified. In general, the inclusion of isomers involved moving the position of halogen substitution to different points on the hydrocarbon backbone. The location of hydroxyl groups was frequently not specified. In general, these groups were confined to the 3rd or 4th carbon unless specific identification was made in the database. In all, 51 saturated haloacids were explored for QSTRs using TOPKAT®.

Characteristics of Saturated Haloacids That Appear to Have the Greatest Potential as Carcinogens

Table 6.4 summarizes the TOPKAT® predictions related to the saturated haloacids. The main point of this figure is to illustrate the range of LOAELs that was predicted for the class. The predicted chronic LOAELs within this group ranged from 0.95 to 902-mg/kg day^{-1}. Those saturated haloacids that fell within the criterion of a predicted chronic LOAEL < 10-mg/kg day^{-1} are identified in bold. All chemicals in the class are displayed to illustrate how applying this criterion excluded compounds from the priority list. Display of subsequent classes of DBPs will only include those chemicals whose chronic LOAEL was predicted to be < 10-mg/kg day^{-1}. The haloacetates have not been included in Table 6.4 as there are data available for these chemicals.

The results from models within TOPKAT® that predicted saturated haloacids to be carcinogenic were analyzed in more detail. Those compounds for which positive responses were predicted in a least one sex in two species survived the next cut resulting in the shorter list of saturated haloacids of interest (Table 6.5). The most potent chemicals in this group, as indicated by the chronic LOAEL prediction, were those containing a halogen substitution alpha to the carboxyl group. The convention of including only those compounds with a LOAEL less than 10 mg/kg was reinstated in this table and subsequent tables.

The smaller group of saturated haloacids that were predicted to be carcinogenic in two species was then subjected to a similarity search among the compounds that were used in making the predictions. The criteria suggested by the TOPKAT® instructions is that there should be at least one compound in the equation that has similarity index < 0.2. Therefore, no saturated haloacids are suggested for characterization as potential carcinogens.

Table 6.4
Summary listing of TOPKAT® predictions of carcinogenicity and chronic LOAEL for members of the saturated haloacid class of disinfectant by-products

Index Number	Chemical	Pred. Ames' mut.	No. pos. cancer pred. (8 poss.)	No. outside OPS*	LOAEL mg/kg	LOAEL lower limit mg/kg	LOAEL upper limit mg/kg
K01C14	2,4-Dichloro-3-hydroxybutanoic acid	+	5	5	0.9	0.2	5.7
K01C16	2,5,5-Trichloro-4-oxopentanoic acid	-	3	2	1.8	0.3	10.5
K01C50	2,4,4-Trichlorobutanoic acid	-	3	1	1.8	0.3	10.7
K01C19	2,6,6,6-Tetrachloro-4,5-dihydroxy-hexanoic acid	+	4	3	2.0	0.3	12.3
K01C46	2,4-Dichlorobutanoic acid	+	5	0	2.0	0.4	10.7
K01C02	2,2,6,6-Tetrachloro-4,5-dihydroxy-hexanoic acid	-	1	0	2.1	0.3	13.3
K01C15	2,5,5,5-Tetrachloro-4-hydroxy-pentanoic acid	+	3	3	2.1	0.3	13.2
K01C18	2,5-Dichloropentanoic acid	+	5	1	2.9	0.5	15.6
K01C17	2,5,5-Trichloropentanoic acid	-	4	2	3.0	0.5	16.5
K01C26	2-Chlorobutanoic acid	+	2	0	3.3	0.6	17.1
K01C20	2,6,6,6-Tetrachloro-4-hydroxy-hexanoic acid	-	5	3	3.3	0.6	19.2
K01C29	2-Chloropentanoic acid	+	2	0	4.0	0.8	21.0
K01C21	2-Chlorohexanoic acid	+	2	0	4.7	0.9	24.6
K01C28	2-Chlorohexanoic acid	+	2	0	4.7	0.9	24.6
K01C27	2-Chloroheptanoic acid	+	1	0	5.3	1.0	28.1
K01C47	2,3,3-Trichloropropanoic acid	-	4	3	5.5	1.0	29.3
K01C45	2-Bromo-3,3-dichloropropanoic acid	-	5	3	5.8	1.1	30.7
K01C44	2-Bromo-4-chlorobutanoic acid	+	5	3	9.0	1.8	45.0
K01C13	2,3,4,4,4-Pentachloro-3-hydroxy-butanoic acid	-	5	3	10.7	2.1	54.8
K01C41	4,4,4-Trichloro-3-hydroxybutanoic acid	-	3	1	11.1	1.8	67.1
K01C40	4,4,4-Trichloro-2-hydroxybutanoic acid	-	5	1	11.2	1.9	67.8
K01C37	5,5-Dibromo-4-oxopentanoic acid	+	1	0	11.5	2.4	54.4
K01C38	3-(Dichloromethyl)-2,5,5,5-tetrahloro-4-oxopentanoic acid	-	4	4	12.1	2.3	63.0
K01C24	2-Bromo-5-chloropentanoic acid	+	5	2	13.8	2.9	65.5
K01C12	2,3,3,5,5,5-Hexachloro-4-hydroxy-pentanoic acid	-	5	4	14.9	2.9	75.0
K01C25	2-Bromobutanoic acid	+	3	1	16.6	3.7	75.6

(continued)

Table 6.4 (Continued)

Index Number	Chemical	Pred. Ames' mut.	No. pos. cancer pred. (8 poss.)	No. outside OPS*	LOAEL mg/kg	LOAEL lower limit mg/kg	LOAEL upper limit mg/kg
K01C42	5,5,5-Trichloro-4-hydroxy-pentanoic acid	-	4	3	17.7	3.2	96.5
K01C01	2,2,6,6,6-Pentachloro-4-hydroxy-hexanoic acid	-	5	3	18.7	3.0	115.8
K01C32	3,3,3-Trichloro-2-hydroxypropanoic acid	-	4	3	36.1	7.9	166.0
K01C48	2,2,3-Trichloropropanoic acid	-	2	1	36.4	8.0	165.5
K01C36	3,3-Dibromo-3-chloropropanoic acid	-	1	0	36.7	8.0	167.4
K01C39	3-Bromo-3,3-dichloropropanoic acid	-	1	0	40.2	8.9	182.1
K01C23	2-Bromo-2-chlorobutanoic acid	-	2	0	40.3	9.0	179.3
K01C06	2,2-Dichloro-3-hydroxypropanoic acid	-	2	0	52.4	11.2	245.5
K01C11	2,2-Dichloropropanoic acid	+	1	0	55.9	12.9	242.9
K01C34	3,3,4,4,4-Pentachloro-2-hydroxy-butanoic acid	-	4	4	61.4	13.6	276.3
K01C04	2,2-Dibromobutanoic acid	-	2	0	61.5	14.1	269.3
K01C08	2,2-Dichlorobutanoic acid	-	1	0	62.9	14.5	273.4
K01C05	2,2-Dibromopentanoic acid	-	3	2	65.1	14.9	285.3
K01C10	2,2-Dichloropentanoic acid	-	3	2	69.7	16.0	303.2
K01C35	3,3,5,5,5-Pentachloro-4-hydroxy-pentanoic acid	-	5	3	72.5	16.3	322.4
K01C33	3,3,3-Trichloropropanoic acid	-	3	0	73.0	16.8	316.6
K01C31	3,3,3-Trichloro-2-hydroxy-2-methylpropanoic acid	-	2	1	75.1	17.2	327.3
K01C49	2,2-Dibromo-4-hydroxybutanoic acid	-	5	2	75.1	16.3	345.6
K01C09	2,2-Dichlorohexanoic acid	-	2	0	76.4	17.6	332.4
K01C51	4,4,4-Trichlorobutanoic acid	-	5	1	83.4	19.3	361.3
K01C43	5,5,5-Trichloropentanoic acid	-	5	1	92.7	21.4	401.4
K01C22	2-Bromo-2-chloro-4-hydroxy-pentanoic acid	-	0	0	370.6	51.8	2700.0
K01C03	2,2-Dibromo-4-hydroxy-pentanoic acid	-	2	0	511.9	72.6	3600.0
K01C07	2,2-Dichloro-4-hydroxy-pentanoic acid	-	1	0	515.3	73.4	3600.0
K01C30	2-Chloropropanoic acid	+	2	0	902.0	76.4	10000.0

* OPS = Optimal predictive space of the model.

Table 6.5
Prioritization of saturated haloacids based upon strength of evidence provided by TOPKAT® predictions

Index Number	Chemical	Pred. Ames' mut.	Male rat	Female rat	Male mice	Female mice	LOAEL mg/kg
K01C46	2,4-Dichlorobutanoic acid	+	pos	pos	pos	pos	2
K01C18	2,5-Dichloropentanoic acid	+	pos	pos	pos	pos*	2.9
K01C26	2-Chlorobutanoic acid	+		pos	pos		3.3
K01C29	2-Chloropentanoic acid	+		pos	pos		4
K01C21	2-Chlorohexanoic acid	+		pos	pos		4.7
K01C28	2-Chlorohexanoic acid	+		pos	pos		4.7
K01C45	2-Bromo-3,3-dichloropropanoic acid	-	pos	pos	pos	pos*	5.8

* Outside the optimal predictive space of the model, but within permissible limits

Characteristics of Saturated Haloacids That Appear to Have the Greatest Potential as Developmental Toxicants

Several saturated haloacids were identified as potential developmental toxicants by TOPKAT® (Table 6.6). It is of interest that these chemicals were largely distinct from those predicted as carcinogens. All eleven of the identified chemicals were within the OPS of the model.

Table 6.6
Members of the saturated haloacid class predicted to be developmental toxicants

Index Name	Chemical	Probability of develop. toxicity	Within OPS*?	LOAEL mg/kg	LOAEL lower limit mg/kg	LOAEL upper limit mg/kg
K01C06	2,2-Dichloro-3-hydroxypropanoic acid	1.0	Yes	52	11	246
K01C31	3,3,3-Trichloro-2-hydroxy-2-methylpropanoic acid	1.0	Yes	75	17	327
K01C48	2,2,3-Trichloropropanoic acid	1.0	Yes	36	8	166
K01C04	2,2-Dibromobutanoic acid	1.0	Yes	62	14	269
K01C08	2,2-Dichlorobutanoic acid	1.0	Yes	63	15	273
K01C11	2,2-Dichloropropanoic acid	1.0	Yes	56	13	243
K01C23	2-Bromo-2-chlorobutanoic acid	0.99	Yes	40	9	179
K01C33	3,3,3-Trichloropropanoic acid	0.99	Yes	73	17	317
K01C36	3,3-Dibromo-3-chloropropanoic acid	0.98	Yes	37	8	167
K01C39	3-Bromo-3,3-dichloropropanoic acid	0.98	Yes	40	8.9	182
K01C51	4,4,4-Trichlorobutanoic acid	0.95	Yes	83	19	361

* OPS = Optimal predictive space of the model

None of the chemicals in the saturated haloacid class identified as developmental toxicants met the criteria for a predicted LOAEL below 10 mg/kg. Therefore, none are carried forward as high priority in this area. Those chemicals in the saturated haloacid group that are predicted to be developmental toxicants are distinguished from those predicted to be carcinogens by multiple substitutions of halogen on a single carbon. All are clearly related to the di- and trihaloacetic acids in terms of polarity. Finally, the chronic LOAEL falls into a narrow band that is close to the actual LOAELs obtained experimentally with the dichloroacetic and trichloroacetic acids. Experimentally, the di- and trihaloacetic acids have been shown to be active as developmental toxicants at or near the LOAEL for chronic toxicities, including carcinogenicity (Smith et al. 1989b).

Unsaturated Haloacids

Unsaturated Haloacids Formed With Disinfection

Unsaturated haloacids were identified in the Richardson et al. (1998) report. As with other groups a number of structural isomers were reported. Consequently, several possible structural isomers were added to round out the TOPKAT® analysis of this group of DBPs.

Unsaturated Haloacids Predicted to Have the Greatest Potential as Carcinogens

A total of 39 unsaturated haloacids were subject to analysis. Chemicals in this class meeting the criterion of a predicted LOAEL < 10-mg/kg day^{-1} and predicted to be carcinogens are listed in Appendix E, Table E.4.

The unsaturated haloacids that were evaluated as carcinogens were generally predicted to be more toxic than the saturated haloacids. Twenty-four chemicals evaluated with TOPKAT® had chronic LOAELs lower than 10-mg/kg day^{-1} and two of these had predicted LOAELs below 1-mg/kg day^{-1}. TOPKAT® did not distinguish between cis and trans forms of the same chemical so the predictions of toxicity are identical where they were listed, reducing the effective number of unsaturated haloacids to meet the criteria to twenty-one. Because of limited time only a few structural variants of the unsaturated haloacids were explored. A potentially useful follow-up exercise would be to generate all possible structural isomers and to repeat the analysis.

The next step was to identify those unsaturated haloacids for which TOPKAT® predicted a positive response in two species. These compounds are identified in Appendix E, Table E.5. A similarity search was made of the database for each model within TOPKAT® for which there was a positive prediction. Instructions for the program indicate that greater confidence can be had in a prediction of carcinogenicity if the similarity index is less than 0.2. Those with similarity indices < 0.3 are identified in Appendix E, Table E.6. A distinction is made between those compounds whose predictions depend upon ranges of similarity indices between 0.2-0.3 in the final prioritization scheme. In any case, these data point strongly to 3,3-dibromopropenoic acid as an important DBP for study. Other unsaturated haloacids for which the positive predictions were most robust include 2,3-dichloro-4-oxopentenoic acid, 3,3-dichloropropenoic acid, and 3-chloro-2-butenoic acid.

Unsaturated Haloacids Predicted to be Developmental Toxicants

Only four of the unsaturated haloacids assessed with TOPKAT® were classified as developmental toxicants. These are identified in Appendix E, Table E.7. The prediction for one of those (2,3,3-trichlorocyclohexadien-5-one carboxylic acid) was outside OPS of the model. It remains of concern because its predicted chronic LOAEL was 0.22-mg/kg day^{-1}. One compound, 3-chloro-butenoic acid, stands out from the others in the group as it has a predicted LOAEL of 2.8-mg/kg day^{-1}. The predictions of the chronic LOAEL for the remaining two compounds were in excess of 10-mg/kg day^{-1}.

Haloalcohols

Haloalcohols Formed With Disinfection

The haloalcohols included in the present analysis were drawn from Richardson et al. (1998). Additional compounds, primarily bromohydrins, were added based on the paper of Collette et al. (1994) that studied production of ozonation by-products. In addition, some alcohols were added to this list based on identification of the corresponding haloaldehydes and haloacids in chlorinated water. The amounts of the alcohol relative to those of the corresponding aldehydes and acids are likely to be a function of how oxidizing the conditions of disinfection are. For example, chloramination might be thought to result in the formation of higher amounts of the corresponding alcohol than break-point chlorination.

Haloalcohols Predicted to Have the Greatest Potential as Carcinogens

Table E.8, Appendix E lists those haloalcohols that were predicted to be carcinogenic in at least one animal model where the TOPKAT® model indicated a chronic LOAEL of less than 10-mg/kg day^{-1}. The list includes five of the twenty chemicals identified with the group. Two of these chemicals are analogs of dihaloacetic acids (bromochloroethanol and dichloroethanol). Their predicted chronic LOAELs are somewhat lower than the actual LOAELs that have been reported for the corresponding acid. However, the chronic LOAELs that TOPKAT® predicted for dichloroacetate and bromochloroacetate (5.9 and 2.7-mg/kg day^{-1}, respectively) were also lower than the actual LOAELs. These latter figures are consistent with the predictions for dichloroethanol (5.2-mg/kg day^{-1}) and bromochloroethanol (5.2-mg/kg day^{-1}). Therefore, the program was consistent in its predictions for these related compounds. The other three haloalcohols were 2-chloro-4-dimethylbutanol, 3-chloro-2-pentanol, and 2,5,5-trichloropentanol.

No haloalcohol was predicted to be carcinogenic in two species. Therefore, no member of this group was carried forward as a priority. It should be pointed out that the corresponding acid to dichloroethanol is in fact carcinogenic in both mice and rats and that dichloroethanol will be converted to dichloroacetic acid in vivo. If this event were necessary for carcinogenicity, the cancer risk would be equal to the sum of the risks for corresponding haloalcohols plus the haloacids.

Haloalcohols Predicted to be Developmental Toxicants

The haloalcohols predicted to be developmental toxicants are listed in Appendix E, Table E.9. Ten of twenty-nine haloalcohols subjected to analysis were predicted to be developmental toxicants. Trihalosubstituted ethanols were predicted to be developmental toxicants, while the dihaloethanols were not. This outcome appears to be influenced by the fact that trichloroacetate is included in the TOPKAT® database. However, none of the haloalcohols predicted to be developmental toxicants were predicted to have a chronic LOAEL less than 10-mg/kg day^{-1} (range from 62.2 – 453-mg/kg day^{-1}).

Haloaldehydes

Haloaldehydes Formed in Disinfection

Although the original analysis separated haloaldehydes into five subgroups (K04, K10, K13, K15, K22), they are combined here for the purposes of simplifying the presentation of the results of the QSTR analysis. The majority of the compounds included in the analysis were obtained from the reviews of Richardson et al. (1998), but selected analogs of identified haloacid DBPs were added to more thoroughly explore the potential health hazards from this class.

Haloaldehydes Predicted to Have the Greatest Potential as Carcinogens

Table E.10, Appendix E lists those aldehydes, by subclass, that met the criteria of being predicted as carcinogens in at least one TOPKAT® model and that were predicted to have a chronic LOAEL below 10 mg/kg day^{-1}. Two classes of haloaldehydes contributed no members to this list, the trihaloacetaldehydes (K13), and aromatic haloacetaldehydes (K22). However, there were two dihaloacetaldehydes, six saturated haloaldehdyes, and 9 unsaturated haloaldehydes that met the initial criterion as potential carcinogens.

Those haloaldehydes for which there were positive predictions of carcinogenic activity that were within the OPS for two species were identified (Table E.11, Appendix E). Dichloroacetaldehyde, 3,3-dichloropropanal and 7 unsaturated haloaldehydes met this criterion.

Dihaloacetaldehydes are inevitable intermediates in the conversion of the corresponding haloalcohols to the haloacids. It is important not to succumb to the temptation of combining these aldehydes with the alcohols and acids. The dihaloacetaldehydes have a much greater genotoxic potential than the dihaloacids (and probably the dihaloalcohols) (Bignami et al. 1988; Guengerich and Persmark 1994; Moore and Harrington-Brock 2000; IARC, 2004). The reason is that the concentrations of haloaldehdyes will be maintained at a very low level as an intermediate in vivo because the alcohol form is favored by the cellular redox potential. Once the aldehyde has been converted to the acid, it cannot be reconverted to the aldehyde. Consequently, virtually all of alcohols that are not glucuronidated and eliminated in the urine, will eventually end up as the corresponding acids. Thus, the peak concentrations of a bolus dose of ingested dihaloacetaldehydes will give rise to higher peak concentrations than would be produced by metabolism of an equivalent dose of the alcohol.

Finally, the similarity indices for each of the positive model predictions were examined. These are displayed in Table E.12, Appendix E in approximate order of the strength of evidence developed against the criteria of similarity indices <0.2 or between 0.2 and 0.3. Elimination of aldehydes that did not meet these criteria left dichloroacetaldehyde, 2,3-dichloropropenal, and 2-chloropropenal as the members of this class that are of a higher research priority.

Haloaldehydes Predicted to be Developmental Toxicants

Among the five classes of aldehydes reported to be DBPs, two contained chemicals that were predicted to be developmental toxicants with a probability of greater than 0.7 (Table E.13, Appendix E). Of a total of 37 aldehydes studied, seven were identified in the saturated aldehyde group and included all four possible brominated and chlorinated analogs of chloral hydrate. None of these chemicals were predicted to have a chronic LOAEL below 10 mg/kg (predictions ranged from 28 to 72 mg/kg day^{-1}). Therefore, no halogenated aldehyde was assigned a high priority for study as developmental toxicants.

Haloketones

Haloketones Produced in Disinfection

The haloketones listed in the reviews of Richardson et al. (1998) fall into two groups, unsaturated haloketones (K05) and saturated haloketones (K12). Six compounds were included in the TOPKAT® analysis in the first group and fourteen in the second.

Haloketones Predicted to Have the Greatest Potential as Carcinogens

Of the 20 haloketones included as query structures, only three were to fit the initial criterion for carcinogens of potential importance (Table E.14, Appendix E). One of these was 1,1,3-tribromo-3-chloropropanone in the class of saturated haloketones. Two were identified in the unsaturated haloketone class, trichlorocyclopentenedione and 1-chloro-3-buten-2-one. These were subjected to analysis for positive predictions in more than one species. Only 1,1,3-tribromo-3-chloropropanone survived the application of this latter criterion.

The similarity indices for each cancer prediction for 1,1,3-tribromo-3-chloropropanone are displayed in Table E.15, Appendix E. Only a single prediction (FDA male rat model) was found to have a similarity index less than 0.3. No prediction had a similarity index of < 0.2. Therefore, the QSTR analyses provide only minimal support for further research on haloketones.

Haloketones Predicted to be Developmental Toxicants

Three unsaturated haloketones and twelve saturated haloketones were found to have a probability > 0.7 for being developmental toxicants (Table E.16, Appendix E). All were found to be within the OPS of the TOPKAT® developmental toxicity potential model. All were also found to have predicted chronic LOAELs > 10-mg/kg day^{-1} (ranged from 29-328-mg/kg day^{-1}). Two of the unsaturated haloketones, trichlorocyclopentadione and 1-chloro-3-buten-2-one had predicted chronic LOAELs of 0.5 and 3-mg/kg day^{-1}, respectively. Both compounds had similarity indices > 0.2 and were eliminated from further consideration as developmental toxicants.

Haloacetonitriles, other Halonitriles, Halopyrroles, and Halonitromethanes

Haloacetonitriles, Other Halonitriles, Halopyrroles, and Halonitromethanes Reported to Occur With Disinfection

Four nitrogen-containing groups of DBPs are combined to simplify the presentation of the results. Seven haloacetonitriles (K17) are considered with fourteen other halonitriles (K30), three halopyrroles (K27), and six halonitromethanes (K26). As in previous groups of by-products, most of these structures were identified in the papers of Richardson et al. (1998, 1999a & b, 2003). As with prior groups discussed a few compounds were added to provide a more robust analysis of these classes of DBPs. Virtually all of the "other halonitriles" fall into the latter category. The authors are not aware of such compounds being reported in drinking water, but some of these compounds are likely by-products based upon predictions of reactions with precursor substructures in NOM. Others were added to explore the potential toxicity of an additional carbon in the background (all had three or four carbon chains) with varying placement of the halogen substitutions.

Prediction of Carcinogenic Activity among Haloacetonitriles, Other Halonitriles, Halopyrroles, and Halonitromethanes

The chemicals in these groups that had predicted chronic LOAELs of less than 10-mg/kg day^{-1} and at least one prediction of a carcinogenic response by the TOPKAT® cancer models are listed in Table E.17 in Appendix E. A large number of halonitriles were found to meet the criteria set to identify compounds as potential carcinogens.

None of the six halopyrroles were found to have a predicted chronic LOAEL below 10-mg/kg day^{-1}. For example, the 2,3,5-tribromopyrrole was predicted to be carcinogenic by the FDA male rat model, but its chronic LOAEL was predicted to be 560-mg/kg day^{-1}. Two of six halonitromethanes met these criteria, with both being dihalogenated rather than trihalogenated analogs that have been found in drinking water. In contrast virtually all of the halonitriles were identified as potential carcinogens and all of the predicted chronic LOAELs were well below the 10-mg/kg day^{-1} criteria. As discussed in prior sections, most of the predicted LOAELs for the halonitriles were outside the OPS of the model, and many exceeded the permissible limits of the model.

Compounds within the halonitrile class were examined to identify those which were predicted to be carcinogenic in at least one model for each of the two species. These chemicals are displayed in Table E.18 of Appendix E.

Three of the haloacetonitriles and eight of the other halonitriles were found to have positive predictions of carcinogenic activity in two species (Table E.19, Appendix E). Two halonitromethanes survived the application of this criterion, dichloronitromethane and dibromonitromethane. As has been pointed out previously, the predicted chronic LOAEL values for the halonitriles were generally below 1 mg/kg, but most exceeded the permissible limits of the OPS as well.

The halonitriles yield the most consistent and defensible predictions of carcinogenic activity among all the classes of halogenated DBPs that were examined. Fully nine of ten chemicals listed in Table E.19, Appendix E have at least one positive prediction based on a similarity index of < 0.2. The only compound found outside this range was cyanogen bromide, which had two predictions with similarity indices < 0.3. Dichloroacetonitrile, trichloroacetonitrile, 2,3-dibromopropionitrile, 3,3-dichloropropenenitrile, 2,3-dichloropropionitrile and 3,3,3-trichloropropenenitrile were consistently predicted to be carcinogenic. 2,2-Dichloropropionitrile was set aside because the three predictions of carcinogenic activity fell between 0.2 and 0.3.

None of the halonitromethanes survived the criterion of a similarity index of < 0.3. The present analysis provides little support for further research related to the carcinogenicity of the halonitromethanes.

As the halonitriles appear to be one of the more important classes of halogenated DBPs based upon TOPKAT® predictions, it is useful to examine the bases of the models that appeared critical in the predictions. The differing predictions are significantly affected by the similarity of chemicals recognized in each of the databases. This can be illustrated by examining the differing bases for predictions for 3-chloropropionitrile and 3-bromopropionitrile, by-products that differ only in the nature of the halogen substitution.

Table 6.7
Contrast in the most similar chemicals identified in the databases for each model used to predict the carcinogenicity of 3-chloropropionitrile and 3-bromopropionitrile

	3-Chloropropionitrile			3-Bromopropionitrile		
Model	Prob. carc.*	Most similar chemical in database	Simil index	Prob. carc.	Most similar chemical in database	Simil. index
NTP – M rat	0.035	1,2-Dichloropropane	0.061	0.054	1,2-Dichloropropane	0.088
FDA – M rat	0.021	Acetamide	0.338	0.001	Acetamide	0.326
NTP – F rat	0.003	Acetonitrile	0.543	1.000	1,2-Dibromoethane	0.140
FDA – F rat	0.433	Ethanol	0.322	0.963	Ethanol	0.280
NTP – M mouse	0.000	n-Butyl chloride	0.256	0.000	Acetonitrile	0.235
FDA – M mouse	0.347	Dichlorvos	0.388	0.903	Iodoform	0.885
NTP – F mouse	0.006	Trichlorofluoromethane	0.100	0.999	1,2-Dibromoethane	0.147
FDA –F mouse	0.408	Chlormadinone	0.397	1.000	Milrinone	0.238

* Probability that the query structure is a carcinogen by the specific cancer model used within TOPKAT®.

Data in Table 6.7 illustrate how differences in the positive predictions arise. In this case the positive prediction of 3-bromopropionitrile was strongly influenced by bromine substitution as opposed to chlorine substitution. It met our criteria as a carcinogen, a fact influenced by the presence of a strong carcinogen with similarity indices < 0.2 for two models. In contrast, 3-chloropropionitrile had no positive predictions of P > 0.7 and the similarity indices in NTP female rat and mouse models reflected very poor matches with the database.

The position of the bromine had a varying impact on the overall ranking in that 2-bromopropionitrile (not shown) did not meet the criteria for carcinogenicity and was predicted to have a higher chronic LOAEL (19.6-mg/kg day^{-1}). On the other hand, 2-chloropropionitrile (not shown) was predicted to have a relatively low chronic LOAEL (3.8-mg/kg day^{-1}) and was predicted to be carcinogenic in only one model, the NTP female mouse.

Prediction of Developmental Toxicity Among Haloacetonitriles, Other Halonitriles, Halopyrroles, and Halonitromethanes

The by-products in these classes that have a high probability of being developmental toxicants are identified in Appendix E, Table E.20. A significant number of the halonitriles were projected to be developmental toxicants. As previously noted, the predicted chronic LOAELs for this group were largely outside the OPS. The limited number of halonitriles for which toxicological data exist suggests that these chronic LOAELs are underestimates, but the experimentally determined LOAELs are within the confidence intervals of the predictions. These differences are too small to discount the consistent prediction of potential developmental toxicity from many of the halonitriles. Toxicological investigation of selected members of this class of DBPs for developmental effects is clearly indicated. A key question is whether or not some of the halonitriles with longer carbon chains are as potent as implied by their predicted chronic LOAELs. Virtually all of the halopropionitriles were predicted to be developmental toxicants as was the one halobutanenitrile examined. While there was significant variation in the predicted chronic LOAEL based on the type of halogen substitution or its position, this did not significantly affect the probability that a halonitrile was predicted to be a developmental toxicant. Compounds with saturated carbon chains were as active as those with unsaturated chains. The final prioritization among the halonitriles should be modified by the likelihood of appropriate precursors and a preliminary assessment of their occurrence and stability in drinking water.

Some halonitromethanes were also predicted to be developmental toxicants. The trihalogenated members of this group were predicted to have chronic LOAELs above 10 mg/kg and would seemingly be of no more concern than the corresponding trihalomethanes. One of the monohalonitromethanes evaluated, bromonitromethane, was predicted to have a chronic LOAEL less than 10 mg/kg. Concerns over this compound should be tempered by occurrence data.

Of the halopyrroles examined only 2,3,5-tribromopyrrole was predicted to be a developmental toxicant. However, to the extent that the predicted chronic LOAEL can be considered indicative of its potency as a developmental toxicant, this chemical should not have a high priority for research without some experimental data to raise the level of concern.

Haloalkanes and Haloalkenes

Haloalkanes and Haloalkenes Formed in Disinfection

The THMs are the classic members of this group. As the THMs received significant toxicological study and are discussed in the section dealing with characterized DBPs they are not included in the TOPKAT® analyses. There have been a number of other halogenated alkanes (n = 13) and alkenes (n = 5) that were included in the reviews of DBPs published by Richardson (1998).

Haloalkanes and Haloalkenes Predicted to Have the Greatest Potential as Carcinogens

Haloalkanes that were predicted to have a chronic LOAELs < 10 mg/kg and to be carcinogenic by at least one TOPKAT® model are included in Table E.21, Appendix E. The haloalkane group included a variety of dihalogenated methanes, ethanes and propanes, but only 1,2,3-trichloropropane was predicted to be a carcinogen and to have a chronic LOAEL below 10 mg/kg. Among the haloalkenes evaluated only hexachlorocyclopentadiene and tetrachlorocyclopropene met these criteria. The inclusion of these compounds as DBPs is suspect. 1,2,3-Trichloropropane is a pesticide and hexachlorocyclopentadiene a pesticide precursor. While these compounds were found in chlorinated water, the rigor with which they were shown to be by-products of disinfection was not described in the reviews and it was difficult to track back to the original observation (Richardson 1998). However, 1,2,3-trichloropropane could conceivably be formed by chlorination of drinking water. It was the only compound that met the criterion of having positive predictions of carcinogenicity that were within the OPS of the models in more than one species. The similarity indices for these predictions are depicted in Table E.22, Appendix E.

Data available on 1,2,3-trichloropropane do provide some insight into how different models in TOPKAT® should be evaluated. This compound was tested in the NTP bioassay program and shown to be carcinogenic and as a result is in the each of the NTP models. Since this compound was not in the FDA databases, predictions with the FDA models were based upon other chemicals in the respective databases. While there was one negative prediction for the female rat in the FDA database, the major difference in the result was that the predictions in female rat, male and female mouse were found to be outside the outside the OPS of these models. This result justifies the use of a positive prediction of carcinogenicity for a given species or gender with only one of the four models/species being considered sufficient to identify the chemical as a probable carcinogen if there are compounds of sufficient similarity in the training set.

Haloalkanes and Haloalkenes Predicted to be Developmental Toxicants

Three haloalkanes (hexachloroethane, dibromomethane, and dichloromethane) and one haloalkene (bromocyclohexene) were predicted to be developmental toxicants (Appendix E, Table E.23). Based upon precursors available and consideration of chlorination chemistry, the three haloalkanes may not be DBPs. The documentation of their formation was not available in the review articles and it was difficult to trace the basis of their identification as such from the review articles searched. In any case, the predicted chronic LOAELs for the haloalkanes were in excess of 10-mg/kg day^{-1} (range 89-492 mg/kg day^{-1}). Bromocyclohexene was also predicted to have a very high chronic LOAEL (664 mg/kg day^{-1}). As a result, expending resources to determine whether these chemicals are developmental toxicants has a low priority.

Haloesters, Haloamides, and Halothiophenes

Haloesters, Haloamides, and Halothiophenes Reported to be Formed with Disinfection

Again, the major source for identifying members of these chemical groups as DBPs was the review of Richardson (1998). As with other groups, a few compounds have been added to the haloamide group, in particular, to explore how variation in structures may affect the predicted toxicological outcomes.

Haloesters, Haloamides and Halothiophenes Predicted to Have the Greatest Potential as Carcinogens

The chemicals in these three classes that were predicted to have chronic LOAELs below 10-mg/kg day^{-1} and to satisfy the initial criterion of at least one TOPKAT® model as carcinogens are listed in Table E.24, Appendix E. One of six haloesters, ten of twenty-four haloamides, and one of seven halothiophenes evaluated were found to meet the dual LOAEL and carcinogenic criteria to be carried forward.

Table E.25, Appendix E identifies those chemicals within these groups that were predicted to be carcinogenic in both mice and rats. All were found to be in the haloamide group and included bromochloroacetamide and dichloroacetamide. These two chemicals can be precursors of the two corresponding dihaloacetic acids commonly observed as chlorination by-products.

The similarity indices for these compounds are presented in Table E.26, Appendix E. Bromochloroacetamide has two predictions with probabilities exceeding 0.9 and similarity indices < 0.3. One of these predictions falls within the similarity index criteria of < 0.2 (NTP female mouse model). Two other positive predictions were obtained, but these were either outside the OPS or had similarity indices > 0.3. 2,2-Dichloroacetamide was predicted to have one predicted probability > 0.9 to be carcinogenic in the FDA male rat model and a similarity index < 0.2. Two additional positive predictions were made for dichloroacetamide (FDA male and female mouse models), but these did not meet the criteria for similarity.

Haloesters, Haloamides, and Halothiophenes Predicted to be Developmental Toxicants

Table E.27, Appendix E lists those haloesters, haloamides and halothiophenes that were predicted by the TOPKAT® model to be developmental toxicants. A large fraction of the members of all three classes were predicted to be developmental toxicants. Most had predicted chronic LOAELs above 10 mg/kg. It is of interest that the haloesters predicted to have developmental toxicity potential were confined to the monohalogenated forms. In contrast, among the haloamides all of the predicted developmental toxicants had two or three halogens substituted on a single carbon. The halothiophenes predicted to be developmental toxicants all had three or more halogens, but in two cases, these substitutions were split among 3 carbons.

Judged by their high probability of being a developmental toxicant and chronic LOAELs < 10-mg/kg day^{-1}, six propamides and two butamides appear to be of concern as potential developmental toxicants. Two trihaloacetamides were among the group as well, but their chronic LOAELs > 10-mg/kg day^{-1}. It is of potential interest that butamides with a single halogen at the 2-carbon appear to be of more concern that propamides with a similar substitution. The probability that propamides were developmental toxicants appeared to increase significantly when the halogen substitutions were > 2. The compounds with the lowest predicted chronic LOAELs were dihalopropamides with halogen substitutions on the third carbon.

Among the halothiophenes, only tetrachlorothiophene was predicted to have a chronic LOAEL below 10 mg/kg. All substitutions of chlorine by bromine in this structure were predicted to decrease potency, as indicated by a progressive increase in the predicted chronic LOAEL.

The similarity indices for the five haloamides and tetrachlorothiophene ranged from 0.257 to 0.593. Therefore, they were not further considered as potential developmental toxicants.

Halodicarboxylic and Halotricarboxylic Acids

Halodicarboxylic and Halotricarboxylic Acids Formed in Disinfection

A wide variety of reported halogenated DBPs possess more than one carboxylic acid function. These chemicals are prominent in the Richardson (1998) review. To expand the analysis, a number of structural isomers were introduced into these classes of DBPs because the isomers were not clearly identified in the primary literature. Additions to identified compounds were also made for exploratory purposes, but no attempt was made to be comprehensive. It is presumed the toxicological properties of these three groups will be similar to the corresponding halogenated monocarboxylic acids and among themselves. Therefore, the predictions for these groups are presented together.

Halodicarboxylic and Halotricarboxylic Acids Predicted to Have the Greatest Potential as Carcinogens

Halogenated members of compounds with multiple carboxyl functions that were predicted to be carcinogenic and to have a chronic LOAEL < 10 mg/kg day^{-1} are listed in Table E.28 of Appendix E. Twelve of twenty-five saturated and seven of seventeen unsaturated halodicarboxylic acids were predicted to be carcinogenic and to have predicted chronic LOAELs < 10-mg/kg day^{-1}. Neither of the two halotricarboxylic acids examined met these criteria.

The extent to which positive predictions of carcinogenic potential were obtained for both mice and rats is displayed in Table E.29 of Appendix E. Only two of the saturated halodicarboxylic acids were predicted to have a high probability of being carcinogenic in both mice and rats. On the other hand, five of the seven unsaturated halodicarboxylic acids were predicted to be carcinogenic in both species.

The similarity indices of these chemicals are provided Table E.30, Appendix E. The similarity indices of the saturated halodicarboxylic acids predicted to be carcinogenic were all found to be > 0.3. Five of the unsaturated halodicarboxylic acids were found to have similarity indices < 0.3, but none were found to have similarity indices for any prediction that was < 0.2. Please note that the values for the cis and trans forms of 2-chlorobutenedioic and 2-bromobutenedioic acids are identical. This is because the TOPKAT® cancer models are not capable of distinguishing differences in the potential activity of geometrical isomers.

Halodicarboxylic and Halotricarboxylic Acids Predicted to be Developmental Toxicants

The multifunctional carboxylic acids predicted to be developmental toxicants are listed in Appendix E, Table E.31. Six of twenty-nine saturated halodicarboxylic and eleven of eighteen unsaturated halodicarboxylic acids were predicted to be developmental toxicants. Neither of the two halotricarboxylic acids examined were predicted to be developmental toxicants. Two saturated (2-chloromethylpropanedioic acid and 2,2-Dichloro-4-hydroxypentadioic acid) and one unsaturated (2-chloro-3,4-dicarboxy-2-pentenedioic acid) halodicarboxylic acids were outside the OPS of the model. If these three chemicals are removed from the list, only the 2-chlorobutendioic and 2-bromobutenedioic acids were predicted to have chronic LOAELs below 10-mg/kg day^{-1}. The similarity indices within the developmental toxicity model for both chemicals was 0.312, so they were dropped from further consideration as developmental toxicants. Note that the developmental toxicity TOPKAT® model does not distinguish between geometrical isomers, so predictions for the trans- and cis- isomers of these compounds were identical.

Halogenated Aromatics

Halogenated Aromatic Compounds Formed in Disinfection

A variety of aromatic by-products were reported in the reviews of Richardson (1998) and Richardson et al. (2003). These include the halophenols (K08), aromatic haloaldehydes (K22), haloaromatic acids (K24), halogenated aromatic amines (K25), and a group of other haloaromatics (K29).

Halogenated Aromatics Predicted to Have the Greatest Potential as Carcinogens

While some of the chemicals in these classes were predicted to be carcinogenic, none were found to have a chronic LOAEL less that 10-mg/kg day^{-1}. Therefore, members of these classes are not considered to be of a high priority for toxicological testing as important carcinogenic DBPs.

Halogenated Aromatics Predicted to be Developmental Toxicants

The chemicals in these groups predicted to be developmental toxicants are listed in Table E.32 of Appendix E. None of the chemicals in these groups were found to have a predicted chronic LOAEL < 10-mg/kg day^{-1}. Two halophenols, trichlorotrihydroxybenzene and pentachlorophenol, were predicted to have LOAELs that just exceeded this criterion at 12-mg/kg day^{-1}.

MX and Analogs

MX and Analogs Reported to be Formed With Disinfection

MX is a halogenated furanone derivative, 3-chloro-4-(dichloromethyl)-5-hydroxy-2(5H)-furanone. Because of the unwieldy nature of the name, it is commonly known as MX. Similarly, the names of analogs of MX are frequently unwieldy. To avoid misunderstandings the chemical names of those analogs and their acronyms are individually provided here. The chemical name of CMCF is 3-chloro-4-(chloromethyl)-5-hydroxy-2[5H]furanone. Red-MX is the acronym for reduced MX, 3-chloro-4-(dichloromethyl)-2-furanone. 4-Bromomethyl-MX is 3-chloro-4-(bromochloromethyl)-5 hydroxy-2[5H]furanone. 4-Dibromomethyl-MX is 3-chloro-4-(dibromomethyl)-5-hydroxy-2[5H]furanone. Tribromomethyl-MX is 3-chloro-4-(tribromomethyl)-5-hydroxy-2[5H]furanone. An open ring form of MX, E-MX, is E-2-chloro-3-(dichloromethyl)-4-oxobutenoic acid and is also referred to as open-MX. MCA is mucochloric acid or 3,4-(dichloro)-5-hydroxy-2(5H)-furanone. Ox-EMX represents an oxidation of the aldehyde group in E-MX to a carboxyl function and is E-2-chloro-3-(dichloromethyl)-1,4-butenedioic acid.

MX Analogs Predicted to Have the Greatest Potential as Carcinogens

The predictions of carcinogenicity for MX and its analogs are listed in Appendix E, Table E.33. A total of eleven analogs (including MX itself) were examined and all met the criteria of at least one positive prediction of carcinogenicity and chronic LOAELs below 10-mg/kg day^{-1}. In fact all but one of the compounds (Ox-EMX) had a predicted chronic LOAEL below 1-mg/kg day^{-1}.

Only one MX analog was predicted to be a carcinogen in two species, CMCF (Table E.34, Appendix E). MX, itself, was only predicted to be carcinogenic only in mice and these predictions were outside of the OPS of the models. As discussed earlier, MX has been tested for carcinogenicity in the rat and found positive with a chronic LOAEL (based on increased follicular adenomas of the thyroid) of 0.4 mg/kg. This value is within the confidence intervals predicted by TOPKAT®. In addition to follicular carcinomas of the thyroid tumors,

MX also produced cholangiomas of the liver of both sexes, and adenocarcinomas and fibroadenomas of the mammary gland in females. The cancer bioassay of MX, conducted in Finland, appears to be of an adequate design (IARC 2004). MX is also one of the most extensively studied bacterial mutagens and is among the most potent identified. Nevertheless, the TOPKAT® generated probability of Ames' mutagenic activity was 0.000. These predicted values are clearly countered by experimental evidence. This reflects the weakness of relying too heavily on training sets formally generated from FDA and NTP data sets rather than conducting a more broadly based literature search. For this reason, all of the analogs of MX with positive predictions in which the chronic LOAEL was predicted to be less than 10 mg/kg were carried forward in the analysis and additional discussions of this issue are to be found in the discussion of novel DBPs produced by reaction with NOM. The positive predictions in the two species are displayed in Table E.34.

Table E.35 (Appendix E) displays the similarity indices for all predictions of carcinogenicity and non-carcinogenicity of CMCF and MX. The most similar chemicals to MX in the rat models were p,p-DDE (NTP – male rat model), lindane (FDA – male rat), dieldrin (NTP – female rat), and chloramphenicol (FDA – female rat). These compounds do not accurately reflect the chemical and physical characteristics of MX, reinforcing the conclusion that the predictions were not reliable. It is of interest that the positive prediction (with a poor similarity index) for CMCF was based a model where dieldrin was also the most similar compound in the database for MX. The different interpretation of positive predictions in the male mouse for the two compounds was largely related to the fact that the MX prediction exceeded the permissible limits of the model. The positive predictions for both compounds in the FDA female mouse model were outside the permissible limits of the model whereas the prediction of CMCF did not exceed these limits. Collectively, these data indicate that the TOPKAT® models do not accurately predict the potential carcinogenicity of halofuranones.

There is a database of carcinogenesis assays on chemicals that are small modifications of the basic furan ring (NTP 1993, NTP 1990b, NTP 1998a, NTP 1999) that may add to a basic QSTR study of the carcinogenic properties of compounds related to MX. Table 6.8 presents estimates of carcinogenic potency of furan and a number of modifications of furan, including MX. Of the compounds listed, furan is the most potent carcinogen. It is a multispecies, multiple target organ carcinogen. It produces cholangiosaracomas, hepatocellular carcinomas, mononuclear leukemia in male and female rats, hepatocellular carcinomas, and pheochromocytomas in male and female mice. However, it is not mutagenic. Modification of the furan structure substantially reduces the carcinogenic potency. The carcinogenic potencies of tetrahydrofuran and furfural are two to three orders of magnitude less than that of furan. Evidence that furfuryl alcohol is carcinogenic has to be considered marginal, largely targeting the nose in an inhalation study. Thus, it may be that the structure-activity studies of mutagenic effects of MX and related compounds (LaLonde and Leo 1994) are unreliable indicators of carcinogenic potency within this class.

Although MX was not shown to produce bladder cancer in a lifetime study in Wistar rats (Komulainen et al. 1997), it was shown to produce hyperplasia in the bladder of Eker rats treated for 12 months at similar daily doses (McDorman et al. 2003), so the lack of effect in the chronic study may be a strain difference. Therefore, MX and related compounds might be considered potential causes for bladder cancer if consumed in large enough amounts.

It may be more profitable to examine an analog that is mutagenic (LaLonde and Leo 1994), but that incorporates consideration of how the 5-hydroxyl group influences carcinogenic potency. The simplest substitution would be to replace the 5-hydroxyl group on MX by hydrogen. This compound may not be observed in chlorinated drinking water, but understanding how this substitution affects carcinogenic activity may be more informative than spending money on characterizing the carcinogenicity of the brominated analogs of MX.

Table 6.8

Comparisons of the carcinogenic properties and potency of furan derivatives

Compound	Rat TD_{50} *	Mouse TD_{50} *	Mutagenicity†
Furan	0.396	2.72	neg
Tetrahydrofuran	407	1300	neg
Furfural	683	197	pos
Furfuryl alcohol‡	Some evidence/equiv	Some/no evidence	neg
MX	1	Not tested	pos
Hexachlorocyclopentadiene‡	No evidence	No evidence	neg

Source: Data from Gold 2005

* TD_{50}s drawn from CPDB except for MX, which was estimated in the present work.

† Ames' test mutagenicity only. Increased sister chromatid exchange observed in CHO cells in vitro, other tests largely negative.

‡ Tested only by inhalation.

The cyclopentene structures of the cyclopentenoic acids, predicted to be by-products of reactions of chlorine with syringic acid, at least superficially resemble the structure of MX and related compounds. The only halogenated cyclopentene derivative that was identified in the CPDB was hexachlorocyclopentadiene. It was found not to be carcinogenic by the inhalation route, but it is important to recognize the doses administered were limited by the very strong respiratory irritant properties of this compound. Therefore, investigating the carcinogenic potency of a chemical such as 3,5-dichloro-1-hydroxy-4-ketocyclopent-2-enoic acid (illustrated in Figure 7.2) may also contribute substantially to an understanding of the QSTR relationships of MX-related compounds found in chlorinated drinking water.

MX and Analogs Predicted to be Developmental Toxicants

Neither MX nor any of the analogs of MX examined were predicted to be developmental toxicants.

NON-HALOGENATED BY-PRODUCTS PREDICTED TO BE CARCINOGENIC OR DEVELOPMENTAL TOXICANTS

Nitrosamines and Nitrosamides

N-Nitrosodimethylamine (NDMA) is the only nitrosamine that has been consistently found as a DBP when adequate analytical methods have been utilized. Its formation depends upon the occurrence of a suitable precursor in the source water, or as has been commonly found, in materials that are utilized in the treatment of drinking water (certain polymers and ion exchange resins). In general, the presence of chloramines in disinfected water will substantially enhance the formation of NDMA relative to that seen in the presence of free chlorine (Mitch and Sedlak 2002).

The extent to which other nitrosamines occur as a result of the disinfection of drinking water remains a matter of conjecture although others have been clearly identified (e.g. N-nitrosomorpholine, N-nitrosopyrrolidine; Charrois et al. 2004). However, provided appropriate precursors, any number of nitrosamines could be formed. Caffeine is commonly found in waters impacted by domestic wastewater. Mononitrosocaffeidine is an example of a nitrosamine that is apparently produced from caffeine when green tea is treated under alkaline conditions in the presence of appropriate salts (Razdan et al. 1994). The potential for formation of novel nitrosamines or nitrosoureas is discussed later in the context of predicted DBPs.

There are a variety of other sources of human exposure to N-nitroso compounds (Tricker 1997). The diet contains both non-volatile and volatile N-nitroso compounds. The major non-volatile compounds are nitrosated amino acid derivatives, including N-nitrosarcosine, N-nitrosoproline, N-nitroso-hydroxyproline, N-nitrosothiazolidine-4-carboxylic acid, N-nitroso-2-methylthiozolidine-4-carboxylic acid, and N-nitroso-2-hydroxymethylthiazoline-4-caboxylic acids. The volatile nitrosamines include NDMA, N-nitrosopyrrolidine, N-nitrosopiperidine, and N-nitrosothiozolidine. Foods that contain these compounds in the highest concentrations are dried and smoked meats, fish, and smoked cheese. A variety of more complex nitrosamines can be formed by condensation reactions during food processing and/or as a result of contact with packaging material.

Endogenous formation of some of these nitroso compounds has been demonstrated in the human body (Tricker 1997). Much of this formation occurs in the acidic environment of the stomach. Reaction of cysteine with formaldehyde or acetaldehyde results in the formation of N-nitrosothiazolidine-4-carboxylic acid and N-nitroso-2-methylthiazolidine-4-carboxylic acid, respectively. The milder acidic conditions required for the formation of these two chemicals will generally give rise to larger yields of these products than N-nitrosoproline. In addition, there are a number of secondary amines present in the body. These give rise to the formation NDMA, N-nitroso-pyridine and N-nitrosopyrrolidine. Many cells produce NO through oxidation of the terminal guanido-nitrogen of the amino acid, arginine. Activated macrophages, in particular, can form nitroso compounds utilizing NO. Estimates of endogenous formation of nitroso compounds amounts to about 40-75% of the average exposure to nitroso compounds. Total exogenous exposure averages about 1.1 μg/day.

Not all nitroso compounds appear to present a carcinogenic risk and there is considerable range in the potency of those nitroso compounds that are carcinogenic. To put this into perspective, the commonly identified nitroso compounds whose carcinogenic activity has been characterized in animals are listed in Table 6.9.

Among N-nitroso compounds found in the diet or formed endogenously, NDMA is clearly the most potent carcinogen. N-Nitrosopyrrolidine and N-nitrosopiperidine are also carcinogenic, but N-nitrosoproline and N-nitrosohydroxyproline do not appear to present a risk. There appear to be no relevant data for the remaining compounds.

As discussed in Chapter 5, there have also been a number of studies of interactions of nitrogen containing natural compounds and pharmaceutical chemicals with HOCl. A potential scenario might be initial interactions with free chlorine that break down ring structures to form products which could be nitrosated by subsequent addition of ammonia to suppress formation of the regulated halogenated DBPs. For example, in neutral to acidic conditions HOCl can react with purines such as guanine or caffeine to break one or more of the rings in this structure (Suzuki et al. 2003). The exocyclic nitrogens can be targets for nitrosation by conventional reactions provided there is a source of nitrogen dioxide radical. Opening of the rings can also free a number of amine groups for reaction by the unsymmetrical dimethylhydrazine oxidation pathway that occurs as the result of ammonia addition (Mitch and Sedlak 2002). Research on substructures of NOM and various purine and pyrimidine model compounds is needed to determine whether or not such reactions may occur under conditions of drinking water treatment.

Table 6.9
Evidence of carcinogenicity with nitroso compounds found in diet that are formed endogenously

Nitroso compound	Rat TD50*, mg/kg day^{-1}	Rat target organs	Mouse TD50*, mg/kg day^{-1}	Mouse target organs
NDMA	0.0959	Kidney, liver, lung, testis, vascular	0.189	Liver, nervous system
N-Nitrosoproline	Negative			
N-nitrosohydroxyproline	Negative			
N-Nitrosopyrrolidine	0.799	Liver, vascular	0.679	Tumors, site not specified
N-Nitrosopiperidine	1.43	Esophagus, liver, nose	1.3	Liver, lung stomach
N-Nitrosothiazolidine	Not in CPDB		Not in CPDB	
N-Nitrososarcosine	Not in CPDB		Not in CPDB	
N-Nitrosothiazolidine-4-carboxylic acid	Not in CPDB		Not in CPDB	
N-Nitroso-2-methylthiozolidine-4 carboxylic acid	Not in CPDB		Not in CPDB	
N-Nitroso-2-hydroxymethylthiazolidine-4-4-carboxylic acid	Not in CPDB		Not in CPDB	

* TD_{50}s all taken from Gold (2005)

The three nitroso compounds that have been identified in drinking water all have TD_{50}s below 1 mg/kg day^{-1}. The third, not identified in Table 6.9, is N-nitrosomorpholine, which has been tested only in rats. It is carcinogenic with a TD_{50} = 0.109 mg/kg day^{-1}. A fourth N-nitroso compound that is likely to be present is N-nitrosodiethylamine (NDEA), but this has not yet been reported in the open literature. The TD_{50} for NDEA in the rat is 0.0265-mg/kg day^{-1}. To the extent they have been evaluated in toxicology studies, the compounds found in drinking water represent a higher risk per unit dose than those formed endogenously or that have been identified in the diet. However, because many of those that are formed endogenously have not been studied, this conclusion is tenuous. The levels of NDMA found in drinking water are generally below 10 ng/L. Studies in France estimated the daily dietary intake of NDMA to be 0.19 µg/day (Biaudet et al. 1994), which is consistent with intakes in other European countries. Therefore, one would have to consume about 10 L of water containing 10 ng/L to equal the intake from the diet. However, the limited data available on the occurrence of N-nitroso compounds other than NDMA in drinking water, do not allow an accurate comparison of nitrosamine exposure from this source relative to those from the diet or as the result of endogenous formation.

Structure-Activity Relationships Among Nitrosamines

The formation of a variety of N-nitroso compounds from a number of natural and synthetic chemicals can be hypothesized. Six such compounds were evaluated using TOPKAT® and compared to predictions of the program for NDMA. Two of these compounds, mononitrosocaffeidine and dinitrosocaffeidine, have been identified in tea boiled in bicarbonate (Kumar et al. 1992) and are derived from caffeine (Kumar et al. 1993). These were included as a potential model for products that might be expected from reactions of disinfectants with other purines that are likely to present in low concentrations in source waters. The results of the QSTR analyses are presented in Table 6.10.

It is instructive to examine TOPKAT® predictions for NDMA. Positive predictions of carcinogenicity were obtained from three of the four NTP models, but not in any of the FDA models. The carcinogenic activity is well established, so one might ask why the FDA models did not predict NDMA to be a carcinogen. It is simply because the FDA databases do not include the established and generally potent carcinogenic nitrosamines. This illustrates the importance of examining whether the training set of a model is adequate for making predictions for a particular class of carcinogens.

In sharp contrast, mononitrosocaffeidine was predicted to be carcinogenic in the NTP female rat model; the similarity index of the closest related compound (procarbazine) was 0.490, but was within the permissible limits of the OPS. The NTP mouse models also predicted carcinogenicity, but were both outside the OPS of the models and exceeded the permissible limits. However, mononitrosocaffeidine was predicted to be carcinogenic in both the male and female mouse FDA model with similarity indices of 0.273 (nitroacetophenetide) and 0.299 (dacarbazine), respectively. Dinitrosocaffeidine was also predicted to be carcinogenic in the FDA male mouse model with a similarity index of 0.256 (dacarbazine). It was predicted to be carcinogenic in the NTP female mouse model, but the prediction was outside the OPS of the model and exceeded permissible limits with a very poor similarity index (0.623 with nitrofurantoin). Not too surprisingly, both compounds were predicted to be Ames' test mutagens with similarity indices of < 0.2.

Three other N-nitroso compounds were generated to further test the ability of TOPKAT® to address potential products with purine, pyrimidine and indole ringed compounds. None of these chemicals have been identified as by-products of disinfection. In all three of these cases (N-nitrosoadenine, N-nitrosothymine, and N-nitrosoindole) the TOPKAT® output indicated that these molecules contained a fragment that was not present in the training sets.

The analyses of higher molecular weight N-nitroso compounds that could be formed by reactions of disinfectants with naturally occurring higher molecular weight biochemicals indicates that deficiencies in the training sets of TOPKAT® prevented reliable evaluation of these nitrosamines. It is probable that this deficiency extends to other higher molecular weight N-nitrosamines so this avenue was not further pursued.

Table 6.10
TOPKAT® generated probabilities of mutagenic activity, carcinogenic activity, and developmental toxicities for selected nitroso compounds

Index number	Compound	Prob. Ames mut	NTP MR	FDA MR	NTP FR	FDA FR
Rats						
K48C01	N-Nitrosoaniline	0*	0.977	0	0.024	0
K48C02	N-Nitroso-N-dimethylamine	1	0.998	0.003	0.497	0.44
K49C03	N-Nitrosoadenine	1	1†‡	0‡	0.985†‡	0.001‡
K49C04	N-Nitrosothymine	1	1†‡	0‡	0.99‡	0*‡
K49C05	N-Nitrosoindole	1	0.908	0.012	1‡	0‡
K49C06	Mononitrosocaffeidine	1	0.247	0	0.933*	0*
K49C07	Dinitrosocaffeidine	1	0.28	0*	0†	0†

		NTP MM	FDA MM	NTP FM	FDA FM
Mice					
K48C01	N-Nitrosoaniline	0.999	0.055	0	0.54
K48C02	N-Nitroso-N-dimethylamine	0.994	0.042	0.825	0.312
K49C03	N-Nitrosoadenine	1†‡	0.276‡	1†‡	0.029‡
K49C04	N-Nitrosothymine	1†‡	0.035‡	0.007‡	0.904‡
K49C05	N-Nitrosoindole	0.962‡	0.903‡	0.642‡	0.001‡
K49C06	Mononitrosocaffeidine	1†	0.996	1†	0.998
K49C07	Dinitrosocaffeidine	0†	0.796	1†	0.064

DTP and LOAEL		DTP	LOAEL	LOAEL lower conf limit	LOAEL upper conf limit
K48C01	N-Nitrosoaniline	0†	25.2	5	127.3
K48C02	N-Nitroso-N-dimethylamine	0.29*	2.7	0.5216	13.6
K49C03	N-Nitrosoadenine	0.139‡	272.4†‡	63.8	1200
K49C04	N-Nitrosothymine	0.542‡	404.9‡	93.7	1700
K49C05	N-Nitrosoindole	0.506	115.9‡	27.5	489
K49C06	Mononitrosocaffeidine	0.989	49.6	12.3	200.9
K49C07	Dinitrosocaffeidine	0.005	111.9	25.2	498.2

* Outside the optimal predictive space of the model, but within permissible limits
† Outside the optimal predictive space of the model and exceeds permissible limits
‡ Fragment in structure not represented in the training set

Aldehydes

Aldehydes are found with virtually all types of chemical disinfection. However, actual occurrence data are rare. Even more critical, comparisons of the concentrations produced using different disinfectant strategies have not been studied as extensively as halogenated by-products.

Formaldehyde and Acetaldehyde

Formaldehyde and acetaldehyde were identified with those DBPs whose toxicology has received some substantive study in Table 1.1. It should be noted that the TD_{50}s reported in Table 1.1 are largely based on inhalation studies. Inhalation exposure can be important for volatile drinking water contaminants. In the case of formaldehyde and acetaldehyde, their TD_{50}s are such, 2.2 and 153 mg/kg day^{-1}, indicating that the inhalation risk at the concentrations found in drinking water would be minimal (see Table 7.5).

Til et al. (1989) conducted a two-year study administering formaldehyde in drinking water to rats. There were signs of severe irritation and resulting hyperplasia in the stomach seen at doses of 82 or 109-mg/kg day^{-1}, in males and females, respectively, but there was no evidence of tumors at any site. NOAELs of 15 to 21-mg/kg day^{-1} were assigned to males and females, respectively. Soffritti et al. (1989) reported a small increase in the incidence of stomach tumors in Sprague-Dawley rats (2%) administered formaldehyde at 1500 mg/L. In addition, an increased incidence of leukemias was reported that might have reached statistical significance at 1500 mg/L. The former tumors would be expected to result from the strong irritation to the gastrointestinal mucosa that was reported by Til et al. (1989). Leukemia is a variable finding in the rat with a high spontaneous rate (i.e. 10% in one of the control groups in this study). As a result, leukemia in the rat is not generally accepted as evidence for carcinogenic activity on its own. For these reasons, EPA has generally not treated formaldehyde as a carcinogenic risk by the oral route of administration (Hernandez et al. 1994). In a similar vein Til et al. (1988) conducted a four-week study of acetaldehyde and formaldehyde in drinking water and found the NOAELs to be 125 and 25-mg/kg day^{-1}, respectively. These data suggest that acetaldehyde would be even less likely to present a carcinogenic hazard than formaldehyde. In any case, the very low incidence of tumors and leukemia observed in the Soffriti et al. (1989) study would require doses well in excess of 100 mg/kg day^{-1}. Therefore, neither compound meets the criteria of being carcinogens with sufficient potency to be of interest in this study.

There have been reproductive and developmental toxicity studies of formaldehyde and acetaldehyde. In the former case, these have been by both the inhalation (Martin 1990) and oral routes (Hurni and Ohder 1973, Marks et al. 1980). In none of the cases with formaldehyde was there evidence of such effects at levels lower than were maternally toxic. Acetaldehyde has received considerable attention as a possible contributor to fetal alcohol syndrome. However, the very high levels of alcohol (or acetaldehyde) used in these studies (usually several percent in drinking water) are largely irrelevant for the exposures that are obtained from drinking water.

Based on these data, formaldehyde and acetaldehyde will not be further considered as they are simply too weak by the oral route to pose a risk at the very low levels observed in disinfected drinking water.

TOPKAT® Analysis of Aliphatic and Aromatic Aldehydes

In addition to formaldehyde and acetaldehyde, Richardson (1998) identified several aliphatic and aromatic aldehydes in disinfected water that have been evaluated using TOPKAT®. Evaluations of these by-products are presented in Table 6.11.

The TOPKAT® analysis indicates that both glyoxal and methylglyoxal should be considered potential carcinogens (i.e. probability of being carcinogenic was calculated to be > 0.7 in two rat and one mouse model with similarity indices < 0.3). Glyoxal was predicted to have a chronic LOAEL < 10-mg/kg day^{-1}, but the chronic LOAEL of methylglyoxal was predicted to be 15-mg/kg day^{-1}. The probability that methylglyoxal was carcinogenic was greater than 0.7 was matched by a similarity index < 0.2, a somewhat more robust prediction that observed with glyoxal. Therefore, both of these compounds fall just short of the cutoff for high priority. No other aliphatic aldehydes approached the criteria for concern established for the project. Glyoxal is, however, a normal metabolic intermediate in biological systems, including humans.

Benzaldehyde was predicted to be carcinogenic in three models, two in which the similarity indices were < 0.3 and one that was < 0.2. Despite these predictions, the authors conclude that benzaldehyde is of little interest because its predicted chronic LOAEL was 170-mg/kg day^{-1}. The other two aromatic aldehydes had similar predicted chronic LOAELs, but less consistent predictions of carcinogenicity.

Table 6.11
TOPKAT® predictions of toxicity for aliphatic and aromatic aldehydes reported to be disinfection by-products.

Index number	Compound	Ames' Mut	NTP MR	FDA MR	NTP FR	FDA FR
K32	**Aliphatic aldehydes**					
K32C01	Formaldehyde	0.886	0.570	1.000	0.000	1*
K32C02	Acetaldehyde	0.040	0.060	0.582	0.000	0.614
K32C03	Glyoxal	0.133	0.014	0.918	0.996	0.770
K32C04	Methylglyoxal	0.941	0.067	0.949	0.000	0.993
K32C05	Propionaldehyde	0.005	0.006	0.093	0.000	0.385
K32C06	2-Hexenal	0.955	0.000	0.000	0.000	0.000
K32C07	Butanedial	0.957	0.845	0.011	0.000	0.064
K38	**Aromatic aldehydes**					
K38C01	Benzaldehyde	0.000	0.030	0.943	0.000	0.997
K38C02	Phenylacetaldehyde	0.000	0.012	0.257	0.000	0.422
K38C03	p-Hydroxyphenylacetaldehyde	0.000	0.031	0.957	0.000	0.867
			NTP MM	FDA MM	NTP FM	FDA FM
K32	**Aliphatic aldehydes**					
K32C01	Formaldehyde		0.722	0.579	0.007	0.000
K32C02	Acetaldehyde		0.000	0.506	1.000	0†
K32C03	Glyoxal		0.000	0.030	1.000	0*
K32C04	Methylglyoxal		0.000	0.252	1.000	0†
K32C05	Propionaldehyde		0.000	0.523	0.967	0†
K32C06	2-Hexenal		0.000	0.120	1†	0†
K32C07	Butanedial		0.000	0.235	0.000	0.0†
K38	**Aromatic Aldehydes**					
K38C01	Benzaldehyde		0.000	0.393	1.000	0†
K38C02	Phenylacetaldehyde		0.000	0.157	0.165	0†
K38C03	p-Hydroxyphenylacetaldehyde		0.000	0.528	0.742	0†
		DTP	LOAEL	LOAEL lower conf limit	LOAEL upper conf limit	
K32	**Aliphatic aldehydes**					
K32C01	Formaldehyde	0.386	60	13.3	270	
K32C02	Acetaldehyde	0.587	114	25	509	
K32C03	Glyoxal	0.985	3.2	0.546	19.1	
K32C04	Methylglyoxal	0.985	15.5	3.0	82	
K32C05	Propionaldehyde	0.009	166	37	744	
K32C06	2-Hexenal	0.000	176	39	796	
K32C07	Butanedial	0.004	284	62	1300	
K38	**Aromatic aldehydes**					
K38C01	Benzaldehyde	0.012	170	33	883	
K38C02	Phenylacetaldehyde	0.101	183	35	954	
K38C03	p-Hydroxyphenylacetaldehyde	0.275	287	52	1600	

* Outside the optimal predictive space of the model, but within permissible limits

† Outside the optimal predictive space of the model and exceeds permissible limits

EVALUATION OF PREDICTED DISINFECTION BY-PRODUCTS

Organic N-Haloamines

Evidence That Organic N-Haloamines are Produced in Disinfection

As discussed in Chapters 4 and 5, organic N-haloamines are formed in the presence of chlorine or chloramine when primary or secondary amines are present in the source water (Mitch and Sedlak 2002). Common potential precursors that are found at low concentrations in water sources are proteins, amino acids, and nucleic acids. These chemicals are known to form N-chloramines. Minimal occurrence data exist in this area, although there are fundamental data that demonstrate that organic chloramines are produced upon chlorination or chloramination of drinking water and wastewater (Nweke and Scully 1989). Amino acid analyses of humic acids, fulvic acids, and XAD-recovered acids obtained from various surface waters in Britain and Ireland (Watt et al. 1996) indicate an amino acid content of samples that ranged from 53 to 839 nmol mg^{-1}. In general, the greater concentration of amino acids was found in the humic acid fractions. The proportion of the humic samples that was comprised of amino acids ranged from 0.5 – 8.7%. However, the amino acids accounted for only 6.1 to 36.1% of the nitrogen in the sample. Certainly, the amino acids serve as precursors for organic N-chloramines. Formation of the N-chloramines, N-chloroglycine, N-chloroleucine, and N-chlorophenylalanine has been demonstrated in stomach fluid from rats (Scully et al. 1990). These authors also found chemicals that had retention times on high pressure liquid chromatography (HPLC) that corresponded to N-chlorovaline and N-chloroserine, but were unable to confirm this by gas chromatography with mass spectrometry (GC/MS). It is probable that some of the other nitrogenous compounds present in source waters would form organic N-chloramines upon chlorination or chloramination.

Toxicological Characteristics of Organic N-Haloamines

There are no data that demonstrate carcinogenic effects (or lack thereof) of organic N-haloamines in experimental animals. However, formation of organic N-haloamines in vivo has been postulated to be responsible for a variety of chronic diseases, most commonly inflammatory disorders and some human cancers (Halliwell et al. 1987, Jesaitis and Dratz 1992, Daumer et al. 2000). These hypotheses generally begin with the formation of HOCl by myeloperoxidase in leukocytes and phagocytic cells that proliferate in response to infection. There is an extensive literature that documents the formation of N-chloramines of nucleic acids and nucleosides (Hawkins and Davies 2001), amino acids (Peskin and Winterbourn 2001), peptides (Armesto et al. 2001), and intact proteins (Pattison et al. 2003). Products are, in turn, produced from N-chloramines. For example, higher levels of dimerized tyrosine are found in human phagocytes (Jacob et al. 1996). This undoubtedly results from high HOCl to carbon ratios within these cells, but the formation of N-chloramines is more rapid and will occur at lower concentrations (Pattison et al. 2003). Because taurine is present in high concentrations in neutrophils, it is frequently used as the model for subsequent reactions of N-chloroamino acids that may occur in vivo (Peskin and Winterbourn 2001).

The formation of N-chloramines in nucleic acids progress to the evolution of nitrogen centered free radicals that results in oxidation of DNA and/or dimers between DNA bases. It is unlikely that HOCl will travel far enough in cells to directly damage DNA. However, N-chloro groups can be transferred between amines. As the N-chloramines are generally less reactive than HOCl, the formation of N-chloroamines is a way to induce a biological response at sites remote to its formation (Peskin and Winterbourn 2001, Englert and Schacter 2002). They are also less polar than the parent amine, which facilitates diffusion through membranes. It is, therefore, not too surprising that mutagenic activity is more easily produced by N-chloramines than with HOCl, itself (Thomas et al. 1987).

N-Chloramines formed on the N-terminal of dipeptides have been shown less readily degraded than those formed with single amino acids (Armesto et al. 2001). These studies provide some insight into how steric factors may contribute to stabilization of both the N-chloramine and to the transition states on degradation. In this particular case, the N-chlorodipeptides were more stable and the intermediate imine structure survived long enough to be measured. The stability of the imine structure was influenced by the size of the side chain in the N-terminal amino acid. Thus, if the activation energy for the degradation of the N-chloramine could be overcome enzymatically, the resulting intermediate could survive to diffuse through the cell and damage macromolecules. This is not to suggest that dipeptides are the best precursors, as their levels in NOM are likely to be low. Rather it illustrates the possibility that higher molecular weight amines may be effective transmitters of damage to macromolecules.

In the absence of empirical data demonstrating carcinogenic activity, it is not possible to determine the extent to which the formation of these by-products in drinking water would pose a carcinogenic risk. Unfortunately, as will be demonstrated below, this lack of data makes it impossible to make judgments related to the hazards posed by these compounds using QSTR analyses.

Structure-Activity Considerations

A group of 13 N-chloramines were selected from reactions of amino acids known to occur in natural waters and a pharmaceutical (acetanilide) was subject to analysis by TOPKAT® along with two compounds that would be additional reaction products of chlorine with isoleucine, N-chloro-2-methylbutryaldimine and 2-methylbutanenitrile. To confirm that the results that were obtained was not related to the lack of data on some structural feature other than the N-chlorogroup, predictions for N-chloroacetanilide were compared to predictions for acetanilide (i.e. most of the other compounds were based on endogenous biochemicals). The results of these analyses are displayed in Table 6.12.

TOPKAT® estimated probabilities for many N-chloramines, but in every case, the program indicated that the structural fragment was not represented in the training sets of the models. As a consequence, there can be no confidence in the negative predictions of carcinogenic activity or developmental toxicity. Positive predictions carry a little more weight as they take into account other aspects of the structure, but still there can be little confidence in the predictions. The chronic LOAEL prediction resulted in a wide distribution of predicted values, but the since the chloramine fragment was not represented in the training set, they cannot be accepted at face value either.

We specifically examined predictions that would arise from a chemical well represented within the database, acetanilide. As this drug has been demonstrated to be carcinogenic in rodents, the FDA database clearly predicts in all four models that it would be carcinogenic. The corresponding N-chloramine of acetanilide was not predicted to be carcinogenic, but as in other

Table 6.12
TOPKAT® generated probabilities of mutagenic activity, carcinogenic activity, and developmental toxicities for selected N-chloramines, precursors, subsequent products, and degradation products

Category	Compound	Ames' mut	NTP MR	FDA MR	NTP FR	FDA FR
Rats						
K49C01	N-Chloroacetanilide	0.000‡	0.119‡	0.011‡	0.000‡	0.001‡
K49C02	Acetanilide	0.000	0.400	0.898	0.000	0.999
K49C03	N-Hydroxyacetanilide	1.000	0.333	0.000	0.530	0.921
K49C04	N-Chloro-N-acetylglycine	0.646	0.019‡	0.001‡	0.000‡	0.000‡
K49C05	N-Chloroserine	0.000	0.000‡	0.004‡	1.000‡	0.037‡
K49C06	N-Chlorohistidine	0	0‡	0‡	0.999‡	0.001‡
K49C07	N-Chlorophenylalanine	0†‡	0‡	0.029‡	1‡	0.149‡
K49C08	N-Chloroglycine ethyl ester	0.003	0.354‡	0.922‡	0.001‡	0.001‡
K49C09	N-Chloroglycine	0.014	0.005‡	0.17‡	0.015‡	0.001‡
N49C10	N-Chloroalanine	0	0‡	0.002‡	0.048‡	0.399‡
N49C11	N-Chloroglcylaldimine	0.986*	0.254‡	0.001‡	0.993‡	0.76‡
N49C12	N,N-Dichloroglycine	0.151	0.134‡	0‡	0‡	0‡
N49C13	N-Chloroadenine	0.026‡	0‡	0.001‡	0‡	0‡
N49C14	N-Chloro-2-methylbutryaldimine	0.789*	0.995‡	0.113†‡	0‡	0.656‡
N49C15	2-Methylbutanenitrile	0	0.063	0.001	0	0.298
N49C16	N-Chloroisoleucine	0	0.032†‡	0.339†‡	0.094‡	0.845‡

Category	Compound	NTP MM	FDA MM	NTP FM	FDA FM
Mice					
K49C01	N-Chloroacetanilide	0‡	0.379‡	0.002‡	0.000‡
K49C02	Acetanilide	0	0.957	0	0.931
K49C03	N-Hydroxyacetanilide	0.344	0.489	0.001	0.035
K49C04	N-Chloro-N-acetylglycine	0‡	0.008‡	0‡	0‡
K49C05	N-Chloroserine	0‡	0.877‡	0‡	0.922‡
K49C06	N-Chlorohistidine	0‡	0.995‡	1‡	1‡
K49C07	N-Chlorophenylalanine	0‡	0.336‡	0‡	0.917‡
K49C08	N-Chloroglycine ethyl ester	0‡	0.998‡	0‡	0.079‡
K49C09	N-Chloroglycine	0‡	0.130‡	0‡	0.576‡
N49C10	N-Chloroalanine	0‡	0.878‡	0‡	0.396‡
N49C11	N-Chloroglcylaldimine	0‡	0.007‡	0.990‡	0.005‡
N49C12	N,N-Dichloroglycine	0‡	0.007‡	0‡	0.022‡
N49C13	N-Chloroadenine	1†‡	0.005‡	1†‡	0.962‡
N49C14	N-Chloro-2-methylbutryaldimine	0‡	0.007‡	0.999‡	0.947‡
N49C15	2-Methylbutanenitrile	0	0.882	0.13	1*
N49C16	N-Chloroisoleucine	0‡	0.808‡	0‡	0.22‡

(continued)

Table 6.12
(Continued)

DTP and LOAEL		DTP	LOAEL	LOAEL lower conf interval	LOAEL upper conf limit
K49C01	N-Chloroacetanilide	0‡	272‡	53	1400
K49C02	Acetanilide	0*	95	18	491
K49C03	N-Hydroxyacetanilide	0	333	61	1800
K49C04	N-Chloro-N-acetylglycine	0‡	259‡	29	2300
K49C05	N-Chloroserine	0.734‡	7.5‡	1.4	40
K49C06	N-Chlorohistidine	1‡	14.5†‡	2.2	94
K49C07	N-Chlorophenylalanine	0.207‡	64†‡	9	460
K49C08	N-Chloroglycine ethyl ester	0.005‡	8.2‡	1.7	39
K49C09	N-Chloroglycine	0.912‡	4.5‡	0.90	23
N49C10	N-Chloroalanine	0.155‡	123*‡	20	748
N49C11	N-Chloroglcylaldimine	0.899‡	0.42‡	0.071	2.5
N49C12	N,N-Dichloroglycine	1†‡	67‡	14	312
N49C13	N-Chloroadenine	0‡	171‡	41	720
N49C14	N-Chloro-2-methylbutryaldimine	0‡	0.20†‡	0.019	2.2
N49C15	2-Methylbutanenitrile	1	0.10†	0.0074	1.4
N49C16	N-Chloroisoleucine	0†‡	2.2*‡	0.35	13

* Outside the optimal predictive space of the model, but within permissible limits
† Outside the optimal predictive space of the model and exceeds permissible limits
‡ Fragment in structure not represented in the training set

cases these predictions are rendered uncertain by the fact that the one fragment that differs in the two compounds is not in the database. We also entertained the likelihood that N-hydroxy-acetanilide might be carcinogenic in rodents; the predictions were largely negative (an exception was the FDA female rat model). Moreover, the predicted chronic LOAEL for N-hydroxyacetanilide was more than three times higher than that of acetanilide.

Since Nweke and Scully (1989) demonstrated that N-chloroisoleucine degrades to a stable aldimine and a halonitrile, these chemicals were also examined with TOPKAT®. The aldimine predictions fell outside the OPS of all models and exceeded the permissible limits. This finding will be discussed further in the next section. The halonitrile, 2-methylbutyrylnitrile, was addressed adequately by the model. As pointed out with other halonitriles, this chemical is predicted to be a developmental toxicant and the predicted chronic LOAEL was well within our criteria (0.1-mg/kg day^{-1}). It was predicted to be carcinogenic in mice, but not in the rat, so does not rise to our criteria as a high priority for research related to its potential carcinogenicity.

Potential Role of N-Haloamines as Intermediates to In Vivo Formation of DNA-Binding Forms

Clearly, the training sets within TOPKAT® have limited the evaluation of the broad class of compounds referred to as organic chloramines. In view of the fact that organic N-haloamines are clearly formed in chlorination and chloramination of drinking water several specific examples will be discussed that suggest that this may be one of the highest priority research areas related to DBPs. The essential point with all of these examples is that stable chloramines or stable products of chloramine decay provide a potential mechanism for delivery of a potentially carcinogenic form to target organs.

N-chloramines and N-chloramides decay by a variety of mechanisms. Research directed at understanding endogenous formation of chloramines has examined the interactions of HOCl with amino acids, proteins, purine and pyrimidine bases, nucleosides, polynucleotides and a variety of drugs, which provide a basis for speculation as to why these chemicals may be important from a health effects standpoint. A selected set of chemicals that have been studied is provided in Table 6.13 to provide some perspective on the types of by-products that might occur and how some of these chemicals may be further activated by in vivo metabolism. With each chemical, the reactions in which these chemicals participate with HOCl are indicated. Most, but not all of these reactions involve formation of an N-chloramine. It will be noted that halogenation reactions at sites other than the nitrogen have been repeatedly demonstrated when HOCl reacts with some of these precursors. Halogenation of these sites is discussed in detail in Chapter 5. The toxicology of those compounds is also discussed later in this section, so they are not included in this table. Formation of the N-chloramine of phosphotidylethanolamine is included in Table 6.13, but those of phosphotidylserine and phosphotidylcholine are not. N-Chlorophosphosphotidylserine is unstable (Pattison et al. 2003) and phosphotidylcholine does not readily form an N-chloramine.

In most of the cases identified in Table 6.13, formation the N-haloamine is usually the first site of attack by HOCl with these chemicals. The critical question is whether or not adverse health effects result from the N-haloamine or from a subsequent metabolite that has sufficient stability to be absorbed and distributed to critical tissues in the body. Subsequently, the "stable" product decays to one of several types of reactive intermediates that are capable of covalently binding to nucleic acid bases. If not repaired, some of the base modifications may be fixed into a mutation when cells divide. Mutations in the critical gene/s may then give rise to a carcinogenic response. The nature of the reactive intermediates from various N-chloramines appears to range from nitrogen or carbon-centered free radicals, to nitrenium ions, cation radicals, imminim ions, or iminoquinones depending upon the nature of the reacting chemical. A unique case of metabolic activation found in the literature was the case of N-hydroxy-N-(2-fluorenyl)acetamide where the halogen, most favorably from HOBr, adds to the N-hydroxyl group, subsequently causing hydrolysis of the acetamide linkage to fluorene. This yields 2-nitrosofluorene as a major product (Ritter and Malejka-Giganti 1989).

The stability of some of the N-chloramines identified in Table 6.13 is unknown. In some cases the stability of the actual compound or an analog has been sufficiently well studied to suggest it would serve as an effective form for delivery of the chemical to the actual target tissue. HOCl (or other oxidants produced locally) can participate in further activation of the molecule to the reactive form. This would be the case with diclofenac and iminostilbene, for example. Thus, one hypothesis is that HOCl in drinking water treatment will produce some N-haloamines of sufficient stability to be delivered to the target tissue. High local concentrations of HOCl, which can be generated by activated neutrophils can complete the activation of the compound (Lampert and Weiss 1983). In some cases this chemistry can be catalyzed by other peroxidases. One example is the activation of N-acetylbenzidine by prostaglandin H synthase to an N'-hydroxylamine (Zenser et al. 1999). Since this enzyme is highly expressed in the bladder epithelium, it is postulated to be responsible for the ability of this chemical to induce bladder cancer. This enzyme has been found responsible for the fact that o-anisidine acts primarily as a bladder carcinogen (Stiborova et al. 2002).

The relative stability of N-chloramine is held responsible for some of the subtle differences between the ease of N-chloramine formation and the subsequent formation of carbon and nitrogen-centered radicals. In the case of purine and pyrimidine bases, cyclic N-chloramines more easily form on the cyclic nitrogens, but radicals develop on the exocyclic nitrogens. It appears that the N-chlorogroup is transferred from the lowest stability site to the higher stability site and the free radical develops on the exocyclic nitrogens (Prutz 1998, Hawkins and Davies 2002).

The half-lives of many N-chloro amino acids are discussed in Chapter 5. They range from 40 minutes for histidine to 19,000 minutes for N-acetyl glycine. These half-lives may not precisely reflect the in vivo half-life of N-chloramino acids. The half-life of chloramines produced through reaction of HOCl with proteins (largely involving free amino nitrogen of lysine) appears to be > 2 hours at 37° C in simple solution (Hawkins and Davies 1999). N-Chlorolysine and other amino acids do react with thiols (e.g. protein thiols, glutathione, cysteine) and ascorbic acid in plasma (Carr et al. 2001), so the half-life of these compounds may be somewhat shorter in vivo. As previously indicated, the work with N-chloro-dipeptides (Armesto et al. 2001) provides a model for how steric factors may stabilize N-chloramines so that they may be absorbed intact. Steric factors also played a role in the stabilization of the intermediate imine group (Armesto et al. 2001) and may well be a determinant of enzymatic processes that are likely to accelerate the release of the imine in vivo.

The major conclusion is that formation of N-haloamines in the disinfection of drinking water is of potential health concern. It is of interest that many of the reactive metabolites of these chloramines resemble those hypothesized responsible for carcinogenesis by aromatic amine bladder carcinogens (e.g. nitrogen radicals, nitrenium ions). Despite the well-documented formation of organic N-chloramines, there has been no substantive effort to gather the descriptive toxicological data that can be used to judge the degree of risk these compounds pose. Therefore, gathering of descriptive toxicological data on representative organic N-haloamines is among the greatest data needs for resolving issues related to potential cancer risk attributed to chlorinated drinking water.

Table 6.13
Production of organic N-chloramines, other stable intermediates and subsequent degradation products that can react with nucleic acid bases

Chemical	Reactants	"Stable" product(s)	"Reactive" forms produced on decay	Reference
Acetamide	HOCl	N-chloramide	C-centered radical	Hawkins & Davies 1998
N-methylacetamide	HOCl	N-chloramide	N-CH_3 – C-centered radical	Hawkins & Davies 1998
N-acetyl-lysine	HOCl	N-chloramide	C5-centered radical	Hawkins & Davies 1998
Isoleucine	HOCl	N-chloramine → N-chloro-2-methylbutyrylaldimine → 2-methylbutyrylnitrile	Not established	Nweke and Scully 1989
Phosphotidyl ethanolamine	HOCl	N-chloramine	N and C-centered radicals	Pattison et al. 2003
Uridine	HOCl	N-chloramine	N-centered radical	Hawkins and Davies 2001
Adenine nucleosides and nucleotides	HOCl	N-chloramine	N-centered radical	Bernofsky 1991
Diclofenac	HOCl	N-chloramine → hydroxydiclofenac	iminoquinone	Miyamoto et al. 1997
Aminopyrine	HOCl	N-chloramine	Iminium ion + cation radical	Uetrecht et al. 1995
Iminostilbene	HOCl	N-chloramine	Carbonium ion → aldehyde	Furst and Uetrecht 1993
Clozapine	HOCl	N-chloramine	Nitrenium ion	Liu and Uetrecht 1995
Benzidine or N-acetylbenzidine	HOCl	N-chloramine	N-centered radical	Lakshmi et al. 2000a&b
N-hydroxy-N-(2-fluorenyl)acetamide	HOX, especially HOBr	2-nitrosofluorene	Not established	Ritter and Malejka-Giganti 1989

Products Predicted to Arise From Substructures Present in Natural Organic Matter

Several novel by-products were identified through predictions of reactions of chlorine with putative precursor structures or models of those structures. As yet, these compounds have not been investigated as DBPs in drinking water. Some of these compounds are intermediates on the way to by-products that have been previously identified. They may be stable with some disinfection strategies and not with others. However, there are precursors of these compounds, including syringic acid and vanillic acid in drinking water sources, as described more fully in Chapter 5. The emphasis here is on the QSTR analyses of the predicted products. Of 42 compounds subjected to QSTR analysis, 16 were found to meet the prioritization criteria. These are listed in Table 6.14 and their structures provided in Figure 6.10.

The predicted by-products identified through QSTR analysis as the most potent (based on the predicted chronic LOAELs are 5-6 membered ring structures. Many are halogenated quinones. Although these reactions are predicted to occur, it is difficult to estimate the actual yield that may be observed under various disinfection scenarios. For example, the chemicals formed by reaction of HOCl with vanillic acid are intermediates in the formation of more widely recognized DBPs. In other words, even if they are formed these chemicals will be subjected to at least partial degradation by further reaction with chlorine. As a group, they remain as high priorities because if the predicted potencies of these chemicals are confirmed experimentally only small amounts of these DBPs would be required to exceed the effects of the regulated DBPs. The reactions with chlorine that were predicted to give rise to some of these potential by-products are discussed in Chapter 5. Formation of other DBPs that have been reported in the literature with monochloramine, dichloramine, or chlorine dioxide disinfection are discussed in more detail later in this section.

For simplicity, the entire suite of TOPKAT® predictions for the sixteen potential by-products is included in Table 6.14 including probabilities, numerical values, and indications of model shortcomings. Similarity indices will be discussed as deemed appropriate in the text below. Additional quinones were identified as potential DBPs, but did not meet the prioritization criteria. Because the names of these compounds are complex, they will be referred to by their assigned index numbers (K50C01-K50C42).

It was noted that the structures for two of these compounds (K50C06 & K50C07) were at least superficially related to the halofuranones. Justification for accepting the predictions of their chronic LOAELs was discussed with Table 6.3. As indicated there, the confidence intervals for the TOPKAT® predictions of the chronic LOAEL for MX included the actual published LOAEL. Therefore, the predictions for these related chemicals could not be easily dismissed. The predictions made by TOPKAT® suggest that these compounds may be very potent, relative to other by-products, as toxicants.

3,5-Dichloro-1-hydroxy-4-keto-cyclopent-2-enoic acid (K50C07) is clearly the most potent of these chemicals with a predicted LOAEL of 250-ng/kg day^{-1}. This prediction exceeds permissible limits of the OPS for the chronic LOAEL model. Five of eight models predict that K50C07 is carcinogenic, but in only two cases are the predictions within the OPS of the cancer models. It is predicted to be mutagenic in the Ames' test. In all cases, the similarity indices are substantially >0.3. Therefore, one must conclude that analogs of this compound are poorly represented in TOPKAT®. K50C07 was not predicted to be a developmental toxicant, but this prediction was also outside the OPS of the model. Nevertheless, the very low predicted chronic LOAEL indicates that determining the occurrence of this compound should be given a high priority in a DBP research program. Even if it only occurs in the ng/L range, these toxicological predictions should be experimentally verified.

The predicted chronic LOAEL for K50C06 (1,2,2,4-tetrachloro-3-hydroxy-cyclopent-4-enoic acid) was 190-µg/kg day^{-1}, but the prediction fell outside of the OPS. It was predicted to be carcinogenic only in the FDA male and female mouse models and the NTP female mouse models. Only the NTP female mouse model was within the OPS with a similarity index for this prediction of 0.558. It was identified as a non-mutagen, was within the OPS, and had an acceptable similarity index for this prediction. K50C06 was predicted to have low probability as a developmental toxicant, but the prediction was outside of the OPS.

Table 6.15 lists the predictions of the chronic LOAELs by all of the sub-models available within TOPKAT® for haloquinone and related chemicals. In general, TOPKAT® auto-selected the alicyclic submodel, which appears appropriate to the basic structure of this group. However, some predictions of the chronic LOAEL fell outside the OPS of the alicyclic model. Moreover, the similarity indices were generally found to be between 0.2 and 0.5. A difficulty was that the closest compounds matched to these quinones within the alicyclic submodel were dieldrin, dimethipin, captan, or lasiocarpine, accounting for the low similarity indices. Although some of the halogenated compounds within this training set had somewhat analogous structures, they were clearly not quinones.

Predictions by the heteroaromatic submodel improved the similarity index for some of the haloquinones but, as might be expected, none of the chemicals outside the OPS of the alicyclic model were within the OPS of the heteroaromatic model.

The single benzene submodel improved the similarity indices with chemicals in the training set, but inspection of the nearest matched chemicals included no quinone structures. As will be discussed later, the available scientific literature clearly indicates that the quinone structure plays an important role in the toxicities produced by this group of chemicals.

The multiple benzene sub-model also improved the similarity indices and in some cases, the nearest analogs in the training set did include quinone structures (e.g. 2-methyl-1-nitroanthraquinone). However, no halogen substitutions on the quinone ring were within the training set.

Finally, the acyclic sub-model relied entirely on aliphatic halogen-substituted compounds in the training set (i.e. bromodichloromethane and related trihalomethanes, trichloroethylene, tetrachloroethylene, and monochloroacetic acid). Even though the similarity indices were improved with this model, the chemicals matched in the training set could not be expected to be appropriate models for the toxicity of the haloquinones.

Our consideration of these sub-models led us to conclude that the auto-selected alicyclic sub-model clearly contains a better training set for predicting the chronic toxicity of the haloquinones. Therefore, those chemicals with predictions that met our prioritization criteria have been carried forward in the analysis.

3,4,5,6-Tetrachloro-o-quinone (K50CO4) is the most potent haloquinone with a predicted chronic LOAEL of 5-µg/kg day^{-1}. Again, the prediction exceeds the permissible limits of the OPS of the chronic LOAEL model. Four carcinogenic models predict that K50C04 is likely to be carcinogenic, but in only one case was this prediction within the permissible limits of the OPS. In this case the similarity index was 0.402. The compound was predicted to be a non-mutagen and to have a low probability of being a developmental toxicant. While the two latter predictions were within the permissible limits of the OPS, the similarity indices were > 0.3 in both cases. Again, a high priority should be given to assessing the occurrence of this compound in drinking water. If it does occur, experimental verification of its toxic properties would also assume a high priority.

1,2,3-Triketo-4,4,5,6-tetrachlorocyclohex-5-ene (K50C05) was predicted to have a chronic LOAEL of 50-µg/kg day^{-1}. This prediction was within the OPS of the model, but the similarity index was 0.335. There were four models that predicted carcinogenic activity, but only one of these predictions fell within the permissible limits of the OPS. In the single case where this criterion was met (the female rat in the NTP database) the similarity index was 0.374.

K50C05 was predicted to have a low probability of being a developmental toxicant; this was within the OPS of the model. However, the similarity index was 0.357. The low predicted chronic LOAEL remains a concern indicating that the occurrence of this compound should be investigated as well.

2,3-Dichloro-4,5-diketobenzylprop-2-enol (K50C41) and 2,3,5-Trichloro-4,5-diketobenzoic acid (K50C42) were identified as products of reactions of chlorine with coniferyl alcohol (Figure 5.44). These ortho-quinone structures were predicted to have chronic LOAELs with the alicyclic submodel in the 10-100 μg/kg day^{-1} range, both of which were within the OPS of the model. The similarity indices were 0.280 and 0.362, respectively. K50C41 marginally meets the criteria as a predicted carcinogen, whereas K50C42 is predicted to have a high probability of being a developmental toxicant. Therefore, research aimed at determining the occurrence of these chemicals in drinking water should be high priority. If they do occur, experimental work should be conducted to confirm the predicted toxicological properties.

Alternative disinfection strategies probably all have the capacity of producing haloquinone by-products. The question is which strategy creates the conditions in which these intermediary structures become stable.

Recent work by Heasley et al. (2004) demonstrated that monochloramine and dichloramine will produce the series of halogenated quinones identified in Table 6.14 and Figure 6.10 as K50C35, K50C36, K50C37, and K50C38, from reactions with phenol or m-cresol. Formation of these compounds is mediated through displacement of a prior substitution of chlorine in the position para to the hydroxy group with an N-chlorimine group. The N-chlorimine group is hydrolyzed to form the corresponding benzoquinone. Compound K50C40 is one of the N-chlorimines that met the criteria for a predicted chronic LOAEL < 1 mg/kg day^{-1}. The products in Figure 6.10 all have predicted LOAELs < 1 mg/kg day^{-1}; the TOPKAT® data are displayed in Table 6.14 The main finding is the very low predicted LOAELs relative to other putative or established DBPs. In only one case were these predictions outside the OPS of the model. However, the similarity indices were all between 0.3 and 0.4 indicating that closely related chemicals are not represented in the database. A pattern of positive predictions of carcinogenicity for these compounds in the NTP female rat, the FDA female rat, the FDA male mouse, the NTP female mouse and the FDA female mouse models was observed for these five compounds. The carcinogenicity predictions generally fell outside of the OPS of the respective models, with only one case including a chemical with a similarity index of < 0.2. None of these chemicals were predicted to be developmental toxicants.

Two chemicals, 2,2,4,6-tetrachlorocyclo-5-ene-1,3-dione (K50C01) and 1,2-dihydroxy-3,4,5,6-tetrachlorobenzene (K50C03) deserve mention although they do not meet the prioritization criteria. They had predicted chronic LOAELs of 2.4 and 11.7 mg/kg day^{-1}, respectively, which were within the OPS of the model. K50C01 had a similarity index 0.237, approaching the strictest criteria.

In the case of K50C01 there were five models that predicted carcinogenic activity, but only the prediction in the NTP female mouse model was within the OPS. That prediction had a similarity index of 0.540. It also had low probability as a mutagen, but this prediction was outside of the OPS. K50C01 was not predicted to be a developmental toxicant, but the similarity index (0.43) was high relative to the criteria used in this project.

1,2-Dihydroxy-3,4,5,6-tetrachlorobenzene (K50C03) was predicted to be a developmental toxicant. The prediction was well within the OPS of the model with a similarity index of 0.000. The predicted chronic LOAEL for K50C03 slightly exceeded our criterion

(11.7-mg/kg day^{-1}). As a developmental toxicant, it is clearly the most interesting halogenated aromatic compound that was examined within TOPKAT®. It stands out in terms of toxicological potency among halogenated aromatic compounds. Therefore, it is likely that the prediction is strongly influenced by the polyphenol substitution pattern.

It is important to note that chlorine dioxide is more likely to produce polyphenol and quinone by-products with NOM than chlorine. Napolitano et al. (2005) recently demonstrated the rapid formation of polyphenol and the somewhat slower production of quinone by-products from tyrosine, N-acetyltyrosine, and 3,4-Dihydroxphenylalanine (dopa). The reaction scheme leads from tyrosine to dopa, to dopaquinone, and finally dopachrome. The TOPKAT® predictions indicated that the chronic LOAEL of dopachrome (16.4-mg/kg day^{-1}) was significantly lower than that predicted for dopa (72.8-mg/kg day^{-1}) or dopaquinone (128.2-mg/kg day^{-1}). QSTR analysis of these structures indicated a low probability that dopa was carcinogenic, but the dopaquinone could be a carcinogen. This projection fell short of the prioritization criteria used in this study. None of these compounds were predicted to be developmental toxicants. Nevertheless, these data suggest that the formation of quinone by-products by chlorine dioxide needs some attention by researchers.

Two pyrimidine derivatives, 2-hydroxy-2-methyl-3-chloro-4,6-diketopyrimidine (K50C17) and alloxan (K50C20) were identified as potential by-products. Alloxan has some well-established toxicological properties that are used in producing an experimental model of diabetes mellitus. These compounds are listed primarily because the predicted chronic LOAELs with the alicyclic sub-model were both below 1-mg/kg/day^{-1}, but both predictions exceeded the permissible limits of the OPS of the model. The alternative multiple benzene submodel includes quinone derivatives in the training set and had acceptable similarity indices, but predicted higher chronic LOAELs. The lower level of halogen substitution for these chemicals relative to some of the haloquinones might make the multiple benzene sub-model the most appropriate. No positive predictions of carcinogenic activity or developmental toxicity were within the OPS of the respective models. Therefore, these two chemicals are considered to be of a lower priority.

Spiroaminodihydantoin (K50C08) and guanidinohydantoin (K50C14) are potential products of reactions between HOCl and guanine, a nucleic acid base that is expected to commonly occur in NOM. They are included in Table 6.14 because they just meet the prioritization criteria based on predicted carcinogenic activity. However, their predicted chronic LOAELs are sufficiently high that they would have to occur in drinking water in the μg/L range to be of interest.

Table 6.14

TOPKAT® generated probabilities of Ames' mutagenic activity, carcinogenic activity, and developmental toxicities for by-products predicted to occur based on disinfectant reaction with precursor substructures in natural organic matter (NOM)

Index number	Putative disinfection by-product	Ames' Mut	NTP MR	FDA MR	NTP FR	FDA FR	NTP MM	FDA MM	NTP FM	FDA FM	DTP	LOAEL
K50C04	3,4,5,6-Tetrachloro-o-quinone	0	0†	0.929†	1*	0†	0	1†	0	1†	0.007*	0.005†
K50C05	1,2,3-Triketo-4,4,5,6-tetrachlorocyclohex-5-ene	0†	0.003	1†	0.748	0†	0	1†	0	1†	0.010	0.05
K50C06	1,2,2,4-Tetrachloro-3-hydroxy-pent-4-enoic acid	0	0.003	0†	0.079†	0.018†	0.008*	1†	1	1†	0†	0.19†
K50C07	3,5-Dichloro-1-hydroxy-4-keto-cyclopent-2-enoic acid	1†	0	0†	1†	0†	1†	1†	1	1†	0†	0.00025†
K50C11	Spiroaminodihydantoin	0	0.007	0.823	0	1†	0	0.991*	0.918	0.003*	1	7.2
K50C12	Guanidinohydantoin	1	0.834	1*	0	0	0	0.998*	0.060	0.998*	1	5.6
K50C17	2-Hydroxy-2-methyl-3-chloro-4,6-diketopyrimidine	0.988†	0	0*	0†	0†	1†	0.995†	0.253	0.001†	0.023†	0.598†
K50C20	Alloxan	0	0.054	1	0†	0	0	1†	0.368	0.85	0.716†	0.305†
K50C26	1,3,3,5,5-Pentachloro-4-keto-5-hydroxycyclohexene	0	0.004	0†	0†	1†	0	1†	1†	1†	0*	0.21*
K50C35	2,6-Dichloro-1,4-benzoquinone	0	0.001‡	1*	0.955*	0.001	0	1	1†	1†	0†	0.049†
K50C36	2,3,6-Trichloro-1,4-benzoquinone	0	0*	0.996*	0.997*	0	0	1†	1	1†	0*	0.033
K50C37	2,3,5,6-Tetrachloro-1,4-benzoquinone	0	0.001†	0.068†	1*	0†	0.006	1†	0	1†	0.048*	0.005
K50C38	2,6-Dichloro-3-methyl-1,4-benzoquinone	0	0	0.996	0.923†	0	0	1	1	1†	0*	0.079
K50C40	2,3,6-Trichloro-1,4-benzoquinone-4-(N-chloro) imine	0.685‡	0.001*‡	0.997*‡	0.998*‡	0‡	0‡	1†‡	1‡	1†‡	0‡	0.073
K50C41	2,3-Dichloro-4,5-diketobenzylprop-2-enol	0.899	0.001	0.71*	0.992†	0	0	1	0.127	1†	0.001	0.064
K50C42	2,3,5-Trichloro-4,5-diketobenzoic acid	0†	0*	0.009†	1†	0†	0	1†	0	0	0.958	0.073

* Outside the optimum predictive space of the model, but within permissible limits
† Outside the optimum predictive space of the model and exceeds permissible limits
‡ Fragment not represented in the training set of the model

Table 6.15
TOPKAT® submodel predictions of chronic LOAELs for haloquinones predicted to be formed by reaction of chlorine or chloramine with natural organic material (NOM) substructures

Index No.	Compound	Alicyclic	Simil Index	Hetero-aromatic	Simil. Index	Single benzene	Simil. Index	Multiple benzene	Simil. Index	Acyclic	Simil. Index
K50C05	1,2,3-Triketo-4,4,5,6-tetra-chlorocyclohex-5-ene	0.05	0.335	0.305	0.326	1500	0.089	39	0.119	21.2	0.028
K50C06	1,2,2,4-Tetrachloro-3-hydroxy-cyclopent-4-enoic acid	0.19†	0.411	0.23†	0.278	481†	0.147	5.6	0.139	17.9*	0.038
K50C20	Alloxan	0.30†	0.262	3.2†	0.153	149	0.047	61	0.095	98.5	0.036
K50C26	1,3,3,5,5-Pentachloro-4-keto-5-hydroxycyclohexene	0.21*	0.232	0.10†	0.336	1400†	0.171	6.6	0.140	32.3	0.023
K50C35	2,6-Dichloro-1,4-benzoquinone	0.049†	0.313	25.7*	0.103	271	0.052	17.9	0.118	26.3	0.119
K50C36	2,3,6-Trichloro-1,4-benzoquinone	0.033	0.342	8.8†	0.172	509	0.038	21.4	0.118	10*	0.064
K50C37	2,3,,5,6-Tetrachloro-1,4-benzoquinone	0.0053†	0.417	2.9†	0.252	228	0.051	24.9	0.118	113	0.086
K50C38	2,6-Dichloro-3-methyl-1,4-benzoquinone	0.079	0.341	29.3*	0.094	393	0.041	14	0.142	22.7	0.024
K50C40	2,3,6-Trichloro-1,4-benzo-quinone-4-(N-chloro)imine	0.073‡	0.328	25†‡	0.201	505‡	0.033	17‡	0.129	5.4‡	0.086
K50C41	2,3-Dichloro-4,5-diketobenzo-prop-2-enol	0.064*	0.280	26*	0.111	2300	0.101	72.6	0.266	28.9	0.070
K50C42	2,3,5-Trichloro-4,5-diketobenzoic acid	0.073	0.362	3.4†	0.236	308	0.104	8.2	0.171	11*	0.066

* Outside the OPS of the indicated model, but within permissible limits
† Outside the OPS of the indicated model and exceeds permissible limits
‡ Fragment not represented in the training set

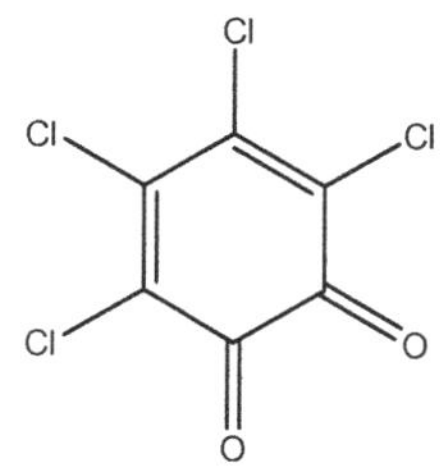

K50C04
2,3,4,5-Tetrachloro-o-quinone

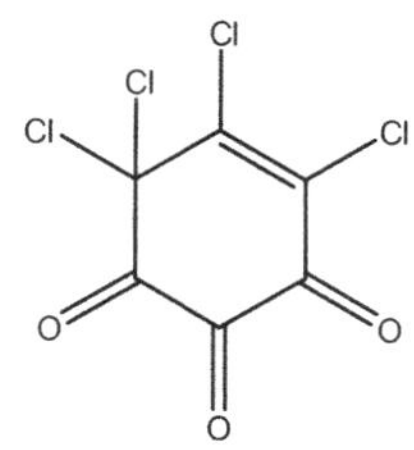

K50C05
1,2,3-Triketo-4,4,5,6-tetrachlorocyclohex-5-ene

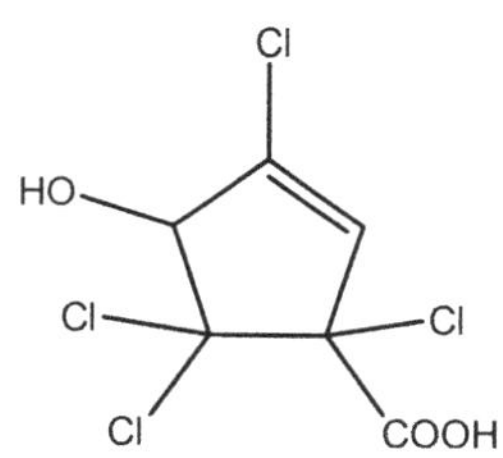

K50C06
1,2,2,4-Tetrachloro-3-hydroxy-cyclopent-4-enoic acid

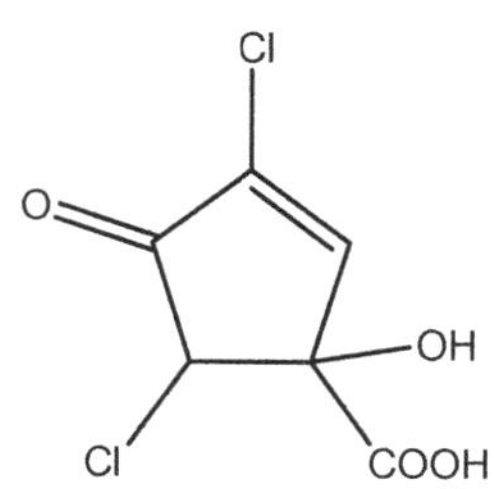

K50C07
3,5-Dichloro-1-hydroxy-4-keto-cyclopent-2-enoic acid

Figure 6.10 Structures of DBPs predicted to be formed from reaction of chlorine with substructures of NOM

(continued)

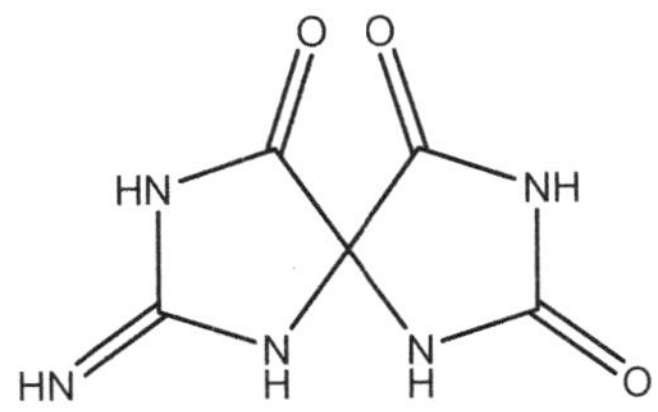

K50C11
Spiroaminodihydantoin

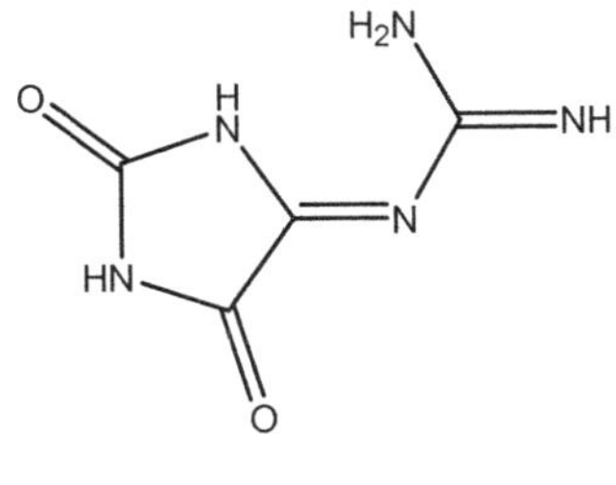

K50C12
Guanindinohydantoin

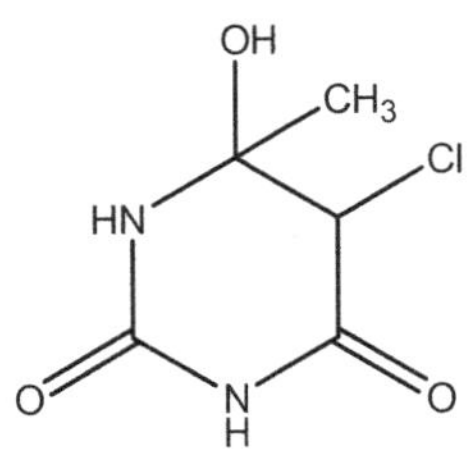

K50C17
2-Hydroxy-2-methyl-3-chloro-4,6-diketopyrimidine

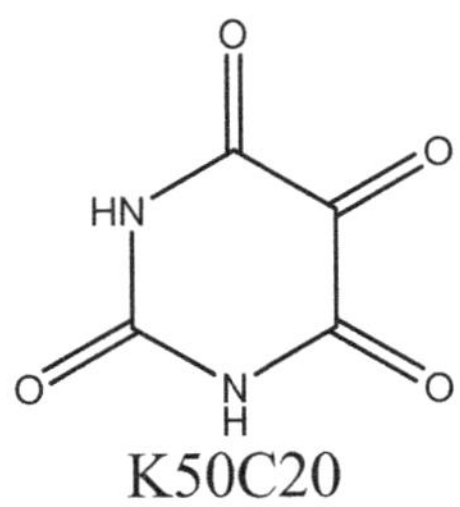

K50C20
Alloxan

Figure 6.10 (Continued)

(continued)

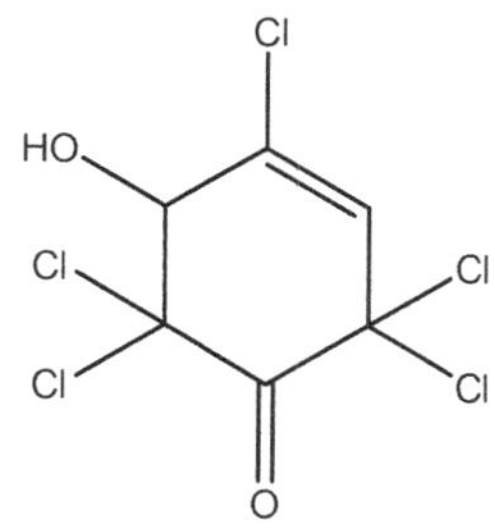

K50C26
1,3,3,5,5-Pentachloro-4-keto-5-hydroxycyclohexene

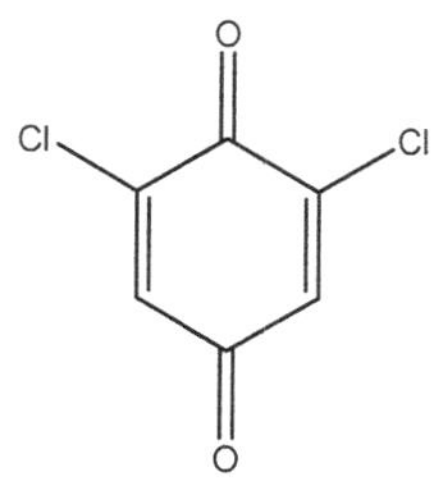

K50C35
2,6-Dichloro-1,4-benzoquinone

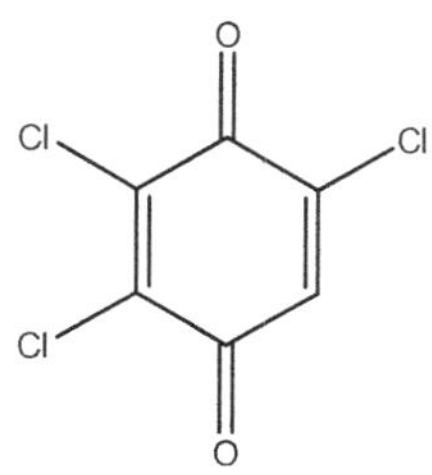

K50C036
2,3,6-Trichloro-1,4-benzoquinone

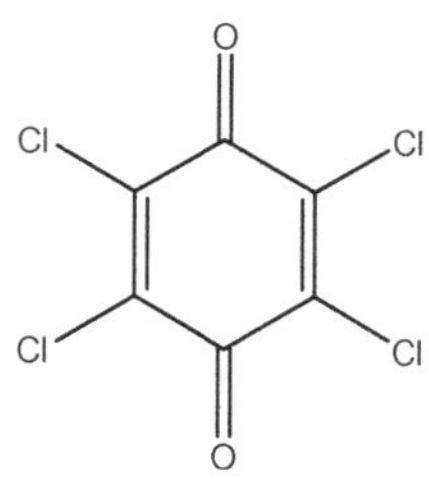

K50C37
2,3,5,6-Tetrachloro-1,4-benzoquinone

Figure 6.10 (Continued)

(continued)

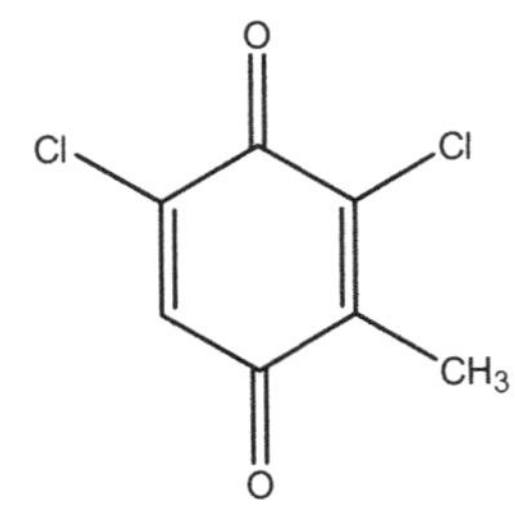

K50C38
2,6-Dichloro-3-methyl-1,4-benzoquinone

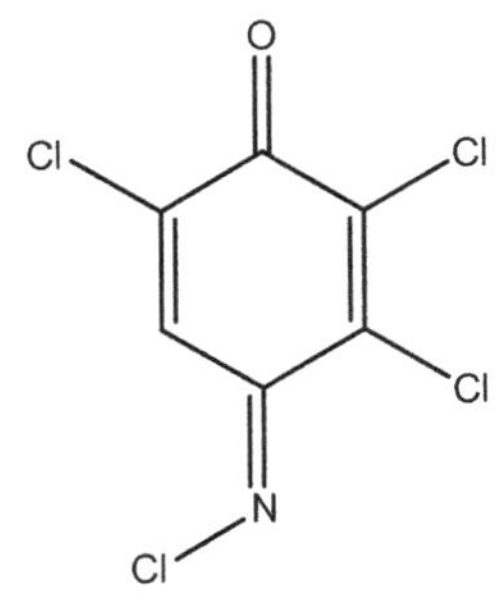

K50C40
2,3,6-Trichloro-1,4-benzoquinone-4-(N-chloro) imine

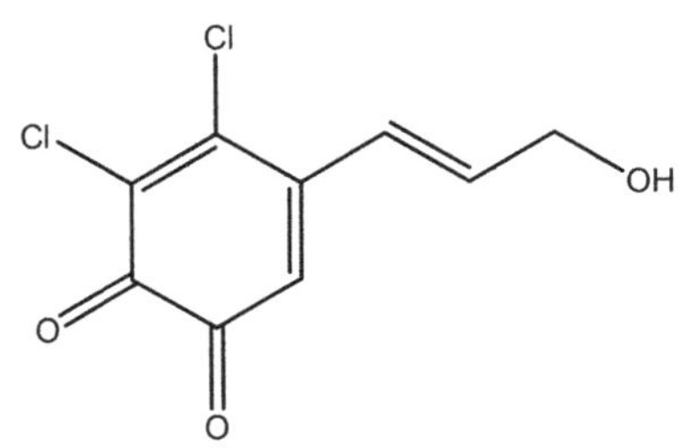

K50C41
2,3-Dichloro-4,5-diketobenzylprop-2-enol

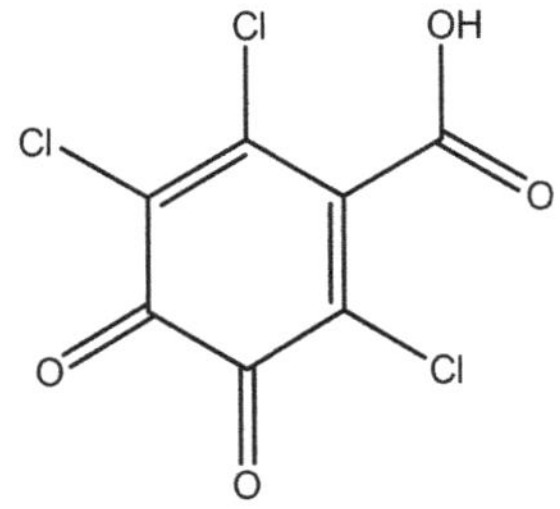

K50C42
2,3,5-Trichloro-4,5-diketobenzoic acid

Figure 6.10 (Continued)

The sequence of the reactions predicted to give rise to K50C03 and K50C14 is discussed in Chapter 5 (see Figure 5.21). This sequence is displayed again in Figure 6.11 where TOPKAT® predictions are identified with each intermediate in the reaction sequence. The reaction sequence emphasizes non-halogenated by-products. Halogenated by-products that result from reactions of guanine and other nucleic acid bases with chlorine are well known and omitted from the sequence. The point is to illustrate how the toxicological properties of the products change substantially as the reactions go through intermediate by-products to completion. The first product (8-oxoguanine) is a mutagenic modification of guanine produced in vivo by hydroxyl radicals and other oxidants. Thus, it is not too surprising that it is predicted to be mutagenic. This product is also predicted to be a developmental toxicant. However, concern over this product is significantly ameliorated by the predicted high chronic LOAEL. Subsequent products in the pathway have predicted chronic LOAELs that are 1/100 to 1/500th that of 8-oxoguanine (although little confidence can be had in the predictions for K50C13 and K50C10 in that at least one molecular fragment is not represented in the TOPKAT® training set). There is some suggestion in the data that the compounds subsequent to 8-oxoguanine may be carcinogenic, however, the similarity indices for these predictions are relatively large. A more consistent finding was that compounds K50C12 and K50C09 have a high probability of being developmental toxicants.

This short exercise illustrates the value of examining the mechanisms by which by-products are formed and the nature of stable intermediates that may arise. In this reaction scheme, developmental toxicants would be more probable with breakpoint chlorination than with preformed chloramine. However, it is important to determine whether or not higher molecular weight N-nitroso compounds will be produced if ammonia is added after free chlorine and whether or not such compounds may be potent carcinogens. Thus, the toxicological risks may vary significantly with each of these commonly utilized disinfection strategies.

	8-Oxoguanine		Spiroamino-dihydantoin	Spiroimino-dihydantoin
		HOCl limited		
Guanine →	K50C14	→→→→→	K50C11 →	K50C08
LOAEL	2500 (0.128†)		7.2 (0.232)	7.2 (0.232
Carcinogen	1(4)		0 (4)	0 (2)
Mutagen	1.000 (0.128)		0.000 (0.436†)	0.000 (0.431)
DTP	1.000 (0.222)		1.000 (0.463)	1.000 (0.459)

Excess HOCl ↓

	N-Carboxyguanidino-hydantoin	Guanidino-hydantoin	Parabanic acid	N,N-Dichloro-parabanic acid
	K50C13 →→→	K50C12 →	K50C09 →	K50C10
LOAEL	16.9 (0.279‡)	5.6 (0.237)	16.3 (0.122)	25.6 (0.124†‡)
Carcinogen	1 (4)	1 (4)	1 (3)	0 (2)
Mutagen	0.999 (0.505)	1.000 (0.435)	0.000 (0.408)	0.000 (0.918)†‡)
DTP	1.000 (0.492)	1.000 (0.323)	1.000 (0.173)	0.000 (0.918†‡)

Figure 6.11 Predicted toxicological properties of degradation products of guanine upon reaction with chlorine Chlorine-limited conditions are envisioned with low chlorine to carbon ratios, or when ammonia limits the reactions of chlorine. The lower line of predicted by-products is more likely to be produced with breakpoint chlorination. The relative amounts of these latter compounds can be substantially different from either of the other two conditions when ammonia is added following chlorine. The ammonia can quench HOCl reactions at some intermediate state if the reactions in excess chlorine have not gone to completion.

FEED-BACK EVALUATION USING ESTABLISHED CHEMICALS AS CAUSES OF ADVERSE HEALTH EFFECTS OF INTEREST

Identification of Model Compounds for Feed-back Evaluation

The feed-back approach to identifying DBPs that are potentially responsible for a given adverse health effect that has been associated with consumption of disinfected water is an inherently more speculative approach to the problem. The general approach is to identify other chemicals that have been demonstrated to be responsible for a given health effect that has been associated with drinking water disinfection and determine whether or not analogous (not identical) compounds have been identified in disinfected water. If none can be identified, the next step is to determine whether or not there are precursor substructures that are likely to exist broadly enough in source waters to allow formation of such compounds. The approach for identifying compounds that target specific organs requires a more sophisticated, expert analysis than is possible with the QSTR tools that are available. In the present work, an example of such an approach was applied to the identification of putative DBPs that have the potential of being bladder carcinogens. This involved three steps: 1) identification of chemicals shown to produce bladder cancer in humans or experimental animals, 2) literature search to identify common metabolic and physiological processes that might be determinants of a carcinogenic response in the bladder, and 3) identification of established or putative DBPs whose activity are likely to be affected by similar metabolic and physiological processes. This is necessary to account for the apparent ubiquity of the association of bladder cancer with chlorination and would suggest that such a precursor(s) would have to occur quite widely in natural waters.

It would have been appropriate to implement a fourth step related to potential modes of action, but this could not be implemented for lack of adequate experimental data. For example, irritation of the bladder is known to contribute to the development of bladder cancer. Therefore, non-genotoxic compounds should be considered as possible causes of bladder cancer as well as genotoxic ones. Compounds in some of the classes discussed here (specifically the quinones) have been shown to be genotoxic in vivo, but that activity has not been apparent in Salmonella testor strains. A non-genotoxic mode of action would typically require that a compound have a lower chronic LOAEL than a genotoxic agent to be important at concentrations likely to occur in drinking water. Despite this bias, the consideration of potency of a potential bladder carcinogen relied strictly upon the predicted chronic LOAEL without addressing differences in mode of action.

To be further considered, it was also necessary that the inherent toxicity of the product must be greater than that of the precursor to be of interest. As pointed out in previous sections chemicals with chronic LOAELs above 1 mg/kg are of marginal interest. Second, the by-product must be produced in sufficient yield to result in a dose that is capable of adversely affecting health. For this reason, most of the work for this project has focused on substructures in NOM. Reactions of hypochlorous acid have also been explored with a number of pharmaceuticals as a means of explaining serious, but idiosyncratic, toxicities that have been associated with their chronic use. Consequently, the third step has been to explore these reactions as well. The reader must recognize that the actual association of these chemicals with bladder cancer and chlorinated drinking water becomes progressively more speculative as the analysis goes from 1) established chlorination by-products, 2) compounds that are postulated to be formed from NOM substructures, to 3) potential model compounds that may or may not be represented widely in the available precursor pool. In the latter two cases, it must also be remembered that very reactive products will not have sufficient bioavailability to produce a systemic health effect such as bladder cancer. Therefore, the authors' analysis focuses on relatively stable products that may be further activated by in vivo metabolism.

Bladder Carcinogens

Chemicals for which experimental data or human studies have demonstrated the ability to produce bladder cancer have been assembled in Appendix B. The database groups these compounds into seven categories: aromatic amines, nitro compounds, azo compounds, heterocyclic amines, nitrosamines and nitrosamides, halogenated hydrocarbons (including alcohols and acids), and a miscellaneous group. The database includes the structure of the studied substances, the level of evidence indicating that it is a carcinogen in the form of an International Agency for Research on Cancer (IARC) classification (if it has been evaluated), and a standardized measure of potency. The TD_{50}s were obtained from the CPDB (Gold 2005). These figures do not necessarily represent the TD_{50} of each chemical that is associated with the production of bladder cancer, as another tumor site may well have been used in calculating the TD_{50}. Since the bioassays from which these data are derived generally span a limited range of doses, this is unlikely to introduce a large error (generally less than a factor of 2). The TD_{50} used in the following analyses will, however, only use TD_{50}s in the species in which bladder cancer was observed as a statistically significant outcome. There are 62 chemicals in the database. Figure 6.12 displays the structures of those bladder carcinogens that have a TD50 $\leq$ 10-mg/kg day^{-1}. All of the potent bladder carcinogens are nitrogenous compounds that fall into the classes of aromatic amines, heterocyclic amines, nitro compounds, azo compounds, and N-nitroso compounds. The single halogenated compound is a nitrogen mustard, cyclophosphamide, which is used in cancer chemotherapy. The general toxicology of each of these classes will be discussed in turn, although among these chemicals only selected members of the nitroso class have been demonstrated to occur after disinfection of drinking water.

Aromatic Amines

Reactions of Aromatic Amines With Chlorine and Other Oxidants That Increase Their Carcinogenic Activity. Aromatic amines are not in and of themselves DBPs. The key questions are whether or not they exist as precursors and whether reactions with disinfectants can substantially increase their carcinogenic potency. Natural organic compounds are aromatic amines to the extent they have extracyclic nitrogens (e.g. histidine, nucleic acid bases and their degradation products such as caffeine). Emerging contaminants of wastewater, including caffeine, pharmaceutical agents, selected pesticides, and industrial chemicals fall into this class. Of interest are certain reaction products that occur with chlorine and chloramine that may result in the formation of potent carcinogens that may be more plausible causes of bladder cancer than the regulated DBPs.

Characteristics of Carcinogenic Aromatic Amines. Aromatic amines have been extensively studied as carcinogens. As illustrated by the collection of aromatic amines highlighted in Figure 6.13, a large number of potent human and experimental bladder carcinogens belong to this class. While aromatic amines are widely recognized as bladder carcinogens, they also induce cancer in other organs. In some cases, the activity of aromatic amines is greater in other organs than in the bladder. In part this is reflected in species differences in responses, but can also reflect within species and strain variations in the metabolism and distribution of the chemical and processing of the adducts formed in DNA.

Mechanisms. The mechanisms of aromatic amine carcinogenicity are better understood than many other classes, at least in a general sense. Oxidation of the amine group to an N-hydroxylated derivative is strongly associated with both mutagenicity and carcinogenicity of this well-studied group of carcinogens. The actual proximate carcinogen is thought to be a nitrenium ion that reacts with nucleic acid bases, although other hypotheses have not been ruled out.

Role of metabolism and kinetics. Metabolic processing is perhaps the key issue in the carcinogenicity of aromatic amines. The qualitative metabolic activity of different organs towards aromatic amines, the compound-specific kinetics of this metabolism, the kinetics of transfer of metabolites between tissues and to organs of elimination are key to understanding both the target organ specificity and potency of aromatic amines.

The metabolism and transport of aromatic amines and their metabolites determines their targeting of the bladder. The first step is activation of the extracyclic amine group by the addition of an N-hydroxyl group, which occurs primarily in the liver and is catalyzed by CYP1A2 or CYP3A4 (Figure 6.13). This reaction competes with acetylation of the amine group by N-acetyltransferase 2 (NAT2) leading to a form that is eliminated in the urine.

The N-hydroxylated aromatic amine is transported directly or as its glucuronide to the bladder via the blood. Much of the glucuronide will be eliminated in the urine, but some will be hydrolyzed and reconverted to the N-hydroxyl amine. The hydroxyl group can either be acetylated to give rise to the N-acetoxy derivative by N-acetyltransferase 1 (NAT1) or to a sulfonated derivative by a sulfotransferase (reaction not shown in Figure 6.13). Hydrolysis of these forms readily occurs in acidic urine to give rise to a nitrenium ion, which binds to DNA in the urothelium. The sulfonate hydrolyzes much more quickly than the O-acetylated form.

The relative activity of these different enzyme systems is the basis of some of the best-established genetic susceptibilities to carcinogens (Ambrosone and Kadlubar 1998). Polymorphisms that reduce the activity of NAT2 (so-called slow acetylators) are susceptible to increased bladder cancer risk by these compounds. Individuals that express high amounts of CYP1A2 or CYP3A4 are also more sensitive to bladder carcinogenesis. Therefore, the sensitivity of the individual will be subtly determined by the relative activity of these enzymes in the liver. The most sensitive individual would have high CYP1A2, low NAT2 and high NAT1 activity. Subtle differences in the metabolic pathways that activate and deactivate aromatic amines can actually shift the target organ from the bladder to the lower colon (Ambrosone and Kadlubar 1998).

Aromatic amines

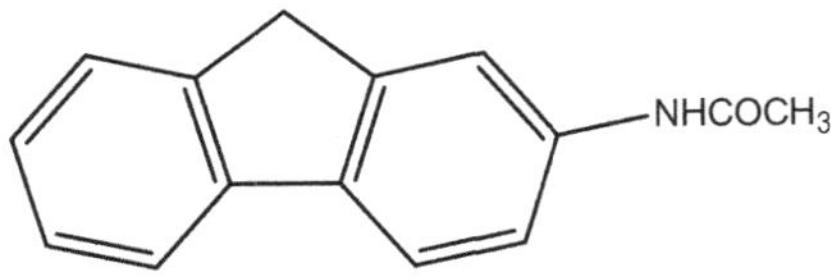

2-Acetylaminofluorene
$TD_{50} = 1.22$

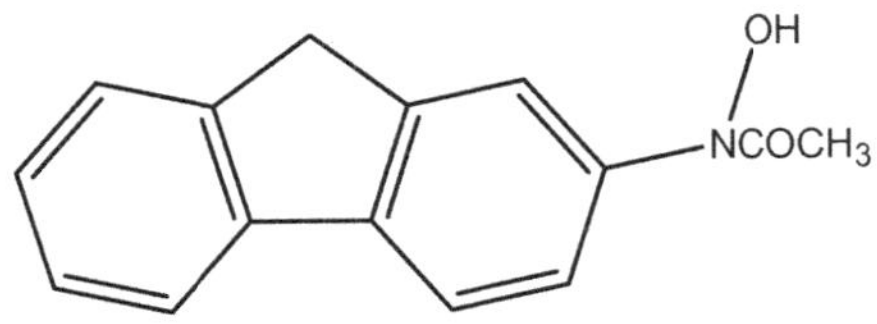

N-Hydroxy-2-Acetylaminofluorene
$TD_{50} = 0.988$

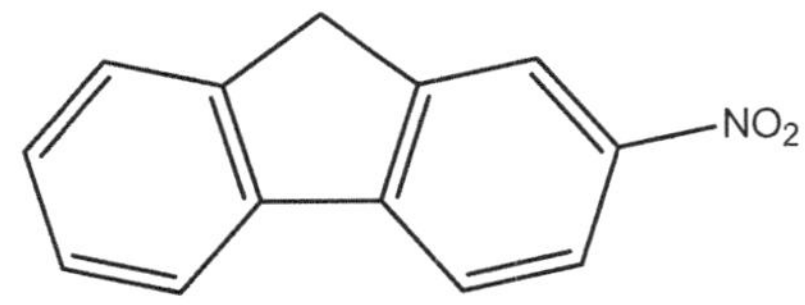

2-Nitrofluorene
$TD_{50} = 0.285$

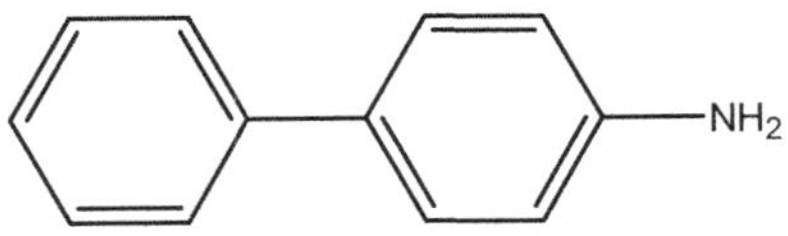

4-Aminobiphenyl
$TD_{50} = 0.98$ (rat)

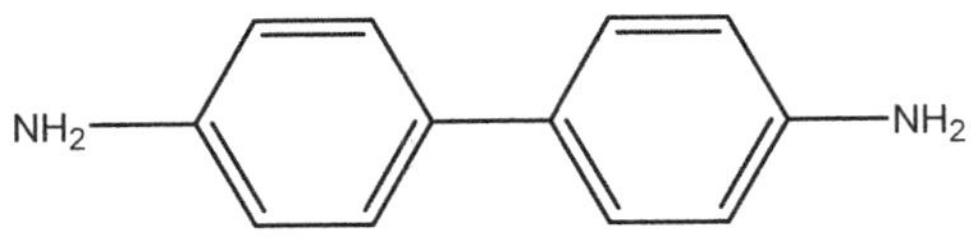

Benzidine
$TD_{50} = 1.73$ (rat)

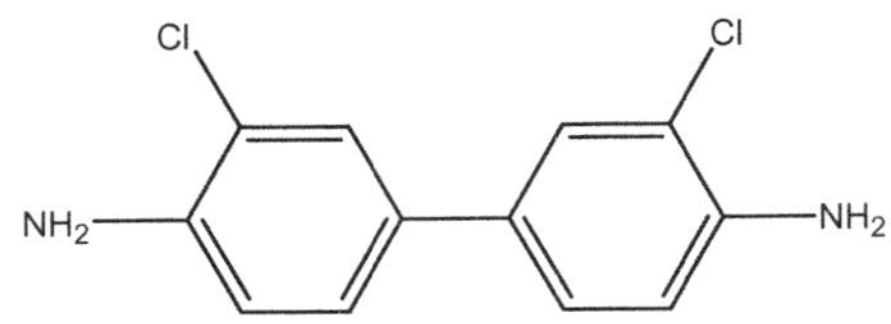

3,3-Dichlorobenzidine
$TD_{50} = 28.1$ (rat)
$= 1.78$ (dog)

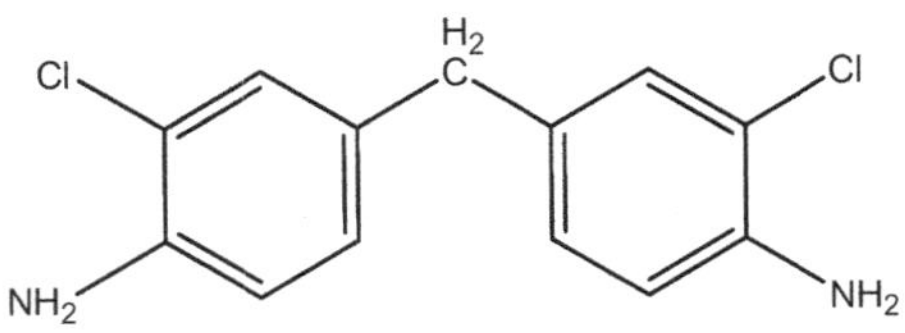

4,4-Methylene-bis(2-chloroaniline)
$TD_{50} = 19.3$ (rat)
$= 2.12$ (dog)

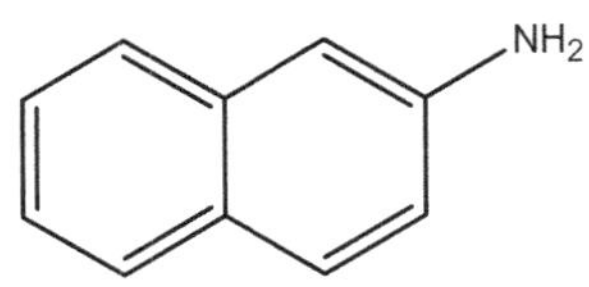

2-Naphthylamine
$TD_{50} = 61.6$ (rat)
$= 5.74$ (Rhesus monkey)

Heterocyclic amines

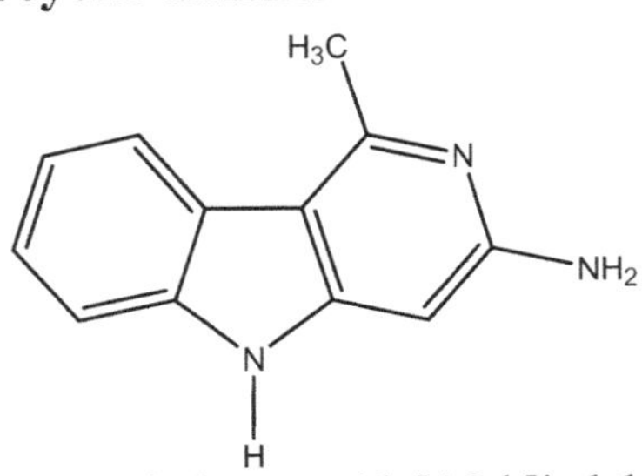

3-Amino-1-methyl-5H-pyrido[4,3-b]indole
(Trp-P-2)
TD50 = 6.66

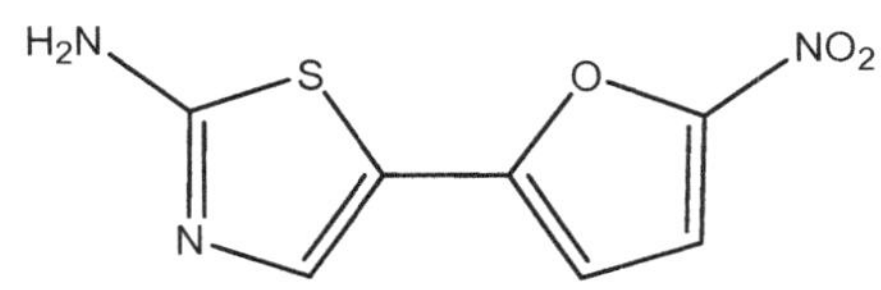

2-Amino-4-(5-nitro-2-furyl)thiazole
$TD_{50} = 5.85$

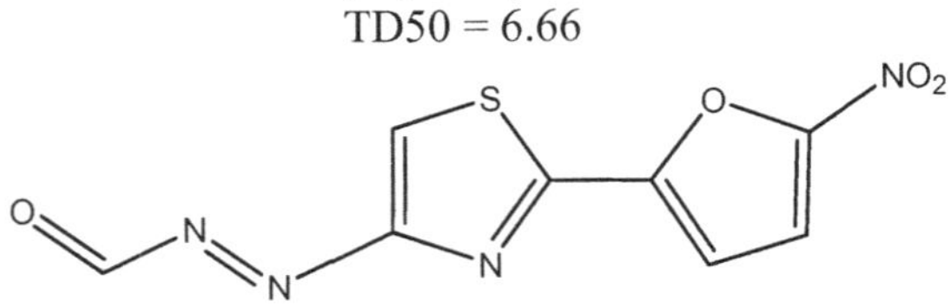

2-(2-Formylhydrazino)-4-(5-nitro-2-furyl)thiazole
$TD_{50} = 5.06$
$= 16.6$ (hamster)

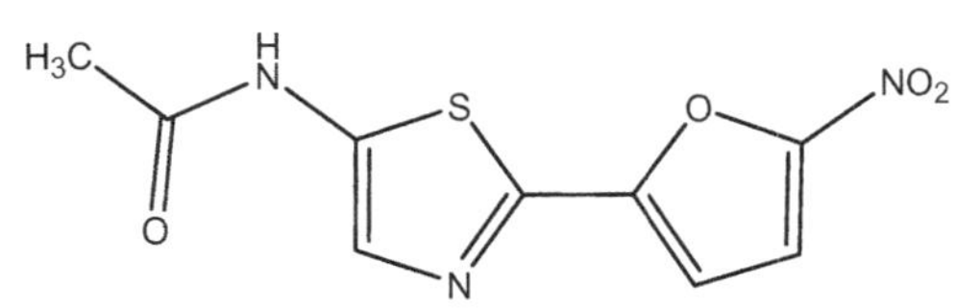

N-[4-(5-nitro-2-furyl)-2-thiazolyl]acetamide
$TD_{50} = 17.8$
$= 8.91$ (hamster)

Figure 6.12 Bladder carcinogens with TD50s below 10-mg/kg (in the rat unless otherwise specified)
(continued)

Nitroaromatic compounds

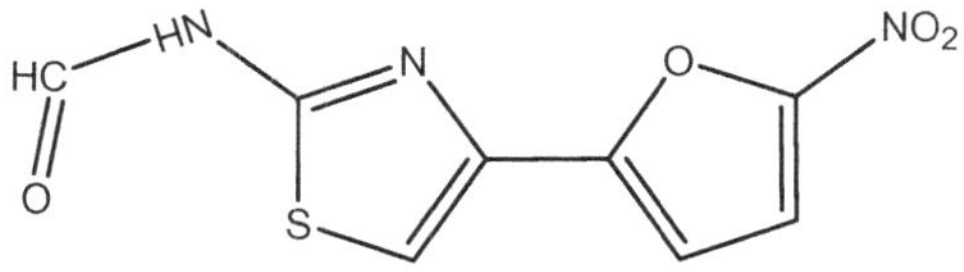

N-[4-(5-nitro-2-furyl0-2-thiazolyl]formamide
(FANFT)
TD_{50} = 4.25

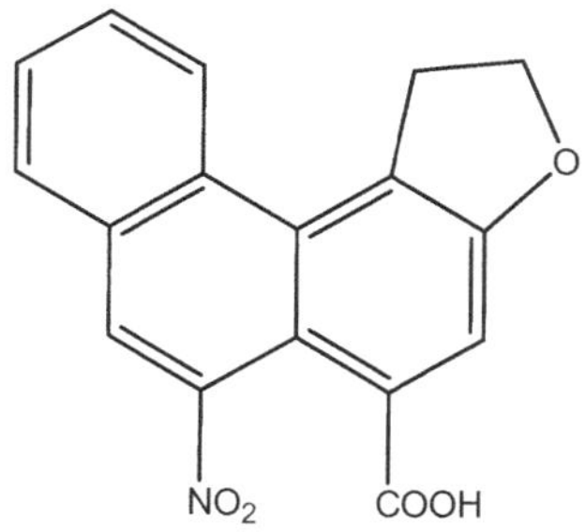

Aristolochic acid
TD_{50} = 0.0141

Azo Compounds

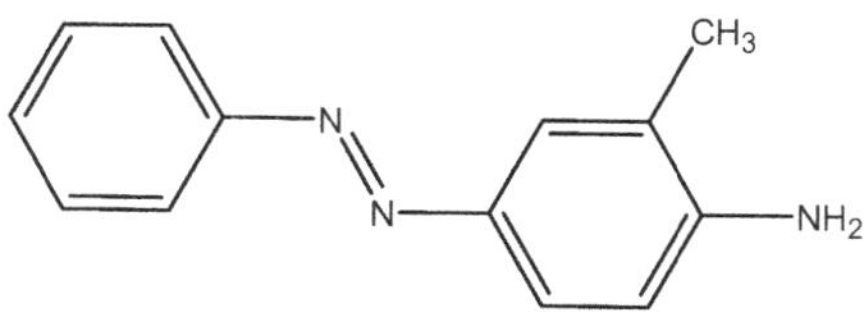

o-Aminoazotoluene
TD_{50} = 4.04

Nitrogen Mustard

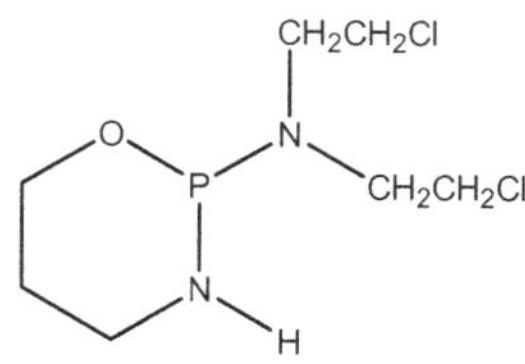

Cyclophosphamide
TD_{50} = 2.21

Nitroso compounds

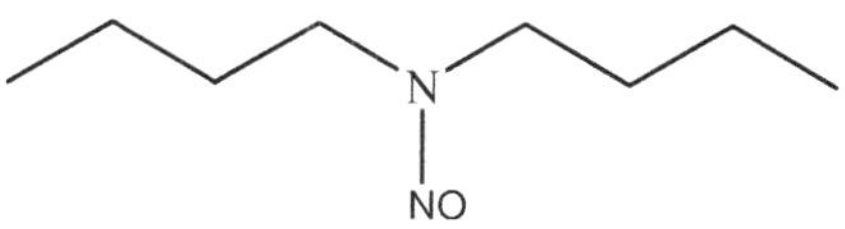

N-Nitroso-di-n-butylamine
TD_{50} = 0.691

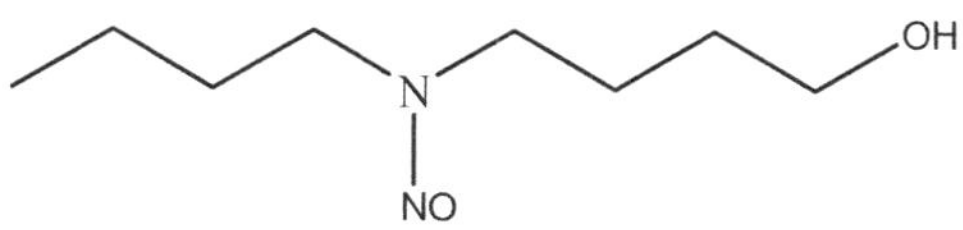

N-Butyl-N-(4-hydroxybutyl)nitrosamine
TD_{50} = 0.457

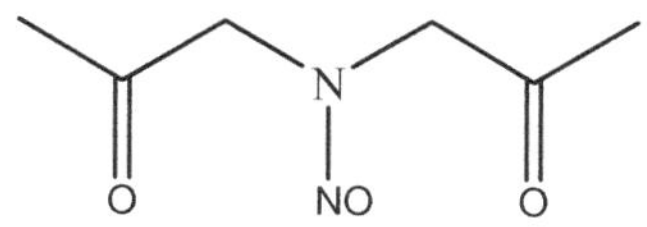

N-Nitroso-(2-oxopropyl)amine
TD_{50} = 0.491

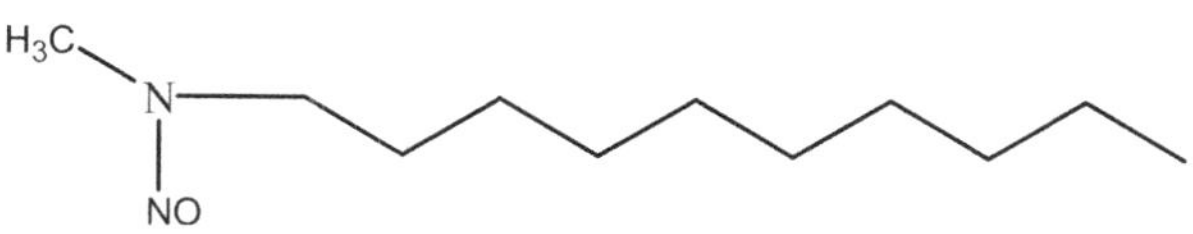

N-Nitroso-N-methyldecylamine
TD_{50}= 1.26

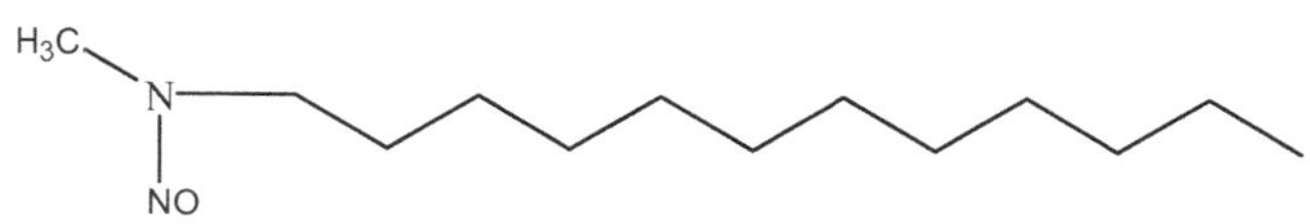

N-Nitroso-N-methyl-N-dodecylamine
TD_{50} = 0.537

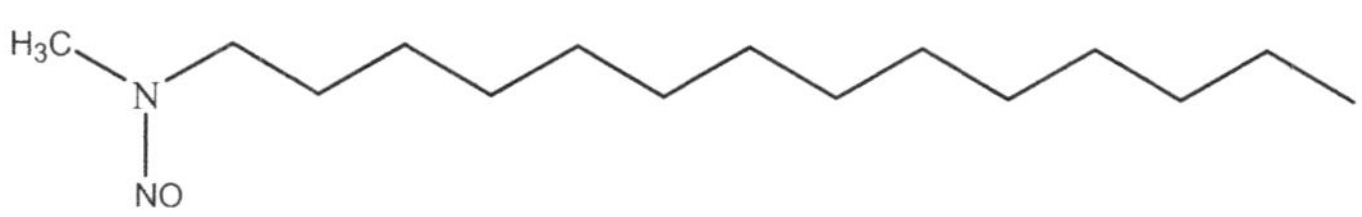

N-Nitroso-N-methyl-N-tetradecylamine
TD_{50} = 1.65

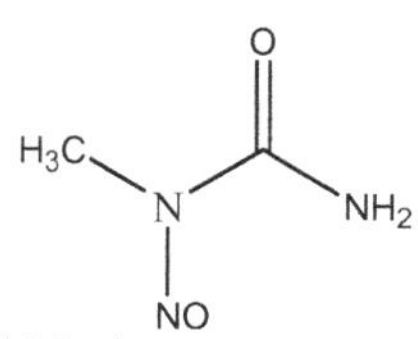

N-Methyl-N-nitrosourea
TD_{50} = 0.0927

Figure 6.12. Bladder carcinogens with TD50s below 10-mg/kg day^{-1} (Continued)

One question is whether or not oxidative reactions of chlorine can introduce an N-hydroxylated group into aromatic amines. Once absorbed, the hydroxylated form of the aromatic amine would distribute to the bladder where it will be acetylated by NAT1 and then activated by hydrolysis to the nitrenium ion that binds to DNA. N-hydroxyl-2-acetylaminofluorene is slightly more potent than 2-acetylaminofluorene when administered to rats, mice, or hamsters (Appendix B). This difference in potency may reflect a more effective delivery of the activated form of the carcinogen as the N-acetylated form is rapidly eliminated in urine. It appears that N-hydroxylation of the N-acetyl aromatic amines does not occur readily with HOCl (Lakshmi et al. 2000a). A more probable interaction of HOCl is with the free amino group to successively form an N-hydroxy, nitroso, and nitro group (Zenser et al. 1999).

Carcinogenic Potency. The carcinogenic potency of aromatic amines is generally consistent with their mutagenic potency in vitro (Benigni et al. 2003b). This conclusion is consistent with a large amount of mechanistic data that has been collected which demonstrates the conversion of aromatic amines to intermediates that react with purine and pyrimidine bases in DNA.

Structure- Activity Considerations. It is of interest that some aromatic amines are sterically prevented from being hydroxylated in biological systems. The best known example of this is the fact that 2-naphthylamine is a well-recognized human carcinogen, whereas 1-naphylamine is essentially inactive as a carcinogen in humans and experimental animals. However, N-hydroxyl-1-napthylamine is a bladder carcinogen (Radomski 1979, Belman et al. 1968, Dooley, et al. 1984). Therefore, oxidation of aromatic amines in drinking water disinfection is a reaction that should be explored as a possible explanation of bladder cancer that has been associated with chlorination of drinking water.

Structural features of aromatic amines that affect their carcinogenic potency of aromatic amines (Benigni et al. 2003) are as follows:

- Number of aromatic rings, - modify the leaving potential of the acyloxy anion. Tends to increase with increasing size (responsible for the number of rings in QSTR models of aromatic amines). Aminobiphenyls are more active than anilines and naphthylamines. A bridge between the rings of aminobiphenyl decreases activity. However, there is also a contribution of the ring structure to log P, which is positively correlated with potency.
- Nature and position of the amine/amine generating groups (hydroxylamino, nitro and nitroso groups), - dimethyl substitution on amine does not affect potency, but longer chain substitutions progressively interfere with N-dealkylation.
- Nature, number and position of other ring substituents, - primarily present steric barriers, especially, if large and in the ortho position. On the contrary, chloro or methyl/methoxy groups ortho to amine function can increase potency.
- Size, shape, and polarity of the molecules

Strictly from a mechanistic point of view, the potency of aromatic amines appears to depend on the ease of forming the N-hydroxylamine, the stability of the nitrenium ion that is produced, and the extent to which epoxides may form on the aromatic ring. The ease of formation of phenols is negatively correlated with carcinogenic activity (Loew et al. 1985).

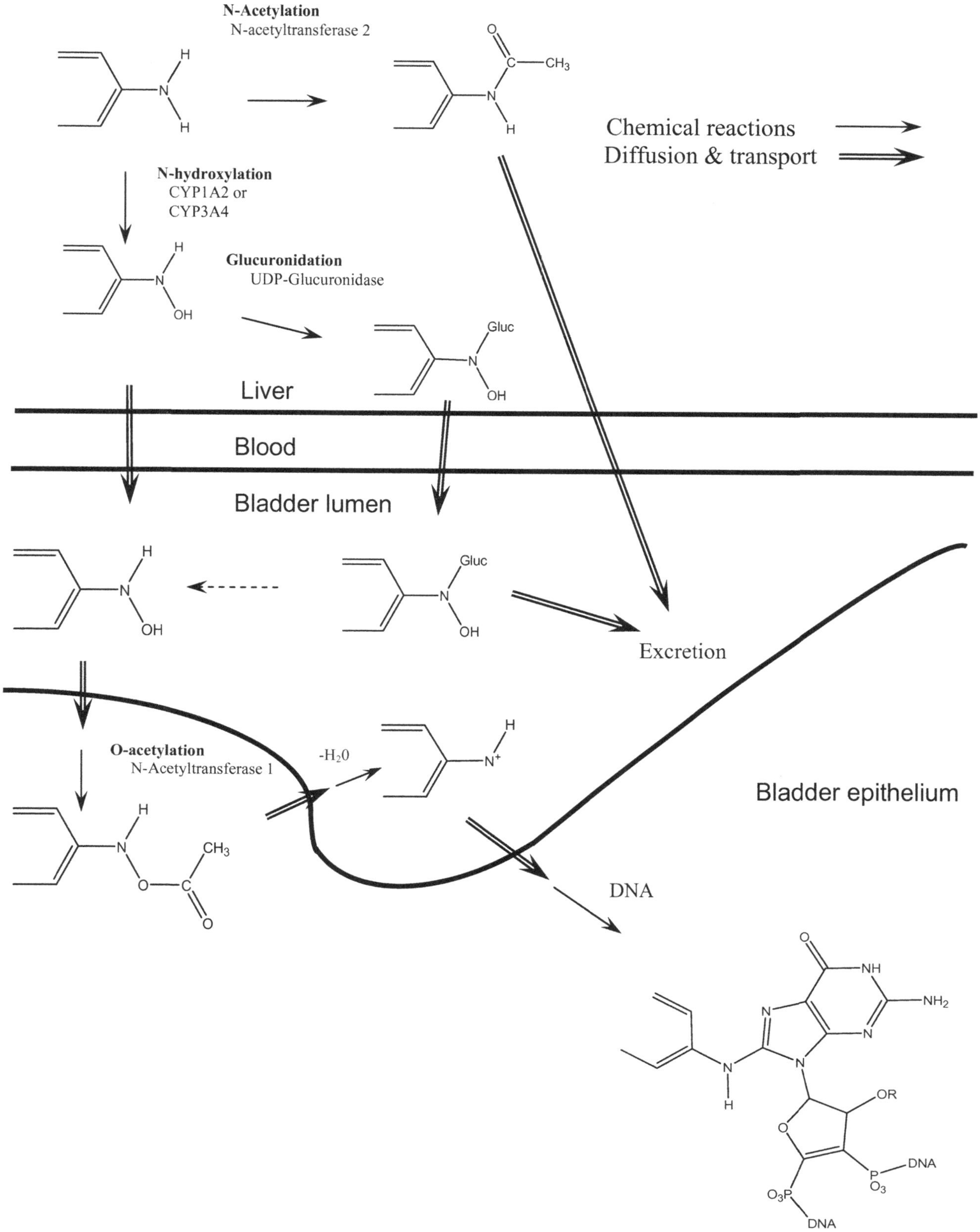

Figure 6.13 Relationship of aromatic amine metabolism to the induction of bladder cancer (see text for details)

This suggests that aromatic amines that are most easily hydroxylated by disinfectants are likely to result in by-products of principal concern from the aromatic amine class of precursors. The actual potency of these derivatives will depend primarily on the stability of the nitrenium ion that will be formed by the hydrolysis of the O-acetylation of these by-products by NAT1.

While evidence strongly supports the role of the nitrenium ion in the carcinogenicity of aromatic amines and related compounds, this may well be an overly simplistic formulation. N-hydroxylation is but one step in the metabolic oxidation of an amino group and oxidation through a nitro compound to a nitrosamine is also possible. The formation of N-chloramines is another way of producing some of the same types of reactive intermediates that have been implicated in bladder cancer (Thomas et al. 1987, Englert and Schacter 2002). However, there are little empirical data to develop QSTR relationships for the carcinogenicity of by-products that may well be produced with chlorine (discussed further in Chapter 7).

Nitroaromatic Compounds

Evidence That Nitroaromatic Compounds are Produced With Chlorination. Nitro compounds can be produced by oxidation of amines or amine precursors. There are few, if any examples of such chemistry in drinking water. There are data that demonstrate that HOCl (or HOBr) can oxidatively hydrolyze acylamides to yield nitroso groups. These are important observations because nitroaromatics that correspond to weakly carcinogenic aromatic amines that are sterically resistant to enzymatic N-hydroxylation can be quite potent as carcinogens. This is illustrated by the finding that N-hydroxy-1-naphthylamine is almost as potent as N-hydroxy-2-naphthylamine when administered to rats though 1-naphthylamine is usually classified as a non-carcinogen (Dooley et al. 1984, Benigni et al. 2003). Of course, 2-naphthylamine is a potent carcinogen. The reduction of the nitro group to produce the N-hydroxyl aromatic amine can utilize a variety of cytosolic and microsomal enzymes that are collectively much less demanding sterically than those that are responsible for the N-hydroxylation of the corresponding amine (Williams and Weisburger 1986, Ritter et al. 2000).

Characteristics of Carcinogenic Nitroaromatic Compounds. There are a large number of nitro compounds represented in the CPDB (Gold 2005). The bladder cancer database in Appendix B formally lists only two compounds in this group. However, the last five of compounds listed in the heterocyclic amine group are also nitro-containing compounds. Three of these compounds produce bladder cancer in all species in which they have been tested.

Mechanisms of Nitroaromatic Compounds. Many, if not all, of the nitroaromatic compounds act as genotoxic carcinogens. Structure-activity relationship studies of the mutagenic activity of the nitroarene group, at least, appear consistent with this interpretation. However, there appears to be limited QSTR study of the carcinogenic activity of these compounds. Therefore, this conclusion should probably be restricted to those compounds with a nitro group directly substituted upon an aromatic ring structure.

It is also important to recognize that there are other non-aromatic nitro compounds that are potent carcinogens (see Appendix C, Table C.1). Examples include 3-nitropentane and 2-nitrobutane. Clearly, the mechanisms involved with these non-aromatic systems are distinct from those of the nitroaromatics.

Role of Metabolism and Kinetics. As indicated above, the mutagenic activity of nitroaromatics appears to be related to their reduction to the corresponding N-hydroxylamines. The resulting hydroxylamines are ultimately converted to electrophiles (postulated to be a nitrenium ion) that react to form mutagenic adducts (Passerini 2003). However, the relative rates of reduction to the N-hydroxylamine and the subsequent reduction to the corresponding amine will greatly affect the carcinogenic potency of the compound (Williams and Weisburger 1986).

Carcinogenic Potency. The dependence of nitroaromatic carcinogens on reductive metabolism leads to the conclusion that their carcinogenic potency will parallel that of the corresponding aromatic amine with the caveat that the rate of reduction to the N-hydroxylamine must be rapid relative to the subsequent reduction to the amine. However, this chemistry should also play a role in the potency of the corresponding amine (i.e. the rate of formation vs. reversal of the N-hydroxylamine formation).

Structure-Activity Considerations. The main QSTR consideration for the nitroaromatics relates to the ability of the nitroaromatic to accept the electrons necessary to reduce it to reactive form (Passarini 2003).

Azo Compounds.

Evidence That Azo Compounds are Produced With Chlorination. Azo compounds are unlikely to be the direct result of reactions of disinfectants. Azo compounds fall into two broad classes, aromatic compounds that are frequently used as commercial dyes and simpler non-aromatic compounds. The dyes can be found in wastewaters. Reactions of HOCl with some members of the class have resulted in the formation of mutagenic 2-phenylbenzotriazole by-products (Nukaya et al. 1997, Shiozawa et al. 1998, Watanabe et al. 2001). The precursors of these compounds appear to be the corresponding dinitrophenylazo dyes. The key reaction involved in the formation of 2-[2-acetylamino)-4-[bis(2-methoxyethyl)-amino]-5-methoxyphenol]-5-amino-7-bromo-4-chloro-2H-benzotriazole (PBTA-1) was shown to be a cyclization reaction of a ring nitro group with the azo group of the dye.

Characteristics of Carcinogenic Aromatic Azo Compounds. The azo group of the aromatic compounds appears to be split to yield the corresponding aromatic amine/s in vivo (Williams and Weisburger 1986). Thus, the carcinogenic activity of some of the members of this class also depends upon the N-hydroxylation of the resulting aromatic amine groups.

Mechanisms. Azo dyes are frequently large molecules (see Appendix B) that can be cleaved metabolically to produce the corresponding aromatic amines; thus, they would generally be considered mutagenic carcinogens. However, the potency of the compound will generally depend heavily on the nature of the released aromatic amine.

Structure-Activity Considerations. There do not appear to be formal QSTR studies of this group of carcinogens. In general, simple azo compounds yield amines with weak or no carcinogenic activity (e.g. azobenzene yields aniline). More complex azo compounds yield amines that are more potent carcinogens (e.g. benzidine)(Williams and Weisburger 1986).

Heterocyclic Amines.

Reactions of Heterocylic Amines With Disinfectants That may Increase Their Carcinogenic Potency. Heterocyclic amines, per se, are not produced by reactions with disinfectants. As they are related to the aromatic amines, these compounds are identified as potential precursors of oxidation products that are likely to be found in nature. Purine and pyrimidine bases found in DNA and RNA are heterocyclic amines; the exocyclic nitrogen would participate in the same types of interactions identified with other aromatic amines. Pharmaceuticals are frequently heterocyclic amines and are increasingly being found as contaminants of domestic wastewaters that may serve as drinking water sources.

Characteristics of Carcinogenic Heterocyclic Amines. Much of the interest in heterocyclic amines has arisen from the identification of highly mutagenic/carcinogenic members of the class in foodstuffs, largely from the cooking of meats (Knize et al. 2002). Several of these chemicals have been identified as potent carcinogens and are being actively studied as potential causes of bladder and colon cancer in humans. Those that have been shown to be bladder carcinogens have been included in Appendix B.

Mechanisms. At least the more potent heterocyclic amines appear to be highly mutagenic. However, there appear to be a larger variety of mechanisms for the biological activation of heterocyclic amines that make assessing the carcinogenic potential of individual compounds much more difficult to evaluate than the simpler aromatic amines.

There are chemicals in this category that are carcinogenic, but for which genotoxicity does not appear to be responsible for the carcinogenic activity. The most notable example of this is sodium saccharin. Saccharin causes bladder cancer by forming insoluble precipitates in the urinary bladder that chronically the irritate the bladder epithelium (Cohen 1999). A similar mechanism is postulated to account for the bladder cancer produced by consuming large doses of uracil, a normal base in DNA.

Role of Metabolism and Kinetics. To the extent that heterocylic amines have an extracyclic amine group, their activation follows the same general pattern described for aromatic amines above (Archer et al. 2000), but with some important variations. The two N-acetyltransferase isozymes generally tend to play somewhat different roles with heterocyclic amines than aromatic amines. Neither enzyme efficiently acetylates the exocyclic nitrogens of heterocyclic amines. However, N-acetyltransferase 2 appears to be more efficient in the O-acetylation reaction that converts it to N-acetoxy metabolites that give rise to the ultimate carcinogenic form. Therefore, while N-acetyltransferase-2 is thought of as a detoxification step for aromatic amines, it is more frequently involved as an activation step with heterocyclic amines. The detoxification of some aromatic amines is sometimes more efficiently accomplished by glutathione transferases depending upon the structure of the aromatic amine.

Reactions of HOCl, $ONOO^-$, and other oxidant species have been shown to contribute to the metabolic activation of some heterocyclic amines as well as aromatic amines (Ritter and Malejka-Giganti 1989, Lakshmi et al. 2000b & 2002). The reactions are catalyzed by a variety of peroxidases, perhaps most notably myeloperoxidase but also other enzymes such as prostaglandin H synthase (Zenser et al. 1999). This latter enzyme is of particular interest because it is expressed at fairly high levels in bladder epithelium.

Carcinogenic Potency of Heterocyclic Amines, Structure-Activity Considerations. Two classes of heterocyclic amines have been studied using QSTR methods, the benzothiopyrano[4,3-b]quinolines and the benzothiopyrano[4,3-b]indoles. The reader is referred to Passerini (2003) for concise summaries of this analysis. Unlike the aromatic amines, the structural requirements for carcinogenicity and mutagenicity activities in this class of chemicals diverge. This suggests that more than one basic mechanism may well be involved within the class and indicates an important departure from the simpler aromatic amine class.

Conclusions about Putative DBPs that Could be Bladder Carcinogens

Three groups of putative DBPs would appear to have potential as bladder carcinogens. These include N-haloamines, haloquinones, and halofuranone-like compounds, including the halocyclopentenoic acids. Haloquinones would appear to be the most attractive candidates from a mechanistic point of view. However, these compounds are likely to occur at higher concentrations in chloraminated water than in the presence of free chlorine. Their actual occurrence has not been documented.

Haloquinones as Potential Bladder Carcinogens. There are parallels in the probable metabolic activation and processing of haloquinones with those of recognized bladder carcinogens from the aromatic amine group. The actual enzymatic bases of metabolic activation and processing of the two types of compounds are different based on literature that is available on quinones in general. Essentially, there is initial metabolism in the liver that tends to be a detoxification step that is followed by transport of the metabolite, concentration in urothelial tissue of the kidney and bladder which, in turn, activates the compounds to forms that alkylate macromolecules and/or generate oxidative stress.

Characterization of the toxicological properties of the haloquinone derivatives predicted to arise from chlorination of syringic acid and coniferyl alcohol brings up additional concerns from a toxicological point of view. There have been numerous studies to implicate phenolic and quinone structures in renal and less frequently bladder toxicity (Bolton et al. 2000). Metabolism of these compounds results in the formation of a series of glutathione adducts in the liver that are processed into reactive cysteinyl conjugates (Figure 6.14). These compounds are reabsorbed and redistributed to cells that express gamma glutamate transpeptidase (γ-GT) or glutathione conjugate transporters. The kidney and bladder are such tissues. In tissues, the conjugates are concentrated in cells expressing these enzymes (e.g. the proximal tubules of the kidney). Quinones are Michel acceptors and are thus capable of binding with nucleophiles such as DNA bases. Catechols have hydroxyl groups ortho to one another and polyaromatic hydrocarbons metabolized to o-quinones, allowing them to participate in redox cycling that produces hydroxyl radicals as a by-product which in turn damage DNA. As indicated earlier, p-phenols can be produced in water and then converted to quinones metabolically. Consequently, there are a multitude of mechanisms by which quinones can participate to produce toxicity and carcinogenicity. One such compound is a metabolite of bromobenzene, 2-bromo-bis(glutathion-S-yl)hydroquinone, which is one thousand times more potent as a renal toxicant than bromobenzene (Monks and Lau 1997). This metabolite produced renal toxicity in guinea pigs given intraperitoneal doses of 10-20 μmol/kg. A metabolite of hydroquinone, 2,3,5-tris(glutathion-S-yl)hydroquinone, given intraperitoneally to Eker rats for 10 months at doses of

2.5 to 3.5-μmol/kg resulted in increased incidence of renal dysplasia, adenomas and carcinomas (Lau et al. 2001). A similar metabolite of 3-tert-butyl-4-hydroxyanisole, 5-(glutathion-S-yl)-tert-butyl-4-anisole, causes cell proliferation and severe hemorrhage in the rat bladder (Peters et al. 1996). 3-tertiary-butyl-4-hydroxyanisole is known to produce preneoplastic and neoplastic lesions in both the rat kidney and bladder (Bolton et al. 2000). Therefore, the halogenated polyphenol/quinone by-products are plausible candidates as a cause of bladder cancer. However, the quinone form is much more toxic.

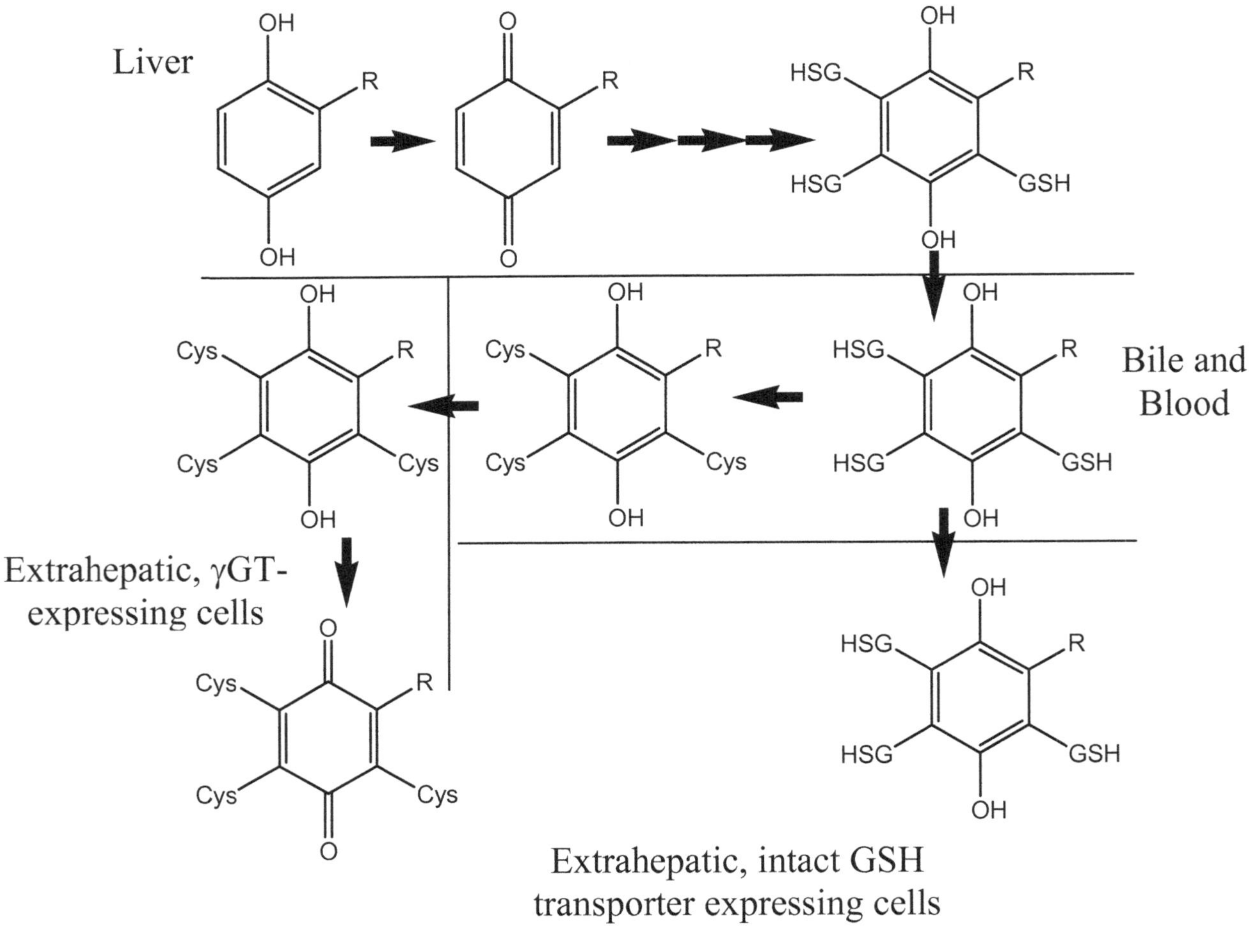

Figure 6.14 Metabolism of quinones animals Quinones are initially conjugated at one or more locations with glutathione (GSH) in the liver. These conjugates are secreted into the bile, reabsorbed into the blood and distributed to internal organs. During their transit in the bile and blood, the glutathione conjugates are degraded to cysteine (Cys) conjugates. The glutathione or cysteine conjugates are transported into tissues that express carriers and concentrate them. The cysteine conjugates are processed by another enzyme to a form that is capable of covalent modification of macromolecules, while the quinone structure participates in cycles of reduction to the corresponding hydroquinone and reoxidation to the quinone. This process leads to the generation of radicals that can also damage macromolecules.

The candidacy of these compounds as potential causes of adverse health effects associated with chlorinated drinking water is supported by examples of bladder carcinogens with polyphenolic or quinone structures. The most potent of these is quercetin with a TD_{50} of 10.1-mg/kg day^{-1} (Gold 2005). At the other end, the TD_{50} for hydroquinone is 82.8-mg/kg day^{-1}. As a consequence, the intrinsic potency of non-halogenated polyphenols and quinones would appear to be marginal relative to our criteria. It has been suggested that o-quinone metabolites may contribute to the carcinogenicity of polyaromatic hydrocarbons, but the formation of diol-epoxides remain a dominant hypothesis to account for the carcinogenicity of these chemicals.

There appear to be no studies of polyhalogenated polyphenols or quinones. Substitution of electron withdrawing groups, such as halogens, on the ring structure results in the quinone behaving as a stronger oxidant (Monks and Lau 1998). Conversely, addition of a nucleophile to the quinone reduces it to the corresponding hydroquinone, which can be reoxidized to the parent quinone. This allows the formation of multiple products capable of damaging DNA. Thus, multiple halogen substitutions, as represented by the two identified products of the chlorination of syringic acid and coniferyl alcohol, are likely to have substantial effects on the toxicological potency of the quinone structure. This conclusion is supported by the unusually low predicted chronic LOAELs for these compounds in the present study.

For these reasons haloquinones should be examined as potential causes of bladder cancer from chlorinated drinking water. This step should not be taken without first determining whether or not these chemicals are produced, whether or not they are at least as stable as MX, and the concentration ranges at which they occur using different disinfection methods in the same water. Despite the uncertainties in the TOPKAT® predictions for these compounds, the very low predicted chronic LOAELs for this class of compounds suggest that that these chemicals need attention. If these compounds are seen to occur in chlorinated drinking water, a set of initial studies should be conducted to prioritize research among the chemicals. In order to conduct chronic studies some of these chemicals will need to be synthesized in quantity. Organic synthesis on the needed scale can add significantly to the cost of toxicological characterization. Therefore, it is suggested that toxicological characterization begin by confirming the extent to which these compounds are tolerated (i.e. a preliminary assessment of the LOAEL in a short-term study), investigations of their mutagenic properties in vitro, and a measure of their potency as DNA-binding chemicals in vivo.

Halofuranones and Halocyclopentenoic Acids as Potential Bladder Carcinogens. There is little theoretical basis in the QSTR analysis for implicating this group as potential bladder carcinogens. However, it is apparent that MX does target the urinary bladder in chronic studies in certain rat strains (McDorman et al. 2003), although this pathology is less apparent in Wistar rats that were utilized in the cancer study (Vaittenen et al. 1995). MX, itself, appears at too low a concentration to account for the risks associated with chlorinated water. However, if its occurrence is paralleled by the formation of a larger group of chemicals with similar modes of action, a significant level of risk might be projected. Consideration of structural analogs of MX that are potent carcinogens (e.g. furan), but not mutagenic, suggest that non-mutagenic modes of action should be considered. The halocyclopentenoic acids are included in this group as one of them, 3,5-Dichloro-1-hydroxy-3-keto-cyclopent-2-enoic acid, had the lowest predicted chronic LOAEL of all compounds examined with TOPKAT®. This prediction was outside the OPS of the model, but the same can be stated for MX, contrary to existing experimental data. These compounds also did not meet the criteria for carcinogenicity, but again, neither did MX, despite the fact it was shown to be a multi-target organ carcinogen in rats (Komulainen et al. 1997).

Organic N-Haloamines as Potential Bladder Carcinogens. Organic N-haloamines are known to occur in both chlorinated and chloraminated drinking water. There is reason to suspect, however, that the form of these organic N-haloamines may differ significantly with the two forms of disinfection.

There is no empirical data for identifying specific organic N-haloamines as likely bladder carcinogens. In part, this is due to a lack of information on the occurrence and concentrations of specific organic N-haloamines. However, study of the activation of certain drugs to toxic intermediates via the formation of organic N-chloramines has demonstrated the occurrence of a number of peroxidases in the urinary bladder that may contribute to the further activation of stable N-haloamines to a variety of reactive intermediates. Such activation may be exacerbated by bladder infections, a known contributor to the risk of bladder cancer in humans. A much better toxicological database is needed for members of this class of DBPs before specific, testable hypotheses for their involvement in inducing human bladder cancer can be formulated.

CHAPTER 7. PRIORITIZATION

This chapter is divided into three parts. First, the results of the quantitative structure-toxicity relationship (QSTR) analyses of disinfection by-products (DBPs) from established classes were examined to determine which of these compounds have high probabilities of contributing to adverse health effects at levels encountered in drinking water. Second, the QSTR results with novel by-products that were predicted to be produced by the reaction of disinfectants with NOM were combined with the established DBP classes. The relative importance of these two groups as health hazards were compared to those posed by regulated DBPs. Finally, reviews of the literature on members of selected classes of DBPs and putative DBPs were conducted to see whether or not they might contribute to the specific health effects associated with chlorinated drinking water. From these assessments a group of research directions are suggested.

PRIORITIZATION OF CHEMICALS SUBJECTED TO QSTR EVALUATIONS.

Disinfection By-products with Chronic LOAELs Predicted to be < 1 mg/kg Day^{-1}

Table 7.1 lists those DBPs from established classes as well as putative DBPs that were predicted to have chronic lowest observed adverse effect levels (LOAELs) below 1 mg/kg and whose prediction did not exceed the permissible limits of the optimal predictive space (OPS) of the model. Exceptions were made to this criterion for 3-Chloro-4-(dichloromethyl)-5-hydroxy-2[5H]furanone (MX) analogs, selected haloquinones and a related haloimine that were predicted to result from chlorination of NOM substructures. These have been retained because the actual LOAELs observed in studies for MX and related compounds were within the confidence interval of the predicted LOAEL. Thus, these chemicals remain on the high priority list. A large number of halonitriles had predicted LOAELs < 1-mg/kg day^{-1}, but exceeded the permissible limits of the OPS, and were eliminated. However, these compounds met the criteria as carcinogens or developmental toxicants and they are identified as high priority under alternative headings.

To put the predicted LOAELs into perspective, they are treated as if they are real and a probable maximum contaminant limit (MCL) is calculated. The lowest chronic LOAELs were for a group of compounds that were predicted to arise from precursor substructures in NOM. The lowest of these LOAELs that was within the OPS of the model and met other criteria was 0.005-mg/kg day^{-1}. Conventional application of uncertainty factors would result in an acceptable drinking water standard of 0.045 μg/L (0.005 X 70/1000 X 2). At the other extreme, the predicted LOAELs of a number of chemicals were just below 1-mg/kg day^{-1}. The analogous calculation for these compounds would result in a MCL of 35 μg/L. These figures assume that all of the exposure is coming from drinking water (i.e. no factor was included for relative source contribution) and no adjustments were made for children, which would generally lower these values. If these compounds were shown to be carcinogenic, the MCLs would be significantly lower, based upon current USEPA policy of linear extrapolation of risk to low doses. Similar calculations can be applied to the other chemicals to further assess their relative priority. DBPs from established classes were of less interest toxicologically than were those that were predicted to be produced by reactions of disinfectants with NOM. Three likely DBPs were three to four orders of magnitude more potent than regulated DBPs, while the remaining 30 compounds are predicted to be 10-100 times more potent. These data illustrate that chemicals may occur as the

Table 7.1
Putative disinfection by-products with LOAELs < 1 mg/kg

Index Number	Chemical	LOAEL (mg/kg day^{-1})
K01C14	2,4-Dichloro-3-hydroxybutanoic acid	0.95
K02C38	2,3,4-Trichlorocyclohexadien-5-one carboxylic acid	0.22
K05C06	Trichlorocyclopentenedione	0.50
K10C07	2,3,4-Trichlorobutanal	0.96
K11C13	2,3-Dichloropentanedioc acid	0.87
K11C23	Chlorobutanedioic acid	0.85
K15C03	2,3-Dichloropropenal	0.97
K16C04	2-Chloro-3-pentenedioc acid	0.22
K17C03	Dibromoacetonitrile	0.95*
K21C01	3-Chloro-4-(bromochloromethyl)5-hdroxy-2[5H]furanone	0.16†
K21C02	3-Chloro-4-(dibromomethyl)-5-hydroxy-2[5H]furanone	0.29†
K21C03	5-Hydroxy-4-trichloromethyl-2-furanone	0.21†
K21C05	3-Chloro-4-(chloromethyl)-5-hydroxy-2[5H]furanone (CMCF)	0.029†
K21C06	E-2-chloro-3-(dichloromethyl)-4-oxobutenoic acid (E-MX)	0.67
K21C07	3,4-(dichloro)-5-hydroxy-2(5H)-furanone (MCA)	0.79*
K21C08	3-Chloro-4-(dichloromethyl)5-hydroxy-2[5H]furanone (MX)	0.10†
K21C11	3-Chloro-4-(dichloromethyl)-2-furanone (Red-MX)	0.074†
K21C12	3-Chloro-4-(tribromomethyl)-5-hydroxy-2-[5H]furanone	0.41†
K23C01	3-Bromo-3-Chloropropamide	0.62
K23C20	2,3-Dichloropropamide	0.51
K30C02	2,3-Dibromopropionitrile	0.15
K50C04	3,4,5,6-Tetrachloro-o-quinone	0.0047†
K50C05	1,2,3-Triketo-4,4,5,6-tetrachlorocyclohex-5-ene	0.050
K50C06	1,2,2,4-Tetrachloro-3-hydroxy-cyclopent-4-enoic acid	0.19†
K50C07	3,5-Dichloro-1-hydroxy-4-ketocyclopent-2-enoic acid	0.00025†
K50C26	1,3,3,5,5-Pentachloro-4-keto-5-hydroxycyclohexene	0.21*
K50C35	2,6-Dichloro-1,4-benzoquinone	0.049†
K50C36	2,3,6-Trichloro-1,4-benzoquinone	0.033
K50C37	2,3,5,6-Tetrachloro-1,4-benzoquinone	0.005
K50C38	2,6-Dichloro-3-methyl-1,4-benzoquinone	0.079
K50C40	2,3,6-Trichloro-1,4-benzoquinone-4-(N-chloro)imine	0.073‡
K50C41	2,3-Dichloro-4,5-diketobenzoprop-2-enol	0.064
K50C42	2,3,5-Trichloro-4,5-diketobenzoic acid	0.073
K50C44	2,3,6-Trichloro-4,5-diketophenylprop-2-enol	0.0065†

* Outside the optimal predictive space of the model, but within permissible limits
† Outside the optimal predictive space of the model and exceeds permissible limits
‡ Query structure contains fragment/s not represented in the training set

result of drinking water disinfection that are capable of producing adverse health effects at much lower concentrations than those of the regulated DBPs.

Disinfection By-products Predicted to be Multispecies Carcinogens with Chronic LOAELs < 10 mg/kg Day^{-1}

DBPs for which the best case of significant carcinogenic potential can be made based on TOPKAT® are displayed in Table 7.2. These chemicals were predicted to be carcinogenic in both mice and rats and had at least one chemical in the training set of the model with a similarity index <0.2. The most potent of these chemicals, as judged by the predicted chronic LOAEL, are listed first and the similarity indices for those predictions are provided. It was noted previously that several of the LOAEL predictions are outside the OPS of this model, especially for the halonitriles. Nevertheless, the halonitriles have been retained in the priority list because LOAELs observed in subchronic or reproductive/developmental studies for selected members of the class that were not included in the training set were within the confidence intervals of the predicted chronic LOAELs.

Table 7.2
Halogenated chemicals from established classes of disinfection by-products predicted to be potent carcinogens (Chemicals are listed in order of priority based on LOAEL and mutagenic potential.)

Index number	Chemical	Pred. Ames' mut.	Similarity index	LOAEL (mg/kg)	Similarity index
K30C03	2,3-Dichloropropionitrile	+	0.109	0.043†	0.205
K30C02	2,3-Dibromopropionitrile	+	0.121	0.15	0.183
K17C05	Dichloroacetonitrile	+	0.000	0.29†	0.170
K30C08	3,3-Dichloropropenenitrile	+	0.230	0.77†	0.144
K30C15	3-Bromopropionitrile	+	0.200	0.89†	0.149
K15C03	2,3-Dichloropropenal	+	0.223	1.0	0.164
K23C08	Bromochloroacetamide	+	0.242	1.5	0.159
K15C09	2-Chloropropenal	+	0.258	1.7	0.158
K02C13	3,3-Dibromopropenoic acid	+	0.191	1.9	0.168
K17C07	Trichloroacetonitrile	+†	0.494	4.5†	0.166
K09C02	1,2,3-Trichloropropane	+	0.156	4.7	0.000
K04C04	Dichloroacetaldehdye	+	0.192	5.8	0.047
K30C05	3,3,3-Trichloropropionitrile	-	0.212	3.9†	0.169
K01C45	2-Bromo-3,3-dichloropropionic acid	-	0.271	5.8	0.068

* Prediction is outside the optimal predictive space of the model, but within permissible limits
† Prediction is outside the optimal predictive space of the model and exceeds permissible limits

Lack of predicted Ames' mutagenic activity in the TOPKAT® analysis resulted in 3,3,3-trichloropropionitrile and 2-bromo-3,3-dichloropropionic acid falling to the lowest priority in this group of DBPs. The similarity indices for the Ames' predictions are provided, but were not used in the prioritization scheme. In general, risk from genotoxic carcinogens is considered to extend linearly into the low dose range. In contrast, the risk from non-genotoxic carcinogens is considered to extend to low doses in a non-linear fashion. Consequently, at concentrations corresponding to 1/1000-1/100,000 the benchmark dose for a 10% response rate (i.e. extrapolating to the 10^{-4} to 10^{-6} range of lifetime risks), the risk from a non-genotoxic carcinogen would be less than the risk predicted for the genotoxic carcinogen.

2,3-Dichloroproprionitrile topped the list of chemicals from established classes of DBPs based on the lowest predicted chronic LOAEL; this prediction exceeded the OPS of the model. The prediction for 2,3-Dibromopropionitrile was within the OPS and is used to illustrate levels of potential importance for predicted carcinogens. The LOAEL would generally be 3-10 times higher than the lowest confidence limit of the benchmark dose (BMDL) that is used as a point of departure in the new cancer risk assessment guidelines. If a factor of five is used to estimate the lowest confidence limit of the benchmark dose that is predicted to produce a response in 10% of the population ($BMDL_{10}$) from the predicted LOAEL, the concentration of this compound in drinking water that would be calculated to increase lifetime risk of 10^{-4} would be in the range of 1 μg/L; the 10^{-6} risk would be predicted to be approximately 0.01 μg/L. Approximations of possible drinking water concentrations that would be derived if the predictions proved to be correct can be made in a similar fashion. These concentrations are within the range at which the dihaloacetonitriles occur in chlorinated drinking water. Although halopropionitriles are known to be produced from amino acids, only the haloacetonitriles have been included in surveys of DBPs in drinking water.

The inclusion of bromochloroacetamide in this list of compounds is of interest because haloacetamides have been observed in chlorinated waters at appreciable concentrations (Weinberg et al. 2002). It is not entirely clear why bromochloroacetamide remained on the priority list when the dibromo and dichloro analogs did not. Dichloroacetamide was eliminated because the similarity indices in the databases on which positive predictions were made did not meet the < 0.2 criteria. Dibromoacetamide was rejected because its chronic LOAEL was predicted to be 10.6-mg/kg day^{-1}. While there were five predictions of carcinogenicity with a probability > 0.7, there was only one model in which a similarity index of < 0.3 was observed. Therefore, dibomoacetamide narrowly missed several of the criteria established at the outset.

It is of interest to note that acetamide, itself, has been shown to be carcinogenic in male and female rats and male mice. The TD_{50}s for rats and mice were 180 and 3010-mg/kg day^{-1}, respectively (Gold 2005), significantly larger than the chronic LOAELs predicted for the halogenated derivatives. Therefore, halogenation is expected to have a substantial effect on the carcinogenic potency of the haloacetamides.

Despite the fact that the present analysis provides a basis for interest in members of some classes of established DBPs, it seems unlikely that the combination of potency and probable concentrations of individual DBPs would be sufficient to account for cancer risks of the magnitude suggested by epidemiological studies of chlorinated water. Collectively, halonitriles might approach such levels of risk, but this would depend upon the occurrence of members outside the haloacetonitrile class at concentrations that are at least as high as observed with the dihaloacetonitriles. This would mean that a substantial portion of the organic nitrogen in chlorinated drinking water would be in the form of halonitriles.

As the authors were unable to show that the probability that established DBPs could account for the level of cancer risk projected from epidemiological studies, it was necessary to explore the formation of novel DBPs by reactions of disinfectants with substructures of NOM.

Disinfection By-products Predicted to be Developmental Toxicants with Chronic LOAELs < 10 mg/kg Day^{-1}

Chemicals from recognized classes of DBPs that TOPKAT® predicted to be developmental toxicants are listed in Table 7.3. Included are those with a predicted chronic LOAEL below 10 mg/kg and a similarity index with at least one chemical within the training set of < 0.2. Halonitriles are prominent among these chemicals. As indicated earlier, many of the chronic LOAEL figures are outside the OPS of the TOPKAT® model. Since there are data on selected halonitriles in less than lifetime studies that fall within the confidence limit of the TOPKAT® chronic LOAEL predictions, these chemicals were retained with the high priority chemicals as potential developmental toxicants. Although only cyanogen chloride, cyanogen bromide, and the haloacetonitriles have been shown to occur in drinking water, the halopropionitriles have been included in this analysis because they are very likely to be formed and simply have not been measured. However, the fact that none of these has really been measured in drinking water means that their significance will depend upon the concentrations at which they occur. The only other compound in the high priority list for developmental toxicants is 3-chloro-2-butenoic acid, an unsaturated haloacid.

Table 7.3
Halogenated disinfection by-products predicted to be developmental toxicants by TOPKAT®

Index number	Chemical	Probability of DTP	Similarity index	LOAEL (mg/kg)
K02C20	3-Chloro-2-butenoic acid	0.854	0.154	2.8
K17C06	Tribromoacetonitrile	1	0.000	4.5†
K17C07	Trichloroacetonitrile	1	0.000	4.5†
K17C02	Bromodichloroacetonitrile	0.998	0.091	1.5†
K17C04	Dibromochloroacetonitrile	0.998	0.091	1.5†
K30C10	3-Bromobutanenitrile	1	0.205	2.5†
K30C05	3,3,3-Trichloropropanenitrile	1	0.029	3.9†
K30C08	3,3-Dichloropropenenitrile	0.975	0.127	0.77†
K30C01	2,3,3-Trichloropropenenitrile	0.923	0.165	1†
K30C13	Cyanogen chloride	0.869	0.157	0.23†
K30C12	Cyanogen bromide	0.869	0.157	0.25†

* Outside the optimal predictive space of the model, but within permissible limits
† Outside the optimal predictive space of the model and exceeds permissible limits

All of the chemicals on the list, with the exception of 3-bromobutenenitrile, had similarity indices well below 0.2. This means that the electrotopological character of the chemicals was well represented in the developmental toxicity database. A limited exploration of non-halogenated analogs of the halonitriles was undertaken (i.e. acetonitriles, propionitrile, and 3-cyanopropanoic acid). All were predicted to be developmental toxicants with similar probabilities as indicated above for the halonitriles. Moreover, the predicted LOAELs were in the same 1-10-mg/kg range (although only one prediction, 3-cyanopropanoic acid, was within the OPS of the model).

Predictions of developmental toxicity appear to be associated with the nitrile group rather than being a property specific to the halonitriles. Compounds that donate cyanide as a by-product of their metabolism are known to have antithyroid activity (by liberation of cyanide which is subsequently converted to thiocyanate, which inhibits iodide uptake). Thyroid dysfunction is a recognized cause of developmental problems. In recent years, concerns have tended to focus on potential effects on brain development. Thyroid dysfunction generally requires substantial compromise of this function before it results in developmental effects in humans (NRC 2005). The probable potency of these chemicals would be related to their relative ability to produce cyanide as a product of their metabolism. It would require substantial concentrations of a halonitrile to result in an amount of cyanide sufficient to significantly affect thyroid function. Therefore, this is not considered a probable mechanism with the low levels of halonitriles that have been observed in drinking water surveys to date. Similarly, the prediction of carcinogenic effects of the halonitriles may reflect comparisons to other cyanophores that cause thyroid cancer in rodents. In general, the induction of cancer of the thyroid by this mechanism is not thought likely with the low doses of exposure to humans (NRC 2005). The haloacetonitriles are known to be mutagenic and to act as tumor initiators in the mouse skin (Bull et al. 1985). As a consequence, the carcinogenic activity of the halonitriles is of more interest than their effects as developmental toxicants.

The National Toxicology Program (NTP) subjected several recognized DBPs to short-term studies of reproductive and developmental toxicity. These studies have utilized a common 35-day treatment period with the chemical included in the drinking water. Parameters of both male and female reproductive functions were independently evaluated. Fetuses were recovered and examined for structural abnormalities. The LOAEL for liver toxicity of bromodichloromethane (53-mg/kg day^{-1}) was lower than the reproductive and developmental toxicity LOAEL (88-mg/kg day^{-1}) (NTP 1998c). Similarly, general toxicological effects were observed with dibromochloromethane at doses as low as 5-mg/kg day^{-1}, with no evidence of reproductive or developmental toxicity at doses as high as 40-mg/kg day^{-1} (NTP 1996). Bromochloroacetic acid affected reproductive function in female rats at doses of 50-mg/kg day^{-1}, a dose that also produced other indications of toxicity in both male and female adult rats (NTP 1998b). In the case of dibromochloroacetic acid, no effects on the reproductive performance of females were observed and only decreased sperm velocity was observed in males at a dose of 112-mg/kg day^{-1}. There was evidence of mitogenic effects in the liver at this same dose level of dibromochloroacetic acid (NTP 2000b). Tribromoacetic acid-treated rats had evidence of kidney involvement at 39 mg/kg day^{-1} without evidence of it being a reproductive toxicant in either male or female rats at this dose (NTP 1998d). Two haloacetonitriles were examined, bromoacetonitrile and dibromoacetonitrile. In the case of bromoacetonitrile the LOAEL for reproductive and developmental toxicities was the same as that observed for general toxicity (7-mg/kg day^{-1})(NTP 1997a). Dibromoacetonitrile produced some decrements in water consumption at a dose of 9.9-mg/kg day^{-1}, the highest dose examined, but found no evidence of reproductive or developmental toxicity (NTP 1997b). The LOAELs for these two compounds are below 10 mg/kg day^{-1}, and are generally compatible with the chronic LOAELs predicted by TOPKAT® for other haloacetonitriles. However, these data suggest that established DBPs within the THM, HAA, or HAN classes are weak developmental toxicants at best. NTP has apparently abandoned the pursuit of these effects in more traditional two-generation studies on the basis of the lack of suggestive findings in studies that included premating and in utero exposures for the duration of the pregnancy. In fact the lack of such effects in these drinking water studies would seem to cast significant doubt on previous results of shorter duration where gavage and non-aqueous vehicles were utilized.

This analysis of chemicals from recognized classes of chlorination by-products failed to identify any chemicals that are likely causes of reproductive effects associated with chlorinated drinking water in some epidemiological studies. Halonitriles would appear to the best candidates, but the median predicted chronic LOAELs within this class were generally above 1-mg/kg day^{-1}. Data obtained with respect to reproductive and developmental effects of the haloacetonitriles certainly suggest that the halonitriles might be active as reproductive toxicants. However, the data from these compounds do not suggest that such effects would be observed at doses lower than the chronic LOAEL.

Comparisons of QSTR Results with Prior Efforts

Moudgal et al. (2000) utilized TOPKAT® to identify DBPs as potential carcinogens or developmental toxicants. The basic form of the predictions made in that publication were the same as those of the present effort. The main difference is that this work extends itself into the realm of prioritization. This involves a systematic combination of data generated from eight individual carcinogen models, a developmental toxicology potential model, an Ames' mutagenicity model, and a chronic LOAEL model within TOPKAT®. The key decision criteria related to developing an estimate of toxicological potency that could be coupled to occurrence data to decide whether or not a chemical is likely to contribute significantly to risks that have been associated with disinfected water. The subsequent prioritization efforts can be broadly described as weight-of-evidence decisions. The effort is biased toward identifying a relatively small number of by-products for which toxicological data should be required. Relaxation of these criteria to be inclusive of all compounds that could possibly produce an effect would greatly increase the number of compounds, but provide little guidance for prioritization of research to clarify causal relationships.

Woo et al. (2002) performed structure-activity analyses of 209 DBPs for which there were data to support the actual occurrence of the chemical in drinking water. The sources of occurrence information for the Woo et al. study were the same reviews that were utilized in the present effort. The application of the occurrence criterion in the Woo et al. study was not well documented except to say that evidence solely based on laboratory or model studies was excluded.

The Woo et al. requirement for demonstrated occurrence was clearly not followed in the present effort. Chemicals produced under conditions that adequately modeled potential formation in situ were included in this study. Moreover, several potential DBPs were included in this study that were postulated to occur based on knowledge of precursor substructures known to occur in NOM present in virtually all drinking water sources. In exploring particular classes of DBPs, a limited number of hypothetical products within specified classes were included to more thoroughly explore the potential impact of changes in the carbon skeleton and other functional groups on predictions that were made by TOPKAT®. Those compounds for which there have been no measurements in water or that are not plausible considering precursor substrates are identified in the final prioritization below. Despite these methodological differences, the intent of the Woo et al. (2002) effort was more of a prioritization effort than the Moudgal (2000) paper. Therefore, it is useful to compare those by-products given higher priority by Woo et al. with those identified as high priority in this project.

Woo et al. (2002) prioritized DBPs using a semi-quantitative six-category scale based on the likelihood that a by-product was a multispecies and multi-target carcinogen. The top three categories are comparable to the criteria utilized in the present study and were described as 1) highly likely to be potent, multispecies/target carcinogen even at low doses (H), 2) highly likely to be a moderately active multispecies/target carcinogen at moderate doses (HM), and 3) likely to be a moderately active multispecies/target carcinogen at high doses or active single species/target carcinogen at low doses (M). The authors identified no DBPs in the H category. The chemicals that were identified in the HM and M categories are listed in Table 7.4 (in some cases, the nomenclature has been changed to match that used in the present study).

Woo et al. (2002) identified three halofuranones in the HM category. These are abbreviated BMX-1, BMX-2, and BMX-3 where the chlorines on MX are replaced by bromine. These were identified as 4-bromomethyl-MX, 4-dibromomethyl-MX, and tribromo-MX in this project. One halofuranone is included in the M category, mucochloric acid. In addition, the M category included five haloalkanes/haloalkenes, six halonitriles, 2 haloketones, one haloaldehyde, one halonitroalkane, and one non-halogenated aldehyde.

Halofuranones were not adequately addressed by TOPKAT®. Eight halofuranones were identified by virtue of low predicted chronic LOAELs (Table 7.1). There were several positive predictions of carcinogenic potential, but these fell outside the OPS of the cancer models. Most notably, TOPKAT® underestimated the carcinogenic and mutagenic properties of MX, itself. This shortcoming of TOPKAT® extends to some haloquinone and halocyclopentenoic acid structures that were predicted to arise from the reaction of chlorine with NOM (3,4,5,6-tetrachloro-o-quinone, 1,2,3-triketo-4,4,5,6-tetrachlorocyclohex-5-ene, 1,2,2,4-tetrachloro-3-hydroxy-cyclopent-4-enoic acid, and 3,5-dichloro-1-hydroxy-4-keto-cyclopent-2-enoic acid [see Table 6.11]). As the prediction of the chronic LOAEL of MX was consistent with experimental data on MX, these compounds were retained through the next step in the prioritization process.

The current study was in essential agreement with the M category of Woo et al. (2002) that the halonitrile group of DBPs requires attention. A difference was that the selection process used here was driven primarily by low predicted chronic LOAELs. Developmental toxicity potential was also included. Woo et al. (2002) focused strictly on carcinogenic properties, thus there are some differences in the actual chemicals identified as high priority. The higher molecular weight halonitriles, added to the previous lists of DBPs to explore the properties of the class, are not seen in Woo et al (2002) because these authors focused strictly upon chemicals that had actually been measured in drinking water.

The haloalkanes and haloalkenes, identified in the M category of Woo et al. (2002), were all predicted to have chronic LOAELs > 10-mg/kg day^{-1}. Therefore, to the extent that the chronic LOAEL is an adequate surrogate for carcinogenic potency, the authors concluded that these compounds are unlikely to contribute significantly to the carcinogenic risk associated with chlorinated water. This is reinforced by the observation that, to the extent these compounds occur, they are found at levels an order of magnitude or more less than the THMs. Many of these are iodinated methanes and ethanes. Since iodine and bromide occur in source waters in a more or less fixed ratio, it is improbable that this group will ever occur at greater concentrations than their brominated analogs.

1,2,3-Trichloropropane is a high priority and well studied carcinogen, primarily because of its TD_{50} of 4.7-mg/kg day^{-1}. It is used as a pesticide, so its occurrence in water is not necessarily attributable to disinfection. However, precursor substructures do exist that make this and some of its brominated analogs potential by-products of chlorination.

Table 7.4
Priority ranking of disinfection by-products by Woo et al. (2002)

Ranking	Halo-furanones	Haloalkanes& haloalkenes	Halonitriles	Halo-ketones	Halo-aldehydes	Halo-nitroalkanes	Non-halogenated aldehydes
High multispecies carcinogens	Tribromo-MX (BMX-3)		Bromochloro-acetonitrile				
	4-Bromochloro-methyl-MX (BMX-1)		2,3-Dichloro-3-bromoprop-ionitrile				
	4-Dibromo-methyl-MX (BMX-2)		2,4-Dichloro-butanenitrile				
Moderately active carcinogens	Mucochloric acid	Bromochloro-iodomethane	2,3,4-Trichloro-butenenitrile (cis & trans isomers)	1,3-Dichloro-propanone	Dichloroacet-aldehyde	Dibromo-nitromethane	Butanedial
		Bromochloro-methane	Trichloropro-penenitrile	1,1,3-Trichloro-propanone			
		Dibromo-methane					
		Dichloroiodo-methane					
		1,1,1,2-Tetrabromo-2-chloroethane					

Source: Data from Woo et al. 2002

Dichloroacetaldehyde was the only halogenated aldehyde that met the prioritization criteria. This conclusion is in essential agreement with Woo et al. (2002).

Dibromonitromethane was also assigned an M ranking by Woo et al. (2002). As with the other halogenated alkanes, TOPKAT® predicted a chronic LOAEL for this compound of 6.4-mg/kg day^{-1} and an even lower LOAEL for dichloronitromethane of 1.9-mg/kg day^{-1}. However, neither compound met the criteria of this study of being a multiple species carcinogen with at least one similarity index of <0.2. Clearly, elimination of these compounds by the criteria used in this study is borderline. If these chemicals were to occur at substantially higher concentrations than have been reported (Plewa et al. 2004), they could be of interest as potential health hazards.

The TOPKAT® prediction of a chronic LOAEL for butanedial was 284-mg/kg day^{-1} with similarity indices of 0.120-0.184 with seven chemicals in the training set. The high predicted chronic LOAEL forced the conclusion that the chemical should be of little health concern at concentrations likely to be involved in drinking water. One of the analogs of butanedial in the TOPKAT® database was sodium malonaldehdye. It was predicted to have chronic LOAEL of 164.4-mg/kg day^{-1}. This was an overprediction of the actual chronic LOAEL for sodium malonaldehyde in the training set of 35.7-mg/kg day^{-1}. Taking this discrepancy into account butanedial would still not meet the criteria used in this study. In contrast, the evaluation of the non-halogenated aldehyde group by TOPKAT® ranked both glyoxal and methylglyoxal as more probable offenders, based not only on predicted chronic LOAELs, but also from their estimated probabilities to be carcinogenic. The TOPKAT® predictions of carcinogenicity were pretty well supported because malonaldehyde was in the training set, which had been studied as a carcinogen in male and female mice and rats. Only the male rat responded with a tumorigenic response. Glyoxal and methylglyoxal just missed the cut-offs for the prioritization criteria used in this study and should, in our estimation, have a higher priority for research than butanedial.

Modification of Research Priorities Based on Probability of Occurrence

Those chemicals assembled from lists of established DBPs, which were combined with more novel DBPs predicted to be formed by reaction of disinfectants and substructures of NOM, are listed in Table 7.5. To be included in Table 7.5 DBPs had to meet one or more of the prioritization criteria described in Chapter 3. As indicated in Chapter 6, several established and novel DBPs have been retained in Table 7.5 because literature searches provided information that tended to validate some predictions of various TOPKAT® submodels that were outside the OPS or did not meet the minimum requirements for similarity indices. Examples of these DBPs include some of the halonitriles, MX and related compounds, and a few haloquinones.

In effect, Table 7.5 summarizes the individual putative DBPs that QSTR analyses suggest may be of regulatory concern if their concentrations in drinking water were to exceed 1 µg/L. Toxicologically characterized DBPs are included in the first third of the table to allow for comparisons with those that have not been characterized. Generally TD_{50} values are available for those DBPs that have been shown to be carcinogenic in experimental animals. The exceptions to this are MX and sodium chlorate. Sodium chlorate has only recently been shown to be carcinogenic in the rat thyroid and mouse pancreas (NTP 2005a) However, the detailed data required for calculating a TD_{50} will not be available until the actual report is published. A predicted chronic LOAEL was generated for selected characterized DBPs so that comparisons can be made among model predictions, thus avoiding the need to compare between TD_{50}s and predicted LOAELs.

Those DBPs that met one or more of the criteria suggesting a high probability of inducing adverse health effects at low concentration were retained as examples of the types of compounds that should be pursued to clarify the health risks associated with disinfection. Obviously, the first priority is to document their actual occurrence in a range of concentrations consistent with the QSTR predictions of their toxicological potency. If the chemicals are found to occur, it will be important to verify the predictions made experimentally and to thoroughly characterize their toxicological properties.

A relatively simple model was used to determine the importance of any given DBP. The mean dose of the DBP that would be expected was calculated based upon its mean concentrations in drinking water (or an approximation based on limited data). This dose was ratioed with the predicted chronic LOAELs and/or TD_{50}s to provide the margin of exposure relative to accepted toxicological measures of response. LOAELs derived from published experiments were used for chlorite and chlorate as TOPKAT® was incapable of estimating a LOAEL for either compound. The USEPA (2003a) regulates chlorite in combination with chlorine dioxide based on a no observed adverse effect level (NOAEL) for a non-carcinogenic endpoint (a two-generation reproduction study). The LOAEL from this study was 5.6-mg/kg day^{-1} as sodium chlorite and 4.2 as chlorite). The LOAEL for sodium chlorate was obtained from the data summarized in the NTP (2005a) abstract of the chronic study. The lowest dose in the study was 5-mg/kg day^{-1}, which produced thyroid, splenic cell, and bone marrow hyperplasia. Increased tumor yields were observed at the highest dose of 75-mg/kg day^{-1}. Expressed as chlorate, the LOAEL was 3.95-mg/kg day^{-1}. The ratio for formaldehyde is observed to be higher than other DBPs, but this is because its potency as a carcinogen in the respiratory tract was used as the basis of the calculation of its TD_{50}. Chronic studies of formaldehyde in the drinking water of rats failed to produce convincing evidence of carcinogenicity (Til et al. 1988). Therefore, formaldehyde is considered to be of minimal to no hazard when ingested at the concentrations produced by the disinfection of drinking water.

Typically, the calculated ratios of regulated carcinogens ranged from 10^{-7} to 10^{-5}. The ratio with the LOAEL of chlorite was much higher, in excess of 10^{-3}, especially if consideration is given to the fact that chlorine dioxide may be expected to occur at similar concentrations. The ratio with chlorate is smaller, but still in excess of 10^{-4}. Thus, these chemicals have a lower margin of exposure (MOE) than most DBPs.

As ratios begin to exceed 10^{-5}, the other carcinogenic by-products become increasingly more important than the THMs. In order of priority, those chemicals predicted to be carcinogens for which some indication of concentration was available, include chlorate, dichloroacetonitrile, bromochloracetamide, dichloroacetaldehyde, and 3,3-dibromopropenoic acid. Based upon its predicted chronic LOAEL, MX would be included within this group; however, it does not appear to be an issue based on its estimated TD_{50}. Cyanogen chloride was found to have a high ratio, but this was based on a prediction of developmental toxicity, so the ratio is more properly compared to that of chlorite.

Table 7.5
Prioritization of toxicants predicted to have a chronic LOAEL < 1 mg/kg day^{-1}, or to be a developmental toxicant or carcinogen with a predicted chronic LOAEL < 10 mg/kg day^{-1}

Disinfection by-product	Index number	Probable carcinogen	Lowest TD_{50} mg/kg day^{-1}	Probable develop. toxicant	Predicted chronic LOAEL mg/kg day^{-1}	Median conc. µg/L	90th ile or max. reported conc.	Dose at median conc./TD_{50}	Dose at mean conc./LOAEL
Established DBPs									
Chloroform	K09C07	Yes	90.3		31.8	23.1**	51**	0.0000073	0.0000208
Bromodichloromethane		Yes	47.7			8.4**	19.0**	0.0000050	
Dibromochloromethane		No	93.9			4.2**	11.8**	0.0000013	
Bromoform		Yes	648			1.4**	3.8**	0.0000001	
Monochloroacetic acid	K19C02	No			14.3	1**	2**		0.0000020
Dichloroacetic acid	K20C01	Yes	102		5.9	8.9**	24.3**	0.0000017	0.0000286
Trichloroacetic acid	K14C04	?	584		60.3	6.8**	24.9**	0.0000003	0.0000032
3-Chloro-4-(dichloromethyl)-5-hydroxy-2(5H)-furanone (MX)	K21C08	Yes	1		0.1	0.06††	0.31††	0.0000017	0.0000170
2,4,6-Trichlorophenol	K08C02	Yes	405		116.3	< 0.4‡‡		0.0000000	0.0000000
Chloral hydrate	K13C04	No	106		72.2	2.8**	11**	0.0000007	0.0000011
Chloropicrin	K26C02	No			14.6	0.26††	2.0††		0.0000005
Chlorite		No	Neg		5.6^{Exp} (4.2 as chlorite)	440†† ClO_2	840†† ClO_2		0.0029932
Chlorate		Yes			5^{Exp} (3.95 as chlorate)	263*** ClO_2	476*** ClO_2		0.0019024
						486§§ OCl^-	1160§§ OCl^-		0.0035154
Formaldehyde		Yes	2.19			ND (<5.0)** O_3	21.8** O_3	0.0000652	
						5.3** ClO_2	9.0** ClO_2	0.0000691	
Acetaldehyde	K32C02	Yes	153		114	< 5** O_3 or ClO_2	8.2** O_3	0.0000009	0.0000013

(continued)

Table 7.5 (Continued)

Benzylaldehyde	K38C01	No	1390		170	ND (<1.0)****	ND (<1.0)****	0.0000000	0.0000000
Bromate		Yes	9.82 (7.53 as BrO_3^-)			2.7** O_3	7.3** O_3	0.0000102	
						0.05** ClO_2	0.22** ClO_2	0.0000002	
						0.94††† OCl^-	2.60††† OCl^-	0.0000036	
N-Nitroso-N-dimethylamine (NDMA)	K48C02	Yes	0.096		2.7	0.0005§§§ Cl	0.0011§§§ Cl	0.0000001	0.0000000
						0.0029§§§ $ClNH_2$	0.0045§§§ $ClNH_2$	0.0000009	0.0000000
1,2,3-Trichloropropane	K09C02	Yes	0.875	No	4.7				
LOAEL < 1 mg/kg									
2,3-Dibromopropionitrile	K30C02	Yes		No	0.15				
2,3,4-Trichlorocyclohexadien-5-one carboxylic acid	K02C38	No		No	0.22				
Trichlorocyclopentenedione	K05C06	No		No	0.50				
3-Bromo-3-Chloropropionamide	K23C01	No		No	0.62				
E-2-chloro-3-(dichloromethyl)-4-oxobutenoic acid (E-MX)	K21C06	No		No	0.67	0.013††	0.10††		0.0000006
3,4-(Dichloro)-5-hydroxy-2(5H)-furanone (MCA)	K21C07	No		No	0.79*	0.084††	0.71††		0.0000030
Chlorobutyryldioic acid	K11C23	No		No	0.85				
2,3-Dichloropentanedioc acid	K11C13	No		No	0.87				
2,4-Dichloro-3-hydroxybutanoic acid	K01C14	No		No	0.95				
2,3,4-Trichlorobutanal	K10C07	No		No	0.96				
2,3-Dichloropropenal	K15C03	Yes		No	0.97				

(continued)

Table 7.5 (Continued)

3,5-Dichloro-1-hydroxy-4-keto-cyclopent-2-enoic acid	K50C07	No	No	0.00025†			
1,3,3,5-Pentachloro-4-keto-5-hydroxycyclohexene	K50C26	No	No	0.21*			
2,6-Dichloro-1,4-benzoquinone	K50C35	No	No	0.049†			
2,3,6-Trichloro-1,4-benzoquinone	K50C36	Yes	No	0.033			
2,3,5,6-Tetrachloro-1,4-benzoquinone	K50C37	No	No	0.005			
2,6-Dichloro-3-methyl-1,4-benzoquinone	K50C38	Yes	No	0.079			
2,3,6-Trichloro-1,4-benzoquinone-4-(N-chloro) imine	K50C40	Yes‡	No	0.073‡			
2,3,5-Trichloro-4,5-diketobenzoic acid	K50C42	No	Yes	0.073			
2,3,6-Trichloro-4,5-diketo-phenylprop-2-enol	K50C44	Yes†	No	0.0065			
Predicted Carcinogens							
Dichloroacetonitrile	K17C05	Yes	No	0.29†	1.3**	4.1**	0.0001281
3,3-Dichloropropenenitrile	K30C08	Yes	No	0.77†			
3-Bromopropionitrile	K30C15	Yes	No	0.89†			
Bromochloroacetamide	K23C08	Yes	No	1.5	(1)	(3)	0.0000003
2-Chloropropenal	K15C09	Yes	No	1.7			
3,3-Dibromopropenoic acid	K02C13	Yes	No	1.9	0.43††	2.7††	0.0000065
Trichloroacetonitrile	K17C07	Yes	No	4.5†	0.03††	0.15††	0.0000002
Dichloroacetaldehdye	K04C04	Yes	No	5.8	2.2††	11.1††	0.0000108
3,3,3-Trichloropropionitrile	K30C05	Yes	No	3.9†			
2-Bromo-3,3-dichloropropionoic acid	K01C45	Yes	No	5.8			

(continued)

Table 7.5 (Continued)

Predicted Developmental Toxicants							
3-Chloro-2-butenoic acid	K02C20	No	Yes	2.8			
Tribromoacetonitrile	K17C06	No	Yes	4.5†	0††	0††	0.0000000
Bromodichloroacetonitrile	K17C02	No	Yes	1.5†	0††	0.4††	0.0000000
Dibromochloroacetonitrile	K17C04	No	Yes	1.5†	0††	0.15††	0.0000000
3-Bromobutyrylnitrile	K30C10	No	Yes	2.5†			
3,3,3-Trichloropropionitrile	K30C05	No	Yes	3.9†			
3,3-Dichloropropenenitrile	K30C08	No	Yes	0.77†			
2,3,3-Trichloropropenenitrile	K30C01	No	Yes	1†			
Cyanogen chloride	K30C13	No	Yes	0.23†	1.8**	7.2**	0.0002236
Cyanogen bromide	K30C12	No	Yes	0.25†			

* TOPKAT® prediction was outside of the optimum predictive space, but within permissible limits

† TOPKAT® prediction was outside of the optimum predictive space and exceeds permissible limits

‡ Molecular fragment of query structure not represented in the models training set

§ Krasner et al. 1989, mean and maximum

** ICR data base, mean and 90th percentile used

†† Weinberg et al. 2002, mean and maximum reported (too few systems to estimate 90th percentile)

‡‡ Bull et al. 2001. Acetaldehyde and formaldehyde in ozone contactor effluent.

§§ Gordon et al. 1995, mean and 90th percentile for systems using hypochlorite

*** Bubnis 1993, mean and 90th percentile for systems using chlorine dioxide

††† Weinberg 2000

‡‡‡ Disinfectant is assumed to be chlorine (either gas or hypochlorite) unless otherwise specified. If designation is important, OCl^- for hypochlorite solution, O_3 for ozonated systems, and ClO_2 indicate chlorine dioxide.

§§§ Valentine et al. 2005

**** Glaze and Weinberg 1993

( ) Roughly estimated based on occurrence of analogs (e.g. chlorinated analog used to estimate mixed bromochloro analog)

Exp Experimentally determined LOAEL.

No occurrence data were identified in the literature for 29 of the potential DBPs included in Table 7.5. Neither was there a substantive basis for predicting their likely concentrations. Therefore, there is no basis for calculating the ratio that suggests their potential importance relative to other DBPs that have been identified. However, all have a potential of being present in disinfected water based on the nature of precursor substructures that are found in NOM and predicted reactions with a disinfectant. In eight of the 29 cases, the prediction of the chronic LOAEL exceeded the permissible limits of the model and in one case the compound contained a molecular fragment that was not represented in the training set. Of the haloquinones that were identified only two fell into this category; six of members of the class were well within permissible limits of the OPS. For this reason, it was concluded that all of the haloquinones studied that met the criterion that the predicted LOAEL < 1-mg/kg day^{-1} would be retained on the high priority list.

Those chemicals for which there are no occurrence data and which have LOAELs substantially less than 1-mg/kg day^{-1} are high priority for occurrence research. In at least three cases, the low predicted chronic LOAEL is accompanied by predictions of carcinogenic activity that met the criteria. These were 2,3,6-trichloro-1,4-benzoquinone, 2,6-dichloro-3-methyl-1,4-benzoquinone, and 2,3-dichloro-4,5-diketobenzoprop-2-enol. In all three cases the predicted chronic LOAEL was < 0.01-mg/kg day^{-1}. One compound in this group (2,3,5-trichloro-4,5-diketobenzoic acid) was predicted to be developmental toxicant and the predicted LOAEL was in the same range. If the concentrations of these chemicals were to exceed 10 ng/L, their predicted potencies are of a magnitude that might account for the epidemiological associations of adverse effects associated with the consumption of chlorinated water.

CLASSES OF DISINFECTION BY-PRODUCTS THAT REQUIRE TOXICOLOGICAL CHARACTERIZATION.

Four classes of established DBPs were identified for which toxicological information are so limited that reliable QSTR predictions could not be made. In general, chemicals in these classes were identified as being potential carcinogens or developmental toxicants in individual models, but the models were incapable of dealing with them with even a modest degree of confidence. Most important, the ability to predict the chronic LOAELs was limited by the lack of data within these four classes. As a consequence, the QSTR program had no means to estimate their potential quantitative contribution to risks associated with the disinfection of drinking water. Therefore, obtaining appropriate toxicological information on selected members of these classes was deemed to be of the highest priority for addressing the DBP issue in a comprehensive way. These classes include haloquinone, halofuranone, halocyclopentenoic acids, the organic N-chloramines, halonitriles, and high molecular weight nitroso compounds, in approximate order of priority.

Haloquinones, Halofuranones, and Cyclopentenoic Acids

As a result of the predicted formation of a diverse set of quinones from the reaction of chlorine and other disinfectants with substructures in NOM, this group has assumed a much greater priority than was anticipated. Compounds in the haloquinone class are conspicuous by their absence in lists of DBPs that have been identified. As is the usual case, it is difficult to determine if these compounds have been pursued and not found or simply have not been recognized because of the lack of appropriate analytical standards. However, the QSTR analysis clearly showed that many chemicals in the haloquinone class may be quite toxic with predicted chronic LOAELs between 0.005 and 0.08 mg/kg day^{-1}.

Figure 7.1 illustrates why it is necessary to understand the toxicity of any intermediates that may be potent. As previously discussed the haloquinone products will be produced by chlorination, but will be substantially reduced in concentration in the presence of excess chlorine as a result of further reaction. Opening of the ring results in degradation to much less potent carcinogens and toxicants than predicted for the haloquinones. This is illustrated by the sharp decrease in the predicted LOAEL of syringic acid as 1,2,3-triketo-4,4,5,6-tetrachlorocyclohexane is formed. However, the LOAEL increases if the ring is cleaved by further reaction with chlorine to produce 2,3,4,4,4-pentachloro-2-butenoic acid.

A more complex scenario may be set by examining the different chemistries of chlorine and chloramine in reactions with phenol. The initial reactions with both disinfectants are the same (although there may be some differences in the kinetics) in that products chlorinated in the ortho and para positions of phenol result. 2,4,6-Trichlorophenol is used in Figure 7.2 to represent this group of initial products. Note that there are only minor differences in the LOAELs of phenol and 2,4,6-trichlorophenol. However, further reaction with chlorine can produce a cyclopentenoic acid derivative that has a very low predicted chronic LOAEL. On the other hand, further reaction with chloramines can also occur; and this results in the formation of a series of 1,4-benzoquinones. These compounds are also quite toxic and are represented by the 2,3,5,6-tetrachloro-1,4-benzoquinone. The relative risks of chlorinated and chloraminated water will be a product of the relative yields of these two products and their relative toxicological potencies. It is very unlikely that the established DBP, 2,4,6-trichlorophenol, contributes to the risk at all as illustrated in Table 7.5. Similar patterns of changing toxicity with disinfection can be illustrated with other substructures of NOM, such as coniferyl alcohol.

MX and its analogs are included in this group of chemicals as there is a certain structural similarity between the two groups of chemicals. MX analogs also have very low predicted LOAELs in the TOPKAT® model. Moreover, there are data available for MX that serve to justify concerns expressed with respect to the haloquinones and the halocyclopentenoic acids.

The question of how much research attention should be paid to MX and its analogs and the direction that research should take will be discussed first. The brominated analogs of MX have recently been documented to occur when source waters contain sufficient bromide (Weinberg et al. 2002). The brominated analogs of MX were suggested to be of high priority for research among identified DBPs (Woo et al. 2002). We believe that the question of the priority given these compounds needs to be considered in a broader strategic context. It is not clear that chronic study of these analogs would add more than just incremental information to the risk picture. We recommend that a broader QSTR approach be taken for MX and related compounds rather than simply beginning with brominated analogs of MX. First, substitution of bromines for chlorines in MX has only a small effect on the mutagenic activity (LaLonde and Leo 1994). Second, it is not clear that increasing mutagenic activity (i.e. reactivity) of such a potent direct acting mutagen would increase in vivo carcinogenic activity. Third, focusing too closely on mutagenic potency of MX compounds may well be a mistake from a QSTR point of view (discussed more fully in Chapter 6). Our analysis provides reasons including investigation of cyclopentenoic acids derivatives, such as 3,5-dichloro-1-hydroxycyclopent-2-enoic acid, as a more logical first step for clarifying the risks that are associated with MX and related chemicals.

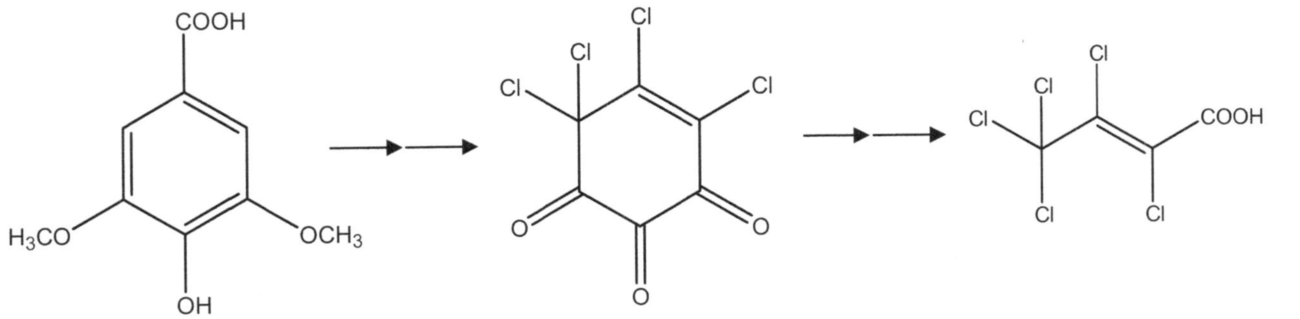

Syringic acid | 1,2,3-Triketo-4,4,5,6-tetrachloro-cyclohexene | 2,3,4,4,4-Pentachloro-2-butenoic acid

LOAEL = 345 | = 0.050 | = 46.5

Note: Predictions of LOAELs made with TOPKAT®

Figure 7.1 Predicted changes in toxic effects that may parallel formation of intermediate vs. terminal by-products of the chlorination of vanillic acid and syringic acid (LOAELs and TD_{50} are given in mg/kg day^{-1})

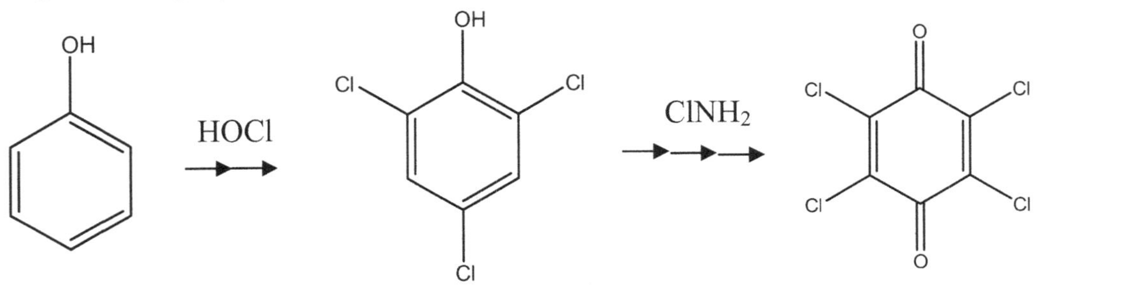

Phenol | 2,4,6-Trichlorophenol | 2,3,5,6-Tetrachloro-1,4-benzoquinone

Actual LOAEL 350 | TD_{50} = 405

Predicted LOAEL = 74.7 | = 116 | = 0.005

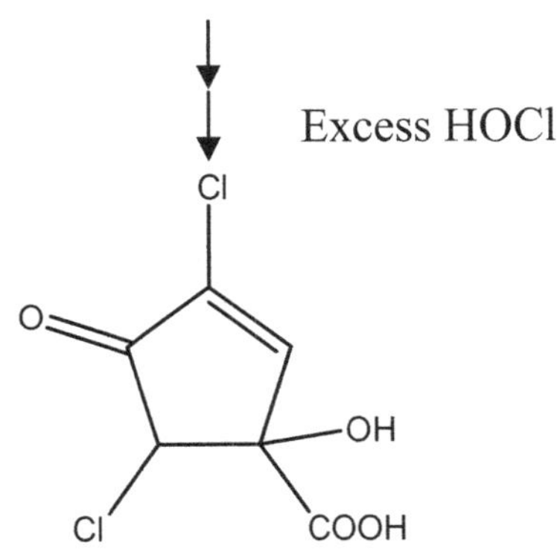

Source: Chloramination reactions Heasley et al. 2004; predictions of LOAELs made with TOPKAT®

3,5-Dichloro-1-hydroxy-4-ketocyclohex-2-enoic acid
Predicted LOAEL = 0.00025†

Figure 7.2 Predicted changes in toxicological potency with chlorination vs. chloramination of phenol (†estimate outside the OPS, and exceeds permissible limits)

Organic N-Haloamines

Organic N-haloamines are formed in chlorination or chloramination of drinking water when appropriate nitrogen-containing precursors are available. It should be recognized that in the presence of bromide in water or in the body, N-bromamines would be formed as well. As only N-chloramines were explored with QSTR, the present discussion will focus on this group. Organic nitrogen compounds increasingly become a concern as the number of drinking water sources that are impacted by municipal wastewater increases.

The organic N-chloramine class was not represented in the TOPKAT® training sets, so these predictions must be considered unreliable. However, concerns over the N-haloamines are not limited to the health effects pursued with the QSTR analysis.

It is of interest that in vivo formation of organic N-chloramines is discussed extensively in the biomedical literature as potential contributors to the development of chronic disease. It must be recognized that these hypotheses were developed strictly on the basis of the potential reactivity of organic N-chloramines with critical macromolecules. Experiments have been limited to studies in cell culture or bacterial systems. While these studies may provide adequate models for local generation of high concentrations of these chemicals in vivo, they do not provide adequate information to quantify risks that may arise from consuming very small doses of the compounds in drinking water. The in vivo formation of these compounds resulting from consuming chlorinated or chloraminated water would yield much more dilute and much less localized concentration of these chemicals than projected to arise from activated leukocytes (Jesaitis and Drutz 1992). Therefore, the determination of risks from drinking water requires sets of descriptive toxicological data based on oral consumption of several of these chemicals. To date, descriptive data of this kind have only been developed for monochloramine.

There are several properties of N-chloramines that suggest they may play a role in the epidemiological association of chlorination and bladder cancer risk is plausible. First, Chapter 5 illustrates that the precursors that will be N-chlorinated are likely to be quite different with excess chlorine than when it is limited. Therefore, different products can be anticipated when chlorine serves as the disinfectant relative to those where contact with free chlorine is limited, such as occurs with chloramination. Evidence is beginning to accumulate that other forms of disinfection reduce the risk of bladder cancer (but not necessarily other cancer risks). While the association of bladder cancer risk is consistently observed with chlorination (Morris et al. 1992, USEPA 1998, King 1999, Villaneuae et al. 2002), case control studies suggest that this risk is reduced, if not eliminated, with chloramination (Zierler et al. 1988, McGeehin et al., 1993) or ozonation (Chevrier et al. 2004). Both the quantity and type of N-chloramines formed with these two alternatives are likely to be different from those produced with chlorination.

Organic N-chloramines are attractive candidates for explaining high cancer risks from a toxicological standpoint, as well. The important properties are as follows:

- The stability of candidate N-chloramines should be sufficient to be absorbed and distributed to target tissues in vivo.
- Some likely metabolic products of N-chloramines resemble proximate carcinogenic forms of well-known bladder carcinogens.
- There are processes that can contribute to further metabolic transformation of organic N-chloramine to potentially active forms in vivo.
- The bladder has local metabolic capability to produce such intermediates.

- The possibility exists that inflammatory processes clinically common in the human bladder may contribute to the risk, particularly in the aged population.

As described in Chapter 5, some organic N-chloramines have very short half-lives while others are stable for hours to days. The N-chloroamines are typically less polar than their parent amine. As a consequence the oral bioavailability of these more stable compounds should be high.

N-Chloramines readily participate in chlorine transfer reactions in the direction of more stable forms. In vivo, the ultimate sink for Cl^+ is likely to be macromolecules. For example, when HOCl is first applied to mixtures of nucleosides, the N-chloro group forms on the ring nitrogens of thymidine. With time, these N-chloro groups migrate to the exocyclic nitrogens of adenosine (Hawkins and Davies 2001). This same behavior is observed in polynucleotides (Hawkins and Davies 2002).

Organic N-chloramines differ significantly in their oxidation reactions relative to HOCl and monochloramine. N-Chlorotaurine and N-Chlorolysine react much more specifically with thiols than with ascorbic acid, whereas the reactions of HOCl are indiscriminate (Carr et al. 2001). Monochloramine also reacts more specifically with thiols than HOCl. The organic N-chloramines react at lower concentrations with thiols than does monochloramine.

The N-chloramine of taurine appears to have assumed some protective role in vivo. It acts to shut down some of the inflammatory processes that are initiated by neutrophils. It has been shown to inhibit collagenase that has been activated by $HOCl/OCl^-$ produced by neutrophils, a property not shared by N-chloroalanine or N-chloroleucine (Davies et al. 1994). It also appears to inhibit nitric oxide synthase and thus limits the production of mediators of inflammation (Barua et al. 2001). Therefore, N-chlorotaurine apparently has some important physiological roles. Nevertheless, the ability of the larger group of N-chloramines to react with thiols to produce oxidative stress indirectly must be kept in mind.

The degradation of N-chloramines formed in plasma proteins (assumed to be primarily formed on lysine side chains) is > 90 minutes at 37° C (Hawkins and Davies 1999). Thus, it appears likely that other organic N-chloramines with stability equal to or greater than N-chlorolysine can be easily distributed to target tissues, including the bladder. A potential qualifier is the influence of first-pass metabolism in the liver. The stability in plasma may, in part, be attributed to the relatively low levels of thiols in plasma relative to cells and tissues (Moriarity et al. 2003). However, liver, as well as other tissues, has thiol concentrations several orders of magnitude greater than plasma, particularly glutathione. The extent of first-pass metabolism would depend very much on the structure of the N-chloramine.

The degradation of organic N-chloramines results in the formation of a variety of carbon and nitrogen-centered free radicals. Evidence of such degradation is observed by crosslinking reactions observed between lysine side chains and nucleosides (Hawkins et al. 2002). As would be expected, these reactions are accentuated by the presence of metal ions (Hawkins and Davies 1998).

It is to be expected that the stability of some, if not most, organic N-chloramines formed in drinking water would parallel those observed with natural products. In part, this is because these natural products are present in NOM. Moreover, the chlorine transfer to ever more stable organic chloramines has been demonstrated in wastewater (Bedner et al. 2004). The framing of the issue, therefore, boils down to the extent to which strong halogenation and oxidative chemistry contributes to a greater diversity of N-chloramine compounds to be formed.

Essentially, there will be differences of products likely to be observed with excess chlorine vs. conditions observed with chloramination. In addition, other man-made amines are likely to be found in some drinking water sources. For example, those drinking water sources impacted by domestic wastewater and agricultural runoff, contain a variety of pharmaceutical agents, many of which are amines. Second, there may be a break in dose-response curves as concentrations of chlorine exceed breakpoint. Examples of N-chloramines formed with some of these chemicals were provided in Table 6.10.

Two endogenous processes can be identified as having the capability of activating the absorbed N-chloramine to reactive forms within the urinary bladder. It has been shown that much of the chemistry of N-chloramine formation and its metabolic processing to intermediates capable of reacting with macromolecules, such as DNA, can occur in vivo. Most of this work has been done in an attempt to explain serious and unpredicted adverse outcomes that have been associated with therapeutic agents. The myeloperoxidase system of activated neutrophils is one such process that has been studied extensively. Activated neutrophils invade tissues in which bacterial infections have occurred and are major players in inflammatory responses. The bladder, either as a result of infection or interstitial cystitis, which has a high prevalence in the aged (Schilling et al. 2001, Kusek and Nyberg 2001), can attract these inflammatory cells. Urinary infections appear to contribute to the cancer of the bladder (La Vecchia and Airoldi 1999). Thus, inflammatory cells in the bladder epithelium can activate N-chloramines to free radicals, nitrenium ions or other DNA-reactive forms that may contribute to the initiation of bladder cancer.

The second factor that may be responsible for local activation of N-chloramines to carcinogenic forms in the bladder is the enzyme prostaglandin H synthase. As previously discussed, this enzyme is expressed in high amounts in the bladder and has been shown capable of activating N-acetylbenzidine and o-anisidine to forms that bind to DNA.

Organic N-chloramines were first suggested as potential carcinogens formed in the chlorination of drinking water in the 1980s (Scully and Bempong 1982). Since that time there has been limited study of the chemistry of these substances and their occurrence. Aside from some early investigation of the mutagenic properties of a few N-chloramines there has been virtually no investigation of their toxicology. Data are so rare for this chemical class that the QSTR models used in this study did not recognize the molecular fragment. Organic N-chloramines are likely to be ubiquitous in drinking water treated with chlorine or chloramines, but study has been limited to the generalized adverse impact these chemicals have on disinfection (Bedner et al. 2004). Due to the apparent lower level of bladder cancer risk associated with chloraminated water (Zierler et al. 1988, McGeehin et al. 1993), a high priority should be assigned to investigating the chronic toxicity of a set of organic N-chloramines that are formed specifically with excess chlorine and which are sufficiently stable in drinking water to be consistently observed at the tap. These studies should be preceded by carefully targeted pharmacokinetic and DNA binding studies that may aid in the prioritization among chemicals in the class to arrive at a few in which the expense of chronic studies would be justified. These studies should be aimed at determining whether or not some of the steps postulated can be shown to occur in vivo, not simply subjecting the chemicals to in vitro screening studies.

Bladder cancer risk has consistently been the highest risk associated with chlorinated drinking water. Therefore, the characterization of organic N-chloramines formed in the chlorination of drinking water and the toxicological characterization of selected members of those that are shown to occur is ranked as a high priority research need in the DBP area.

Halonitriles and Haloamides

Halonitriles and haloamides are two classes of identified DBPs that were predicted to be of the broadest interest from a health effects standpoint. Because adding to the carbon chain may extend these classes, the impact of the pattern and extent of halogen substitution on prediction of toxic effects within these classes was explored. Figure 7.3 projects the impact of varying chlorine substitution on the propionitrile and propionamide members of this class. There is a small trend for a reduction in the predicted chronic LOAEL (i.e. an increase in potency) for both classes with increasing substitution of chlorine in the 2-position. The effect was quite modest with the chronic LOAEL decreasing by approximately 50%. On the other hand increasing substitution from 0, 1, and 2 chlorines on the 3-position decreased the predicted chronic LOAEL by 1/25th to 1/100th of that seen with the non-halogenated forms. Adding a third chlorine at the 3-position sharply increased the predicted chronic LOAEL of the propionitriles (ca. 100-fold). In the case of the halopropionamides, the predicted chronic LOAEL was increased by less than four-fold.

Adding chlorines to both the 2- and 3-positions had an even greater effect. For example, chlorine added to the 2- and 3-positions of the propionamide group resulted in a predicted chronic LOAEL (0.5 mg/kg day^{-1}) that was about 50% of that of 2,2-dichloropropionamide. The predicted chronic LOAEL of 2,3-dichloropropionitrile was 0.04 mg/kg day^{-1}, or about 20% of that of 2,2-dichloropropionitrile. A substitution pattern likely to be found in the propionamides and propionitriles is the 2,2,3-trichloro analogs. This addition of the second chlorine in the 2-position is predicted to increase the chronic LOAELs to 14.7 and 1.7-mg/kg day^{-1}, respectively. These are testable hypotheses that should be investigated.

Examination of the pattern and extent of chlorine substitution had little effect on the predicted probability of members of these two classes as developmental toxicants. All propionamides and propionitriles examined, including the non-halogenated forms, had high probabilities of being developmental toxicants (0.974-1.000). Therefore, the predicted developmental toxicity associated with these classes of DBPs is a function of the amide and nitrile functional groups and is not a function of halogenation. This finding is attributable to the fact that propionitrile occurs in the developmental toxicity database as a positive developmental toxicant and acetamide and acrylamide were the major comparators for the propionamides.

On the other hand, prediction of carcinogenic properties was influenced by the location and extent of chlorine substitution on the propionitrile backbone. No positive predictions were made of carcinogenicity for propionitrile and only a single positive prediction was made for propionamide, based on a single positive prediction that arose from an acetamide bioassay in the Food and Drug Administration (FDA) database for the male rat. Increasing the chlorine substitution in the 2- or 3-positions for the propionamides had very little impact on the predictions of carcinogenicity (one to two positive predictions out of eight possible). In contrast, two chlorines in the 2-position of the propionitriles resulted in positive predictions in the male rat, the male mouse and female mouse. The similarity indices were all between 0.2 and 0.3. Simple substitution of chlorine on the 3-position of propionitrile did not result in a higher probability of carcinogenicity. In contrast, a substitution of chlorine on both positions 2 and 3 resulted in predictions of carcinogenic activity in all four species and both genders for both the propionitriles and propionamides. In both cases, at least one prediction had a similarity index <0.2 and the remaining predictions had indices that were <0.3. In summary, there is a high priority need for better characterization of the chronic and developmental toxicities of the halonitrile class. The class has been so poorly characterized that many of the QSTR predictions are suspect, especially the chronic LOAEL projection. On the other hand, consistent predictions of both carcinogenic activity and developmental toxicity for members of the class indicate that appropriate data need to be developed with a few members of the class to allow more accurate predictions of potency to be made. Finally, the analysis provided above indicates that there is a QSTR basis for prioritization among members of this class. Investigation of these predictions will provide a basis for a broader analysis of risks potentially attributable to this class of DBPs.

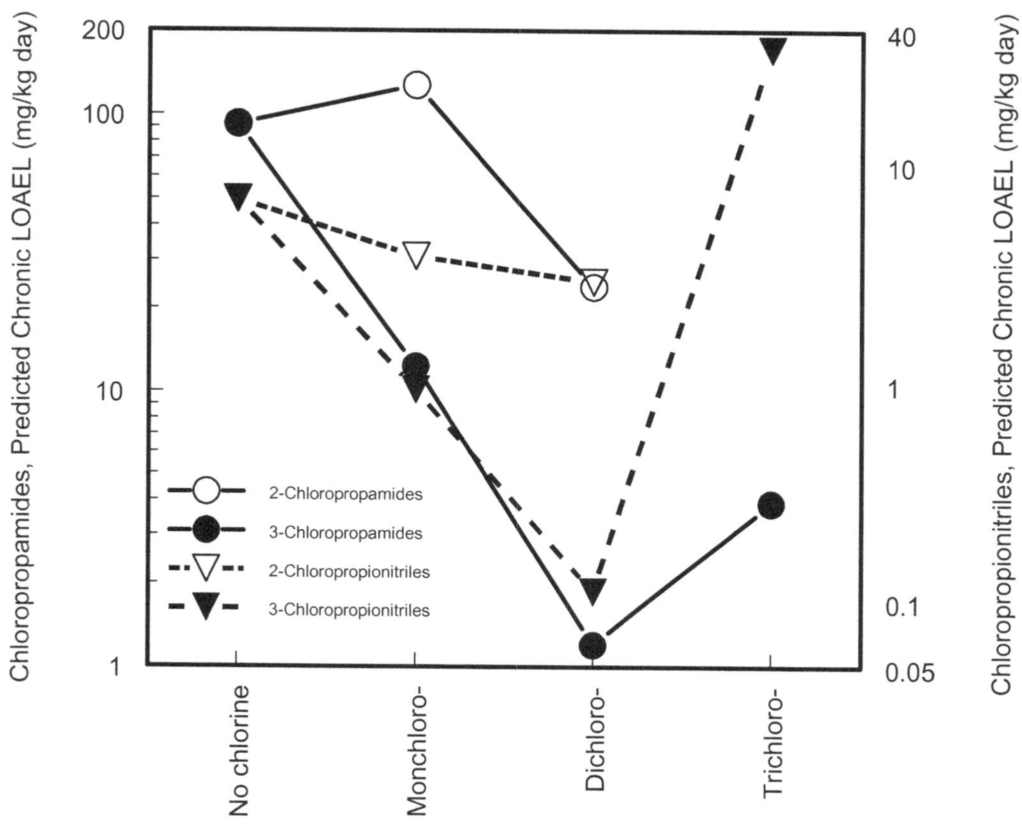

Figure 7.3 The effect of the degree of chlorine substitution at the 2 or 3-positions of propamide and propionitrile on the predicted chronic LOAELs Compounds substituted in both positions were predicted to be more potent, with the predicted chronic LOAEL of 2,3-dichloropropamide being 0.50 and that of 2,3-dichloropropionitrile 0.04 mg/kg day^{-1}.

Large Molecular Weight Nitroso Compounds

The ability of chlorine, and particularly chloramine, to form low molecular weight nitrosamines suggests that this class may be of interest with respect to the risk that may be associated with disinfection of drinking water. Certain of these compounds are extremely potent and well characterized. The most critical issue with low molecular weight nitroso compounds is the extent to which they occur and the concentration ranges in which they are found.

An issue of some specific interest is the potential formation of nitroso compounds with higher molecular weight secondary amines. The possibility that larger molecular weight unsymmetrical nitrosamines formed by chloramine may increase carcinogenic risk from drinking water requires exploration. Since TOPKAT® was not robust in predicting carcinogenic activity of nitrosamines and nitrosamides recent literature on this class of chemicals was reviewed. Current research on nitrosamine products of nicotine provides some insight into why these compounds should be of concern. Several nitrosamine derivatives of nicotine have been shown to be carcinogenic, the N-nitrosonornicotine, N-nitrosoanabasine, N-nitrosonornicotine N-oxide, 4-(methylnitrosamino)-1-(3-pyridyl-(1-butanone), and 4-(methylnitrosamino)-1-(pyridyl N-oxide) (Hecht 1998). It is of interest that these compounds are potent carcinogens in the lung even when administered in the drinking water. The first three compounds represent nitroso formation on ring nitrogens, the latter two nitrosamines formed when the ring is opened. As suggested earlier, similar products may arise by the reaction of excess chlorine with purine and pyrimidine bases in NOM. The case of nitrosamine formation from caffeidine provides an example. Unfortunately, it is not possible to be more specific about the types of larger molecular weight nitrosamines that may form. The potency of N-nitrosonornicotine N-oxide is provided in the CPDB as 0.876-mg/kg day^{-1}. 4-(Methylnitrosamino)-1-(3-pyridyl-(1-butanone) has a total lifetime dose TD_{50} of 0.3-mmol/kg. Therefore, cancer risk from the formation of unsymmetrical nitrosamines cannot be dismissed. However, these nitrosamines may not be bladder carcinogens. The above compounds tend to target the lung, nasal cavities, and esophagus.

Therefore, the highest priority for research is to develop methods for detecting the formation of non-volatile nitrosamines with disinfection, particularly with chloramination. If they are formed, they should be included in epidemiological studies to assess the types of human cancers most likely to be associated with these carcinogens. Only then will it be possible to obtain an accurate estimate of whether or not cancer risk as whole is reduced by chloramination relative to chlorination. In the meantime toxicological research should be initiated to characterize their carcinogenic potency.

PRIORITIZATION OF RESEARCH TO PREDICT AND CONFIRM BY-PRODUCT FORMATION

1. Determine whether or not haloquinone and halocyclopentenoic acids can occur in drinking water.
   - Analytical methodology is required for measuring haloquinone and halocyclopentenoic acid structures that is sensitive into the low ng/L range.
   - The kinetics of the formation and destruction of these compounds should be determined with free chlorine and varying conditions of chloramine use. There is a particular

- ➢ concern for whether or not health risk may be increased with a period of free chlorine contact prior to the addition of ammonia.

2. Characterize the differences in organic N-chloramines that are formed with NOM treated above and below breakpoint concentrations of chlorine (e.g. chloramine vs. excess chlorine).
3. Further explore the types of DBPs that can be formed by:
   - ➢ Identifying additional substructures in NOM.
   - ➢ Developing approaches for predicting products of the interactions between the oxidative and halogenation chemistry of disinfectants with NOM.
4. Determine the potential for forming N-nitroso compounds with substructures in NOM.
5. Develop semi-quantitative models for DBP formation based on known NOM structures and known yields from those structures.
   - ➢ Consider impacts of development in watersheds and other drivers rooted in watershed management.
   - ➢ Consider geographical differences in NOM substructure.

PRIORITIZATION OF HEALTH RESEARCH

1. Study halogenated quinone, and furan derivatives to determine structure-activity characteristics that will facilitate genotoxic and cytotoxic damage to the urinary bladder at low doses (i.e. do not simply extend Ames's mutagenic activity, include in vivo measures of these types of damage to the bladder). This effort should come subsequent to the identification of such compounds as chlorination by-products in actual drinking water (see below).
2. Experimentally identify the important determinants of N-haloamine toxicity and carcinogenicity. Studies should be targeted at key differences in by-products formed with chloramine and with breakpoint chlorination.
3. Experimentally verify characteristics of halogenated quinone DBPs predicted to contribute to the development of bladder cancer based upon metabolic profiles and projected structure-activity relationships.
4. Characterize the toxicological properties of selected members of higher molecular weight (i.e. > haloacetonitriles) dihalonitriles projected to be carcinogenic. Choice of halonitriles for study should be made based on occurrence measures as well as structure-activity predictions.
5. Develop a QSTR methodology for predicting health hazards associated with DBPs. Such methodologies may be based in part upon commercial computer programs, but it must be recognized that these are based on chemicals in commerce and that several important classes of DBPs are frequently poorly represented in their training sets. Therefore, the lack of systematized information in these classes (e.g. N-haloamines, halonitriles, haloquinones, furan derivative, and higher molecular weight nitrosamines) make it necessary to organize data within these classes as data are developed that define the critical issues related to their toxicities.

PRIORITIZATION OF BY-PRODUCTS FOR OCCURRENCE RESEARCH

1. Determine the extent to which haloquinones occur in drinking water. The toxic potential of these compounds suggest that the sensitivity of the methods should extend into the low ng/L range.

2. Extend halonitrile analyses to include halopropionitriles (saturated and unsaturated) and perhaps higher molecular weight compounds. Suggested sensitivity would be 100 ng/L.
3. Develop a means for measuring the extent of total nitrosamine/nitrosamide in drinking water with a sensitivity of 10 ng/L or lower.
4. Explore methods for characterizing variations in the geographical occurrence of DBPs that better characterize the variation in the natural composition of NOM.
5. Develop methods for detecting products of nitrogen addition resulting from chloramination.

CHAPTER 8. SUMMARY AND CONCLUSIONS

CONCLUSIONS SUMMARIZED

Aside from methodological issues, there were five general conclusions drawn from the present work. Embedded in these general findings are a number of specific findings. These are provided here in summary form and are discussed in more detail in the subsequent text.

- Quantitative structure-toxicity relationship (QSTR) analyses of chemicals identified as DBPs in the literature failed to identify candidates with sufficient predicted potency to be credible as potential causes of the epidemiological association between chlorinated drinking water and carcinogenic or reproductive/developmental toxicities.
- Several reaction pathways of chlorine were described with reasonably anticipated precursor substructures in NOM that have the potential for yielding by-products of sufficient predicted potency (and perhaps specificity) to contribute to a bladder cancer risk of the magnitude associated with chlorinated drinking water.
  - Reaction products of disinfectants with substructures in NOM were found to have predicted chronic LOAELs in the low μg/kg day^{-1} range. Representatives within these groups do have sufficient potency, if not the selectivity, to account for epidemiological associations of chlorinated drinking water with adverse health effects, but only if these chemicals occur in drinking water. Thus, several classes of putative DBPs were identified for future research. These include:
    - Haloquinones
    - Halocyclopentenoic acids
    - Organic N-haloamines
    - Nitrosamines and nitrosamides
    - Halonitriles and haloamides
  - Among groups identified as potential by-products of reactions of chlorine or chloramine, the haloquinone group appears the most capable of inducing bladder cancer, as well as other cancers.
    - 2,6-Dichloro-3-methyl-1,4-benzoquinone was identified as the putative chlorination by-product with the greatest toxicological potential be a bladder carcinogen. However, this compound has not been shown to occur in drinking water. It is more likely to be stable in chloraminated water than in the presence of free chlorine.
  - Halocyclopentenoic acids are structurally related to halofuranones, such as 3-chloro-4-(dichloromethyl)-5-hydroxy-2[5H]furanone (MX). By itself, MX, does not appear sufficiently potent to be the sole cause of bladder cancer at the concentrations found in drinking water. However, a group of compounds with similar modes of action, present at concentrations comparable to those found for MX in more recent surveys, may be important collectively.
    - 3,5-Dichloro-1-hydroxylcyclopent-2-enoic acid is a representative of this group of putative DBPs.
  - Members of the N-haloamine class may also be important. There are insufficient toxicological data related to organic N-haloamines to inform judgments for individual members of the class. Chemicals in this class are known to exist in

chlorinated and chloraminated drinking water, but concentrations have not been systematically determined.

- Non-linear relationships should be anticipated for yields of different classes of DBPs. The non-linearity will be exacerbated with the emergence of mixed use of disinfectants.
  - Some of these by-products may be intermediates in the formation of some terminal by-products that have been associated with chlorinated drinking water (THMs and HAAs). Therefore, reduction of risk by alternate methods of disinfection may not parallel the reduction in the terminal by-products produced by excess chlorine. This has particular implication for the use of chloramines for disinfection of drinking water. The concentrations of these compounds at the tap, and thus the risk they pose, may differ if chloramine is used as the primary disinfectant or if ammonia is applied to water after it has been subjected to a period of treatment with free chlorine.
- A different conceptual framework for visualizing the potential health hazards associated with DBPs needs to be developed. A general research approach is suggested to address the safety issues raised by the ever evolving use of reactive chemicals for purposes of disinfecting drinking water. Research efforts in this area need to specifically incorporate:
  - A more organized approach to developing a broad perspective of the types of by-products that can be formed with any specific disinfectant strategy.
  - A much better understanding of the formation and degradation of intermediate products of reactions of disinfectants with NOM is needed at varying disinfectant/carbon and nitrogen ratios. Some intermediates are orders of magnitude more toxic than the end-products of disinfection. Therefore, efforts must be directed at the development of kinetic models of DBP formation that include these intermediates. Such information will be essential for understanding the probable risks of modifying disinfectant treatment (e.g. using a quenching agent, such as ammonia, following addition of a primary disinfectant may minimize regulated DBPs, but is likely to increase the concentrations of more highly toxic intermediates at the tap).
  - Health assessments need to more aggressively pursue studies of the health effects of DBPs beyond the regulated compounds. Toxicological research in the past decade has cast significant doubt on the risk associated with the THMs and HAAs that are subject to regulation.
  - Selection of surrogates for monitoring purposes must be shown to be meaningfully related to risk. This is best done by developing a model that reliably reflects the relationship between the formation of different groups of DBPs.
- The Candidate Contaminant List (CCL) process is poorly suited for solving the problem of health risks associated with the disinfection of drinking water (or other complex processes of concern).

METHODOLOGICAL ISSUES

From a toxicological standpoint, the very low doses predicted to be chronic lowest observed adverse effect levels (LOAELs) for some putative DBPs indicate the distinct possibility that specific DBPs may occur that are of sufficient potency to be of health concern. Some of these chemicals are predicted to be of sufficient potency to qualify as candidates for causing the outcomes that have beeen identified in epidemiological associations with chlorinated drinking water. The chemicals possessing the lowest predicted chronic LOAELs were not found among lists of established DBPs, but were probable products of reactions of chlorine or chloramine with substructures of NOM. As these chemicals have not been identified or measured in drinking water, their occurrence must be considered only a plausible hypothesis. Similarly, predictions of a high level of toxicity must be considered hypotheses, not established facts.

It is important to emphasize that the QSTR program utilized in this effort was selected for reasons largely related to convenience. Thus, these results do not imply endorsement of this particular QSTR tool. The program has many failings, as do other programs that are available for this purpose. Properly applied and aided by further research results, it seems clear that such tools can be used if their inherent shortcomings are recognized and explicitly taken into account in any analyses. These programs will certainly be improved as the role of different mechanisms in chemical carcinogenesis become better understood and descriptive data are developed to aid in their calibration.

Most important, QSTR has limited value in and of itself. Its strength is realized when its predictions are tested experimentally. It can be of real public health importance only if it is coupled with a well organized research program. The lack of appropriate toxicological data to construct training sets containing similar chemicals to DBPs limited the value of many of the predictions in the present project. This further reinforces the need to follow up prediction with experimentation. The short coming is not simply a property of the program used, but due to the fact that little systematic toxicological work has been done on chemicals related to by-products.

FAILURE TO IDENTIFY LIKELY CAUSES FOR EPIDEMIOLOGICAL ASSOCIATIONS BETWEEN CHLORINATED DRINKING WATER AND ESTABLISHED DISINFECTION BY-PRODUCTS

Individual By-Products

Among chemicals that have been firmly established as DBPs none were identified as reasonable explanations for epidemiological associations between chlorinated drinking water and adverse health effects. This statement is based upon the finding that none of the by-products identified and measured in drinking water have ratios of the dose obtained from drinking water to a measured TD_{50} or predicted LOAEL that was appreciably greater than that of dichloroacetic acid. The ratio would have to be 10-100 times larger to reflect the carcinogenic potency implied by epidemiological data, if it were to occur at concentrations similar to those of dichloroacetic acid. A small number of these chemicals was found to occur in sufficient concentrations or had sufficiently high predicted potencies to be of concern as potential health hazards, but likely represent much smaller risks. In addition, some chemicals related to established DBPs were found to be of potential interest, however, in these cases there were no quantitative data relating to frequency of occurrence or probable concentrations.

A second criterion was that the by-product may be of regulatory interest when the predicted dose obtained from drinking water divided by the TD_{50} or predicted LOAEL exceeded 0.00001. This figure approximates the dose-to-effect ratios that are observed among regulated DBPs that are carcinogens. The chemicals that exceed this criterion are bromochloroacetamide, bromate, chlorate, dichloroacetaldehyde, dichloroacetonitrile, formaldehyde, and 3-chloro-4-

(dichloromethyl)-5-hydroxy-2(5H)-furanone (MX) (Table 7.5). Only cyanogen chloride met this criterion among predicted developmental toxicants. The ratio of dose from drinking water to predicted chronic LOAEL for cyanogen chloride was 0.0002236. If the smaller margin of exposure that is generally allowed for non-carcinogens is taken into account, cyanogen chloride is only of minor interest. Guidance for this statement is taken from the ratio calculated for dose of chlorite from drinking water to its actual LOAEL. Chlorite has been well studied for reproductive and developmental toxicity in rodents; its ratio is 0.0029932, more than 10 times the ratio calculated for cyanogen chloride.

As a single member of a potentially larger group of DBPs, it is important to indicate that 3,3-dibromopropenoic acid at 0.0000065 approaches the criterion. It has been included in a limited number of occurrence studies.

Several chemicals were identified with lower predicted LOAELs that may be of regulatory interest if occurrence data become available. Many of these chemicals were halonitriles from outside the well established haloacetonitrile group. The QSTR predictions of toxic potency stand out as potentially significant contributors to health hazards associated with chlorinated drinking water, but it was decided to handle these compounds as classes in the next section because the QSTR predictions were generally of low reliability.

Classes of Poorly Characterized Established Disinfection By-Products of Potential Importance

Since its discovery as a potent mutagen, MX has raised concern as a chlorination by-product. A cancer bioassay in rats demonstrated that MX was also carcinogenic (Komulainen et al. 1997), but the very low concentrations originally found in drinking water suggested that MX posed little risk related to other DBPs. More recent analytical data (Weinberg et al. 2002) have shown that the concentration of MX and related furanones found in chlorinated drinking water is greater than originally thought. By itself, MX is still unlikely to contribute significantly to the cancer risk associated with chlorinated drinking water. However, the finding of higher concentrations of MX raises issues related to the contribution to risk that would be made by a total of MX-related compounds. Interest in this group arose from an emerging characterization of the carcinogenic activity of furan derivatives and by the predicted formation of the cyclopentenoic acid derivatives discussed in the next section. While there have been structure-activity relationships for mutagenic activity among MX-congeners, such an effort has not been taken with respect to the characteristics of members of this class that are important to carcinogenic activity. The fact that furan, itself, is a reasonably potent carcinogen, but a poor mutagen, suggests a need to better characterize the properties of this group of compounds that contribute to carcinogenic risk.

QSTR predictions for the halonitriles also raise concern. These two classes are known to occur in disinfected drinking water, but aside from the haloacetonitriles their occurrence has received little attention. On the other hand, precursor substructures that will give rise to many of these by-products are known to exist in NOM. Therefore, the concern should focus on the occurrence of the entire group of halonitriles that act by similar mechanisms (not a total halonitrile concentration). The QSTR predictions made in this report were based on a set of molecular rules that are derived by examining the carcinogenic activity of chemicals that have been subjected to appropriate testing in the National Toxicology Program (NTP) and Food and Drug Administration (FDA) databases. While predictions cannot be entirely relied upon individually, they are rationally derived from a large set of existing data on other chemicals; very conservative criteria were applied to identify those compounds that should be of the most concern. Therefore, it is recommended that selected classes of putative DBPs be seriously considered in future research efforts.

By way of example, the limited reliability of the chronic LOAEL model for the haloacetonitriles made it difficult to recommend which members of this class should be subject to toxicological characterization. Therefore, it is recommended that a more holistic experimental characterization of the halonitriles should await the development of a better understanding of which halonitriles (beyond the haloacetonitriles) actually occur as a result of differing forms of disinfection. The halonitriles can be viewed as end products of chlorination when conditions are make them reasonably stable (Reckhow et al. 2001), but they can also be viewed more as intermediates when breakpoint chlorination is used (Nweke and Scully 1989). The data that are available on chloraminated systems are not sufficient to determine whether or not the halonitriles would occur at higher concentrations. Therefore, occurrence information is of the highest priority for halonitrile research. The predictions of high potency for the halonitriles are of limited reliability. For this reason it is suggested that toxicological characterization should begin with those members of this class that may occur at concentrations higher than 1 μg/L

The haloamides appear to be of somewhat less concern than the halonitriles. Like the halonitriles, they can be considered either as end products or as intermediates. Their predicted potencies were made with greater reliability than the halonitriles and were not substantially different from the predicted potency of more terminal by-products, such as the corresponding haloacetic acids. Nevertheless, the toxicological predictions made by TOPKAT® for this class do not depend on their similarity to the HAAs in the training set, but on a variety of other chemicals. These included 2[fluoromethyl]-DL-ornithine, 1,2-dichloropropane, methyl carbamate, and bromodichloromethane (BDCM) for predicting the chronic LOAEL, acetamide and carbromal in predictions of carcinogenicity, and acrylamide for reproductive toxicity. For this reason, it was difficult to simply view these predictions in the context of the HAAs. Therefore, research recommendations for this group of chemicals are the same as those for the halonitriles, but with a lower priority.

POTENTIAL IMPORTANCE OF PUTATIVE DISINFECTION BY-PRODUCTS PREDICTED TO BE FORMED FROM NOM

The development of adverse health effects from DBPs will depend upon both the nature and amount of the chemicals that are produced and survive to the tap to be consumed. The extent and nature of the disinfectant residual that remains in the water as it is transmitted via the distribution system is an important determinant of what is observed at the tap. Residual disinfectant increases the amounts of some terminal by-products but its ability to destroy intermediates is not as widely recognized.

A combination of three factors led to the conclusion that there are predicted DBPs that should be more thoroughly investigated relative to their occurrence in drinking water. These are: 1) the chemical was fairly reliably predicted to occur based on probable reactions with established substructures in NOM, 2) there has been no concerted effort to determine whether or not such by-products occur, and/or 3) TOPKAT® predicted a high probability of hazard. Within this group of chemicals, it was deemed prudent to include those compounds for which a high probability of adverse health effect was predicted. Some of these predictions were made with low confidence, but related DBPs were identified that fell within the acceptable similarity bounds or for which experimental data exist to support a higher priority. It is on this basis that these classes of putative DBPs should be given high priority in a research program.

Haloquinones

The haloquinones are considered to be of the highest priority for future research. As illustrated in Chapter 5, precursors of these chemicals include phenolic, polyphenolic, and poly methoxy natural organic chemicals. Haloquinones were predicted to have exceptionally low chronic LOAELs, as much as 1/10,000 of those of the regulated DBPs. Extensive work has been done on the metabolism and toxicology of quinones in general, but limited data are available on halogenated derivatives (Bolton et al. 2000). Quinones can be toxic in a number of organs, but there is a certain affinity for producing effects in the kidney, and in some cases, in the urinary bladder. In part, chemicals in the polyphenol or quinone classes are metabolized in a way that seems to parallel the processes identified earlier for the aromatic amines that are recognized bladder carcinogens. The quinones form glutathione conjugates in the liver, which then are transported to distant organs. Because of the transporters in the kidney these metabolites concentrate in that organ, where they are metabolized to reactive forms that produce renal toxicity and cancer. Moreover, these chemicals frequently act by two different mechanisms, oxidative stress and alkylation of macromolecules.

In the case of the haloquinones, the research priorities are clear. The first priority is to verify whether or not the predictions of formation can be documented, particularly in chlorinated and chloraminated waters. Second, there is a lack of critical toxicological information on halogenated quinone derivatives. The haloquinones are viewed as having a mode of action and a level of toxicological potency that may account for risks of the magnitude that have been epidemiologically associated with chlorinated water, if they are found to occur in drinking water at the tap. The stability of the higher halogen-substituted members of the haloquinone group would be low in the presence of free chlorine. As a result, they may be more of a problem in chloraminated water. There is very limited information on the influence of halogenation on the carcinogenic potency of quinones. Most of the toxicological research has focused on non-halogenated analogs. Therefore, it is suggested that initial research to characterize these compounds toxicologically focus on the toxicological properties of haloquinones.

Haloquinones and Bladder Carcinogenesis

2,6-Dichloro-3-methyl-1,4-benzoquinone. Among haloquinones that have been predicted to be formed on reaction of chlorine or chloramine with substructures in NOM, 2,6-Dichloro-3-methyl-1,4-benzoquinone (DCMBQ) possesses properties that suggest it may be a bladder carcinogen. DCMBQ is only one of several halogenated quinones that were predicted to be formed by reaction of chlorine or chloramine with substructures of NOM. Its selection should not be viewed as indicating that haloquinones are likely to simply be bladder carcinogens. The more important point is the toxicological potency imputed to this group by QSTR analyses. This chemical was highlighted to identify the structural components of the haloquinones that are likely to increase both their carcinogenic potency and their likely specificity for urothelial tissue (i.e. kidney and bladder). Further exploration of substructures of NOM may result in finding other compounds of similar structure that might be formed.

The predicted chronic LOAEL of DCMBQ was found to be low (79 μg/kg day^{-1}). The metabolic activation of this compound would be predicted to involve glutathione conjugation with processes to a cysteine conjugate. The formation of these metabolites will be facilitated by the substitution of halogens on the ring. Electron withdrawing groups, like halogens, would tend to hold the chemical in the more toxic quinone vs. hydroquinone form if they are retained on the quinone ring. It is probable that both glutathionyl and cysteinyl conjugates will be concentrated in the urothelial cells of the kidney and bladder. The cysteinyl conjugates can be converted to forms that alkylate macromolecules with the quinone structure contributing a significant oxidative stress within these tissues. The methyl group is predicted to slow cyclization reactions of the cysteine conjugates that tend to detoxify the metabolite. For these reasons, haloquinones are predicted to be significantly more toxic than simple benzoquinones and the methyl group may provide specificity for inducing renal cancer. These properties may also contribute to the development of bladder cancer (Peters et al. 1997).

Halocyclopentenoic Acids

Two halocyclopentenoic acid derivatives were also predicted to be very toxic. These compounds have a superficial resemblance to MX, so they have been included in the discussion of these chemicals in Chapters 6 and 7. 3,5-Dichloro-1-hydroxycyclopent-2-enoic acid is representative of what could be a larger group of such compounds.

The predictions made by the chronic LOAEL model of TOPKAT® for the two haloacyclopentenoic acids evaluated were both outside the OPS of the model. Neither met the criteria as carcinogens or mutagens. Therefore, these compounds cannot be assigned a high research priority based upon the QSTR analysis.

Similar difficulties were encountered when applying TOPKAT® to furanone derivatives because of the lack of appropriate structures of sufficient similarity in the training sets. The clear evidence that MX is carcinogenic brings these predictions into question. It would be profitable to more broadly explore available data on the carcinogenic activities of furan and cyclopentene derivatives together. One step in this process would be to determine if some of the QSTR predictions were accurate. Such information would provide invaluable guidance to a more comprehensive in vivo toxicological characterization that is important for addressing health risks from this group of DBPs.

N-Haloamines

One large set of almost universal DBPs for which little toxicological data are available are the organic N-haloamines. They are inevitable by-products of HOCl and/or HOBr if amines are present in NOM. N-Haloamines have repeatedly been produced experimentally with precursors that are commonly found in sources of drinking water. Nevertheless, their occurrence has received little serious study beyond the pioneering efforts of Scully and coworkers. This group of by-products was assigned a high priority for future research because the likely qualitative and quantitative differences in their occurrence may well provide an explanation for why increased bladder cancer risk is observed with chlorination, but not with chloramination (Zierler et al. 1988, McGeehin et al. 1993). Chlorination and chloramination may result in different end products due to the increased diversity of N-haloamines that are likely to occur as the oxidative and halogenation reactions of chlorine more thoroughly degrade NOM. The combination of these reactions results in a wider variety of amine structures upon which N-haloamines may be formed.

A small number of N-chloramines formed from the reaction of chlorine with identified constituents of NOM (amino acids, purines, and pyrimidines) were subjected to QSTR analysis. A number of adverse effects were predicted to occur with high probability. However, most of these predictions were of low reliability because the N-chloramine fragment is not represented in the training sets for the TOPKAT models (the one exception is the Ames' mutagenicity model). In addition, the similarity indices were often low.

The level of concern related to N-chloramines is raised by active investigation of their role in the production of chronic disease secondary to their formation in vivo as a result of inflammatory processes. Unfortunately, there has been no attempt to characterize the toxicology of these compounds when they are consumed in drinking water. Several reasons were discussed in Chapters 6 and 7 to indicate why the properties of N-haloamines are such that they may serve as an effective system for delivering a toxic insult systemically and hence generate active forms of toxicants at a variety of organ sites. However, in the absence of appropriate in vivo data, this remains a hypothesis.

It is strongly recommended that research be undertaken to more fully characterize the nature of N-haloamines that are produced in water treated with a variety of disinfectants. An attempt needs to be made to identify the more stable and abundant members of this group. Once the nature of these DBPs is identified a program of characterizing the toxicological properties of this class of DBPs should be developed.

N-Nitrosamines and Nitrosamides

The third broad group of DBPs needing extensive work relative to their occurrence is the N-nitroso compounds that are formed at highest concentrations with chloramination. It has been clearly established that nitrosation can occur from the extensive research of the Valentine (Choi and Valentine 2002 & 2003) and Sedlak (Mitch and Sedlak 2002) groups. What has not been evaluated is whether or not precursors found in NOM may also lead to nitroso compounds with more novel structures. Because of the analytical difficulties of achieving detection limits appropriate to the very high carcinogenic potency of members of this class, the authors recommend first characterizing reactions in model systems. A question that should be addressed early in this work is the formation of N-nitrosamines with substructures of NOM where chloramine is the primary disinfectant vs. conditions where ammonia is added after some period of contact with free chlorine. Because of the diversification of nitrogenous compounds that may be predicted to occur with free chlorine, the nitroso compounds potentially formed under these two conditions may be quite different. If reasonable yields of some novel nitrosamines are demonstrated, then occurrence studies should follow. Finally, given information on occurrence, toxicological characterization should be undertaken on any novel nitroso compounds that are found to occur. The initial prioritization among nitroso compounds may be based upon whether or not the compound is produced at higher concentrations than the mean concentration of N-nitroso-N-dimethylamine (NDMA) that has been observed in chloraminated drinking water, as NDMA is clearly one of the most potent members of the nitrosamine group of carcinogens.

NEED TO CONSIDER NON-LINEAR RELATIONSHIPS IN THE FORMATION OF DISINFECTION BY-PRODUCTS

A major conclusion from this study is that a more sophisticated examination of disinfection chemistry is necessary to address the broad problem of DBPs. Disinfection practice varies widely and the variation must be viewed systematically. At the top of Figure 8.1 a general model of the types of chemical reactions that have been associated with disinfection is provided. The general model is followed by three illustrations that depict broad differences in the character of the three commonly utilized disinfection strategies, chlorination, chloramination, or ozonation. The use of chlorine dioxide has been omitted for simplicity's sake because its participation in the formation of organic by-products is likely to parallel that of ozone.

It is first necessary to understand the kinetics of DBP formation and degradation that occurs with individual disinfectants. The complexity of using combinations of disinfectants cannot be understood without first considering how variations in practice with the commonly used disinfectants affect DBP formation qualitatively as well as quantitatively. All three strategies participate in halogenation and oxidation reactions to varying degrees. All forms of chemical disinfection and some physical methods (e.g. ultraviolet light) utilize one or the other of these properties to inactivate microorganisms. Because of the dominance of oxidative chemistry that occurs with chlorine, chlorine dioxide, and ozone, many of the products that are formed are similar. However, adverse health effects are unlikely to result from the large number of carboxylic acid compounds that are produced as a result of simple oxidative chemistry (e.g. non-halogenated) or the smaller amounts of aldehydes that will be converted to carboxylic acids when absorbed.

The extent to which each strategy contributes halogenated vs. oxidation products varies widely. The varying extent of these basic reactions contributes to the formation of a substantial number of different by-products both on the oxidation and halogenation pathways. The difference in these by-products represents one of the main tools that may be used to explore the differing health hazards that may be associated with each disinfection strategy.

In addition to the bulk reactions of oxidation and halogenation, there are sets of reactions that introduce by-products unique to that disinfectant. These are termed side reactions. They do not necessarily contribute to disinfection. The concern with side reactions is related to compounds that have the potential of contributing substantially to the health hazards of the disinfectant. The examples provided are nitrosamine formation with chloramines and bromate with ozone. Their formation may also be dependent on the occurrence of particular substrates in the source water, e.g. bromate formation by ozone depends upon the occurrence of bromide. Therefore, their formation is likely to vary geographically. There may well be other reactions of this type, but if so, they have not received systematic study.

This project has reinforced the idea that health risks from DBPs are unlikely to be characterized by the major by-products that are formed. In the case of breakpoint chlorination we have shown the possibility that some compounds were formed primarily as intermediates and are destroyed by excess chlorine. However, if a utility makes an attempt to minimize the formation of terminal disinfection by-products of relatively low toxicological potency (e.g. THMs and HAAs) by decreasing chlorine dose or addition of ammonia, there is a risk that some much more toxic intermediates that are destroyed by excess chlorine might be preserved.

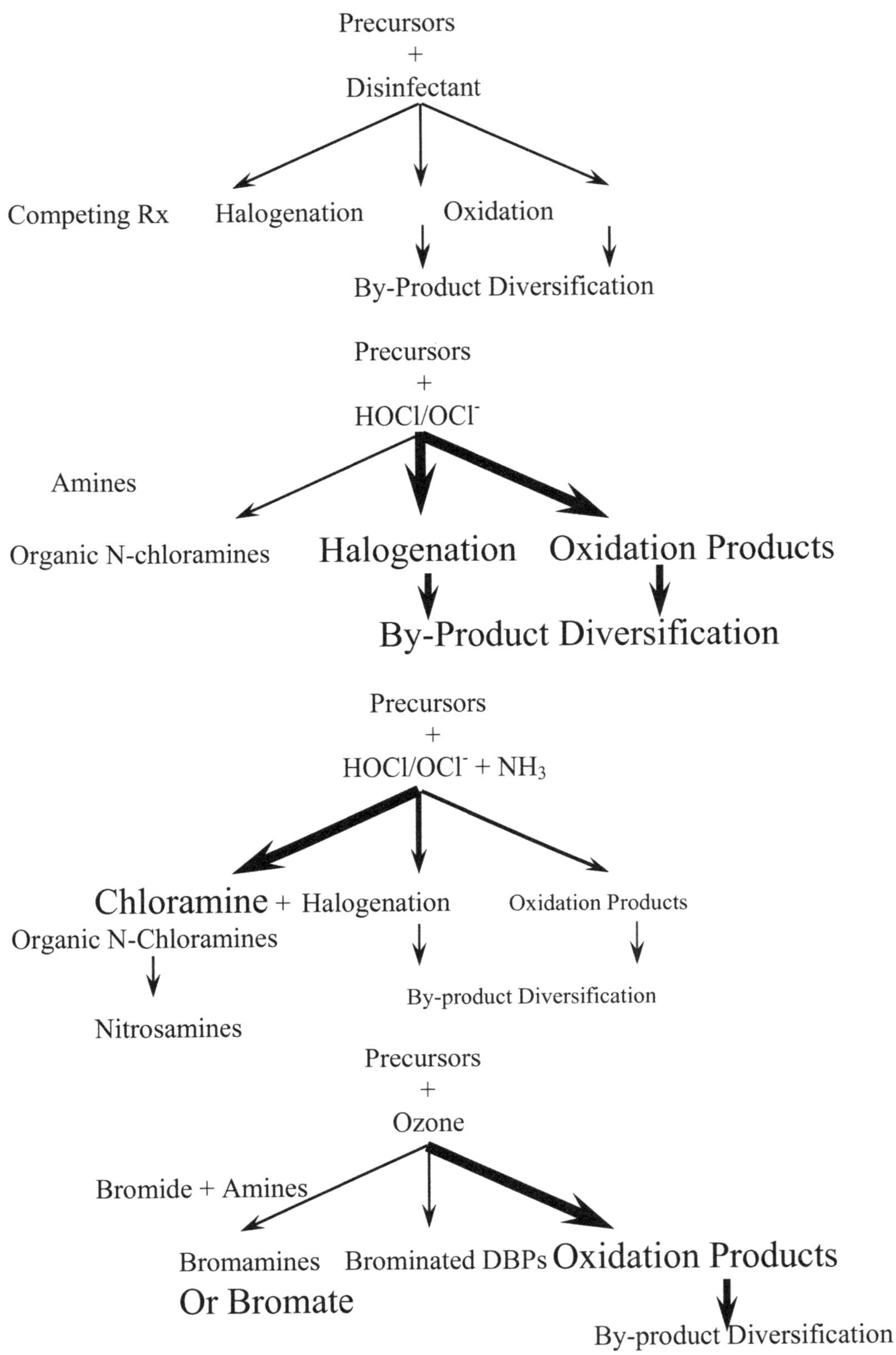

Figure 8.1 Framework for considering potential health hazards from differing disinfection strategies The bolder arrow or font size reflects the reactions that are most favored with a particular disinfection strategy.

There are differences in how each of the disinfectants participates in oxidation and halogenation reactions that are not reflected in simple surrogate measures of by-product formation such as the THMs. One is that oxidation products expected from chloramines will have a substantially different character from those of chlorine and ozone (Kaminski et al. 1976). The products that are observed with strong oxidants will be found at lower yield with chloramine. It does not follow, however, that reaction products that are produced uniquely by chloramine or are produced as intermediates in reactions of ozone or chlorine will be present at lower concentrations just because the THM yield is lower. The stronger oxidants drive oxidation much further and may actually reduce the concentrations of some intermediate species. In the case of chlorine, both halogenation and oxidation reactions are not only strongly involved in producing by-products, but also produce them in interactive ways. Excessive substitution of halogen destabilizes ring structures, resulting in diversification in the same way that oxidation does. Ozone can contribute to both types of reaction, but its participation in halogenation reactions is limited by the bromide concentration in the treated water. As a consequence, the by-products produced by chlorine are much more diverse than those of other disinfectants.

This project developed a structure for predicting by-products that may reasonably be assumed to be part of the uncharacterized by-products that can be inferred by "unknown total organic halogen (TOX)". Additional by-products can be inferred by the fact that measured products of reactions of disinfectants with certain precursor substructures in NOM frequently account for only a small fraction of the disinfectant demand. The first component of this predictive structure is consideration of the nature of the organic precursor material that is present in sources of drinking water. To make the prediction of the formation of novel DBPs tractable, the reactants were considered to be individual substructures of NOM that is assumed to behave independently of the rest of the organic matrix. Second, the chemistry of the disinfectant needs to be considered. Third, the nature of reactions that compete for the disinfectant need to be addressed (e.g. bromine or ammonia). Fourth, by-products that might result from extensive halogenation and oxidation reactions and interactions between these two processes to diversify the by-products also need to be investigated. Limits on resources and the overwhelming complexity of the problem prevented exhaustive exploration of all these factors in the present project. However, the effort provides direction to a more efficient way of continuing the research effort needed to resolve a very important public health problem.

A DIFFERENT FRAMEWORK FOR RESEARCH ON DISINFECTION BY-PRODUCTS

The focus on regulated DBPs (e.g.. the THMs) in recent epidemiological studies has little scientific basis and primarily appears to support a regulatory strategy. Key toxicological studies of these DBPs have either failed to produce effects when given in drinking water or have toxicological potencies much too low to make their activity at the low concentrations found in drinking water scientifically plausible. Mode of action studies have further undermined the supposition that these compounds are causally related to the health effects identified in epidemiological studies. Beyond the inevitable reinforcement of a predetermined regulatory strategy, the main reason that remains for measuring the common DBPs for regulation is that they are easily measured. The fact that the members of these groups cannot plausibly account for the projected risks is widely ignored.

The major conclusions of this report can be viewed in terms of the framework provided in Figure 8.1. As indicated in Chapter 7, these conclusions can be cast directly into detailed objectives for research. Thus, the framework can continue to inform a multi-year research plan that adapts to knowledge that is developed along the way. The information currently available is insufficient to fully implement the framework to the point of making definitive decisions. Implementing a research plan based on the conclusions of this report provides an opportunity to direct research in a reactive way that will lead to much more rapid resolution of this long-standing problem.

For practical purposes, the yield of DBPs is frequently assumed to be a more or less linear function of the total organic carbon (TOC) concentration up until enough disinfectant has been added to completely satisfy the demand presented by that carbon. While it is often true that larger amounts of certain DBPs will be observed as the amount of disinfectant added to a fixed amount of organic carbon increases, this finding is restricted to those by-products that are incapable of further reaction with the disinfectant and to the point that the precursors that produce these by-products have been exhausted. Such a formulation does not take into account the competitive nature of different reactions that occur, particularly at low disinfectant/carbon or nitrogen rations. Second, it does not consider the fact that initial products of disinfection can be destroyed by excess disinfectant. This means that modifying disinfection practice to minimize formation of terminal by-products, such as the THMs and HAAs, runs the risk that intermediates may be stabilized as by-products if the reactions are quenched short of totally satisfying the demand. The use of ammonia to quench chlorine reactions after a period of exposure to free chlorine stands out as an obvious modification of disinfection practice that may lead to the stabilization of some of the haloquinone compounds identified in this report as intermediates.

Moreover, one cannot ignore the likelihood that the concentrations of some by-products may continue to increase even though the yield of other by-products has reached maximum concentrations. There has been little attempt to understand the nature of by-products that may be formed after marker by-product formation has reached a plateau. The result is that it is simply not known whether or not adding more or less chlorine would increase or decrease potential health hazards. There is even less confidence if disinfection strategies other than chlorine are utilized.

A similar fallacy can be identified with respect to side reactions. The question here is, are they are linearly related to the bulk chemistry that is set in motion by the addition of disinfectant? Alternatively, is their formation more dependent on the presence of particular precursors that vary by geographical location? It is probable that the yield of such by-products will vary with the nature of the TOC and the nature of competing reactants for the disinfectant in the water. This chemistry can be expected to vary significantly among sources.

As disinfection of drinking water has moved from classical breakpoint chlorination to the more common use of combined chlorine or combinations of disinfectants, the discontinuities in by-product formation introduced by such practices have simply not been explored. These problems essentially become intractable when utilities move to the use of mixed disinfectants to maintain control of microbial risks while attempting to minimize the hazards resulting from the production of only a small number of DBPs (i.e. those which are regulated).

The body of literature relevant to disinfection chemistry is large, but to these authors' knowledge no prior attempt has been made to organize this information in a way that focuses on the potential importance of intermediate products that may be formed and be of health concern. Hopefully, the approach utilized this project will provide some impetus for research that can more clearly and broadly inform decisions on how disinfection practice can be modified to reduce risks associated with DBPs.

In summary, by-product yield continues to be treated as a more or less linear process that is only modified by degradation of selected by-products in the distribution system by microbial action. A warning of the fallacy of this type of thinking has been the realization that the use of chloramine introduces a new set of by-products with recognized toxicological properties, the nitrosamines, while it clearly reduces by-products that present lower levels of risk, the THMs and HAAs. This does not necessarily mean that formation of small amounts of nitrosamines represent an unacceptable hazard, but it does indicate that the industry needs a more comprehensive model for examining the risks of alternative disinfection strategies.

IMPLICATIONS FOR REGULATION

The CCL has become the primary tool for the Office of Ground Water and Drinking Water (OGWDW) of the U.S. Environmental Protection Agency (USEPA) to identify contaminants of drinking water for regulation. The process recommended by the National Academy of Sciences (NAS) is generally followed, so it is not described in detail here (NRC, 2001). The focus of the CCL is on individual contaminants, which limits the ability to examine broader public health issues such as those introduced by DBPs produced by alternate forms of chemical disinfection. Rather than addressing the broad problem, current regulation of DBPs focuses on establishing MCLs for individual by-products drawn from a limited number of by-product groups. There is no objective evidence, independent of epidemiological associations, that these "surrogate" measures are proportional to risks. The relationships of these surrogates have odds ratios that are not different from those seen with consumption of chlorinated drinking water. Therefore, they have contributed nothing to the resolution of the problem. As data are emerging that show that the THMs are themselves not carcinogenic in animals when administered in drinking water (Jorgenson et al. 1985, NTP 2005b, USEPA 2003a), the inadequacy of this approach has become more apparent. The articulation of benefits from regulation of DBPs was falsely based on attributing benefits to the reduction of inadequate surrogates; this calls the whole regulatory approach into question. It would be far better to focus current regulation on toxicities associated with the regulated by-products, avoiding extravagant claims of benefits. As a second step, it should be acknowledged that these compounds do not account for the effects implied by epidemiological studies relating cancer risk or reproductive effects with the disinfection of drinking water. This illustrates the fact that the CCL, as now implemented, is poorly suited to guiding scientific studies for establishing causal relationships between variables in drinking water and observed adverse health effects.

The research needed to evaluate health risks generated by alternative methods of disinfection is even more complex and will be more protracted than that addressing the potential health effects of a single contaminant of drinking water. It is first necessary to identify DBPs that have a high probability of producing the effect. The rate-limiting step will be testing these chemicals to determine whether or not they have the appropriate toxicological characteristics. The present project has demonstrated that focus can be brought to this effort by simultaneous use of feed-back and feed-forward approaches for identifying candidate by-products. The feed-forward process simply looks at the nature of precursors available and predicts the types by-products that may be formed. The feed-back approach examines chemicals that are known to produce the effect and then examines the possibility that precursors may be present in water that give rise to chemicals of analogous structural and toxicological properties. It is important that this research is not constrained by a set list of chemicals that are examined for potential regulation, but rather that an exploratory effort continues until causal relationships between the use of a particular disinfectant and the existence of adverse health outcomes in humans is established. The linkage with epidemiological studies needs to eventually come from controlled studies of targeted DBPs based upon the classical scientific process of testing alternative hypotheses. Undoubtedly, this will be an iterative process, but it is almost certain to yield definitive answers much more quickly than the empirical processes that have been employed to date.

Once potentially important by-products have been identified, the CCL can operate as it was designed and implemented following the passage of the 1996 Amendments to the Safe Drinking Water Act. Generally, this would follow a process of determining the occurrence and concentrations that are found in drinking water, performing experiments to identify the hazards associated with the chemical, obtaining the quantitative and mode of action data necessary to estimate risks, and finally determining whether or not the compound should be regulated. This activity need not be limited to chemicals with a high probability of accounting for epidemiological observations. It is possible that health hazards will be identified based on toxicological studies that have not been identified in epidemiological studies.

The focus of this project is on accounting for high levels of cancer risk suggested by epidemiological studies of chlorinated water. In this context, it is important to ensure that a chemical occurs broadly before significant resources are devoted to experiments designed to determine the extent to which it represents a risk for cancer. However, one can begin to form a profile of the biological properties that might be expected from a chemical before expensive tests are undertaken. Two tools can be important at this stage, a comparison of the probable toxicological properties of the chemical with chemicals for which there are data followed by the conduct of some preliminary toxicological studies to confirm the predictions. The first step can be accomplished by utilizing QSTR.

This is a logical approach, but some of the limitations of any QSTR tool must be recognized and addressed during this process. Clearly, one should not depend on predictions that are unreliable. Consequently, an examination of the performance of the model used, the adequacy of the training set, and expert evaluation of the outcome of any predictions are essential parts of a process for identifying a chemical for more thorough evaluation by listing it on the CCL.

A second major issue is the development of a rationale for prioritizing chemicals suggested to be of importance by QSTR analysis. This is not simply a matter of accepting what a QSTR program states as the probability of adverse effect, but a critical examination of that output. Within this project a relatively high bar was set for a chemical before it was assigned a high priority because such criteria are necessary if the goal is to identify those compounds <u>most likely, based on current evidence,</u> to contribute to a given risk. It was modeled after approaches to weight of evidence that have traditionally been used in environmental health assessments of carcinogens. However, it is very important to examine the overall output of an effort as complex as that which was undertaken and explore the possibility that other prioritization schemes might be more appropriate.

Unlike prior efforts, the highest priorities established in this study were for 1) chemicals that were predicted to occur from precursor substructures in NOM (quinone and furanone-like compounds), 2) groups of chemicals that are well known to be produced in disinfection, but which have been so poorly characterized that they were essentially not recognized by the QSTR model (e.g. N-chloramines), and 3) a variety of by-products that might be predicted if the character of precursor material was more thoughtfully considered (e.g. higher molecular weight nitrosamines). These first level priorities are supplemented by identifying deficiencies in the database for another well-known group of DBPs (i.e. halonitriles) whose occurrence and toxicology have received only limited attention. Finally, it was recognized that there is a need to more critically assess how small differences in disinfection strategies that are meant to reduce regulated DBPs may lead to increased concentrations of potentially more toxic DBPs. A more scientific approach that utilizes a framework for comparing all disinfection strategies can be used to identify where the most important unanswered questions lie and ensure that limited resources are appropriately focused.

APPENDIX A

SUMMARY OF DATA ON DISINFECTION BY-PRODUCTS THAT HAVE BEEN EXAMINED FOR CARCINOGENIC, REPRODUCTIVE, OR DEVELOPMENTAL TOXICITIES

Table A.1
Results of cancer bioassays of putative disinfection by-products

Disinfection By-product	Experimentally identified target organs*	Rat TD_{50}	Mouse TD_{50}
Halogenated			
Chloroform	kid, liv	262	90.3
Bromodichloromethane	kid, lgi, liv	72.5	47.7
Dibromochloromethane	liv	neg	139
Bromoform	lgi	648	neg
Iodoform	neg		
1,2,3 - Trichloropropane	ezy, kid, orc, pan, pre, sto, cli, mgl, hag, liv ute	1.35	0.88
Hexachlorocyclopentadiene	neg	neg	neg
2-Bromoethanol	sto	not tested	70.1
Trichloroacetaldehyde	liv	neg	106
Trichloroacetic acid	liv	neg	584
Dichloroacetic acid	liv	not in CPDB	102
Chloroacetic acid	neg	neg	neg
3-Chloro-4-(dichloromethyl)-5-hydroxy-2(5H)-furanone (MX)*	thy, liv, mgl	not in CPDB	
2,4,6-Trichlorophenol	hmo, liv	405	1070
2,4-Dichlorophenol	neg	neg	neg
Pentachlorophenol (tech. grade)	liv, adr, vsc	13.1	23
1,2-Dichlorobenzene	neg	neg	neg
1,4-Dichlorobenzene	kid, liv	644	398
Chlorobenzene	liv	247	neg
Non-halogenated			
Acetaldehyde	nas	153	not tested
Formaldehyde	hmo, nas	2.19	43.9
Benzaldehyde	sto	neg	1490
Benzoic acid	neg	neg	neg

* kid = kidney, liv = liver, lgi = large intestine, ezy = ear/Zymbal's gland , orc = oral cavity, pan = pancreas, pre = preputial gland, sto = stomach, cli = clitorial gland, mgl = mammary gland , hag = harderian gland, ute = uterus, thy = thyroid, hmo = hemopoietic system, vsc = vascular system, nas = nasal cavity

Table A.2
Reproductive and developmental toxicities of disinfectants and disinfection by-products

	LOAEL mg/kg		
Disinfectant/DBP	Maternal Repro	Male Repro	Developmental
Chlorine	ns* (5)	†	ns (5)
Chlorine dioxide	ns (10)	ns (10)	ns
Chlorite	ns (300)	ns (300)	ns
Chloramine	ns (10)		ns
Chloroform	50		ns
Bromoform	150		--‡
Bromodichloromethane	50	50	--
Chloroacetic acid	140		140
Dichloroacetic acid	140	62	140
Trichloroacetic acid	800	300	330
Bromoacetic acid	100	ns (100)	100
Dibromoacetic acid	250	90	?
Bromochloroacetic acid		8	
Chloroacetonitrile	55		--
Dichloroacetonitrile	45		25
Trichloroacetonitrile	55		--
Dibromoacetonitrile	50		--
Bromochloroacetonitrile	ns (55)		--
1,1-Dichloropropanone	100		--
Chloral Hydrate	ns (470)	150	ns
Chloropicrin	40		--
1,2-Dichloropropane	300		--
1,2,3-Trichloropropane	50		--

* ns (dose) = no effect was observed at the indicated dose
† blank cell indicates lack of relevant data
‡ -- tested, but negative at dose indicated under maternal reproductive effects

APPENDIX B

BLADDER CANCER DATABASE

Table B.1
Bladder Carcinogen Database

Chemical name	Structure	IARC Class other source	Rat TD50	Mouse TD50	Other TD50
Aliphatic Amines					
11-Aminoundecanoic acid	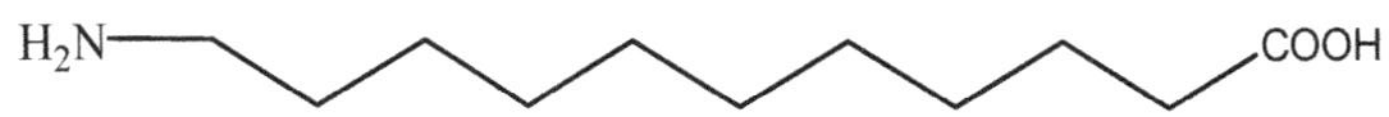	Gold	309 (ubl+)	Inade-quate	
Aromatic Amines					
Melamine	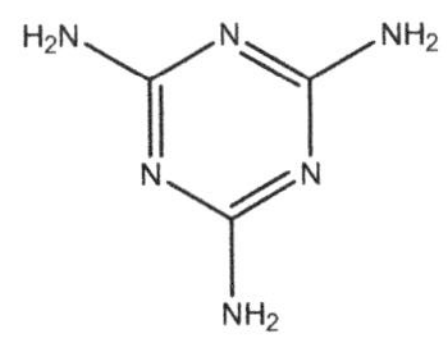	Gold	735 (ubl)	Inade-quate	
o-Toluidine	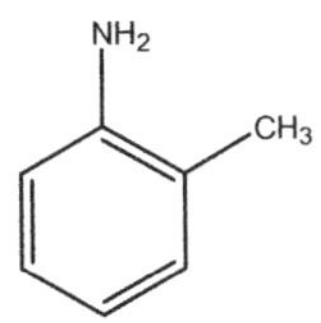	Gold	43.6 (ubl+)	840 (no ubl +)	
o-Anisidine	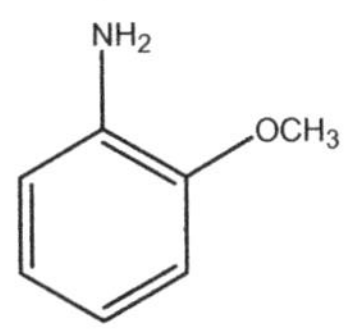	2B	29.7 (ubl+)	966 (ubl)	
4-Chloro-o-phenylenediamine	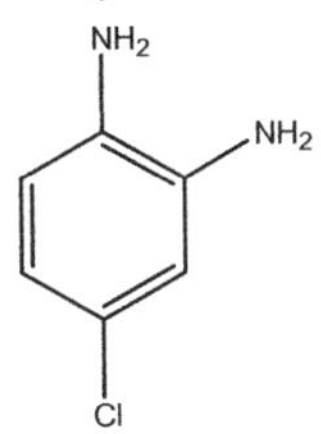	2B	214 (ubl+)	1230 (no ubl +)	

(continued)

Table B.1 (Continued)

p-Chloro-o-toluidine	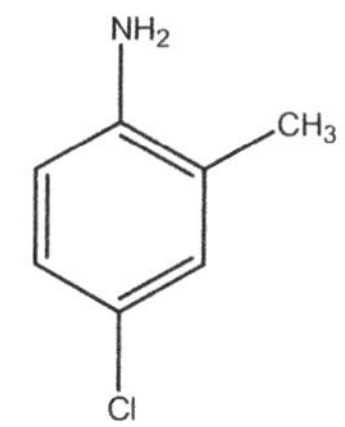	2A	NT	25.8 (no ubl +)
p-Cresidine	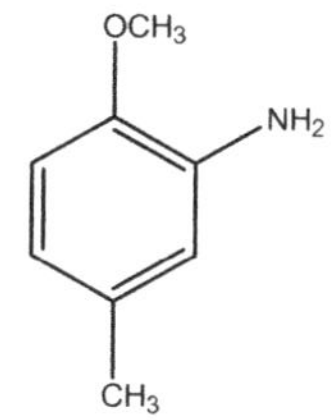	2A	98 (ubl+)	54.3 (ubl+)
m-Cresidine	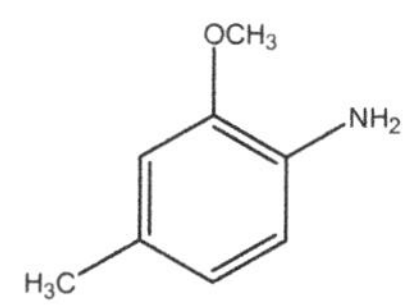	2B	470 (ubl)	Inade-quate
Phenacetin	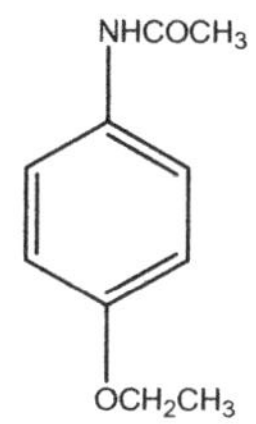	1	1250 (ubl+)	2140 (ubl+)
Acetaminophen	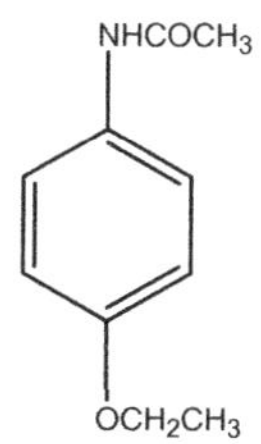	Gold	495 (ubl+)	1620 (no ubl+)

(continued)

Table B.1 (Continued)

4-amino-2-nitrophenol	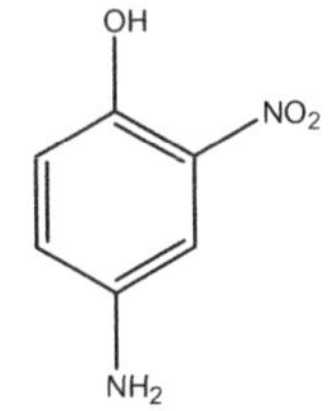	Gold	309 (ubl)	Inade-quate	
4-Aminobiphenyl	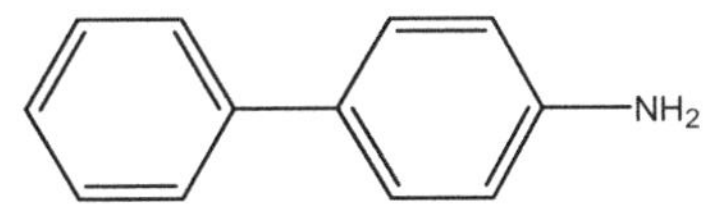	1	0.98 (no ubl+)l	2.1 (ubl+)	
Benzidine	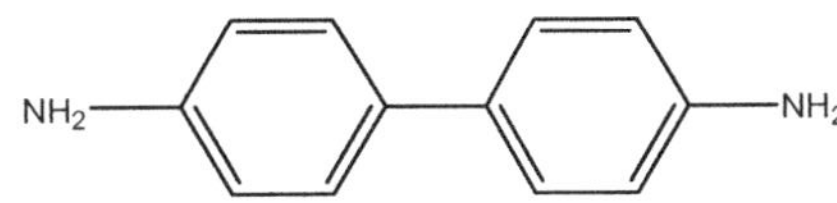	2B	1.73 (ubl+)	19.7 (no ubl+)	
3,3'-Dichlorobenzidine	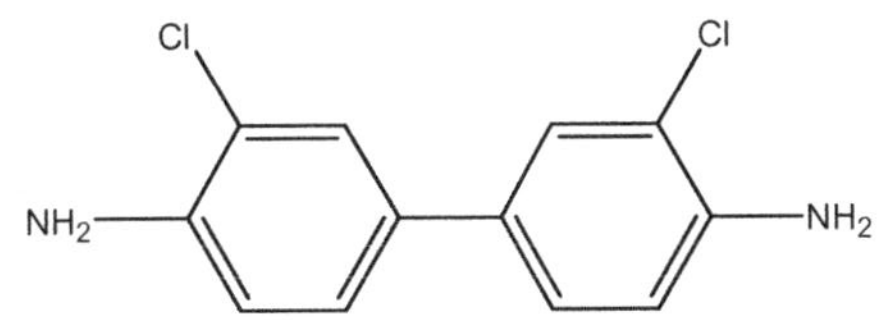	2B	28.1 (no ubl+)	NT	1.78 (ubl+) - Dog
3,3'-Dimethoxybenzidine	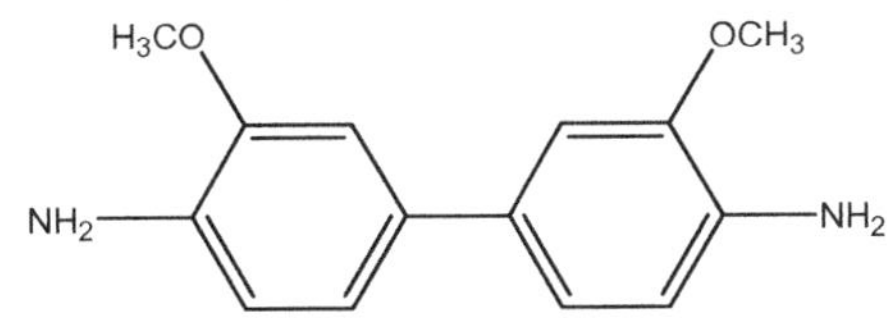	2B	diisocy anate - 1630 (no ubl+) 2HCl 1.4 (no ubl+)	Inade-quate	
4,4-Methylene-bis(2-chloroaniline)	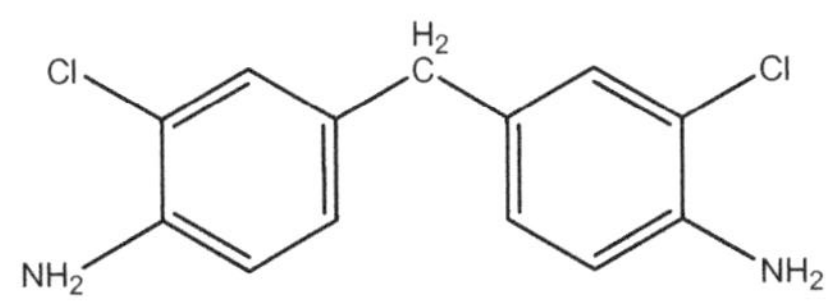	2A	base - 19.3 (no ubl+)	2 HCl - 66.6 (no ubl+)	2.12 (ubl+) Dog
4-Chloro-4'-aminodiphenylether	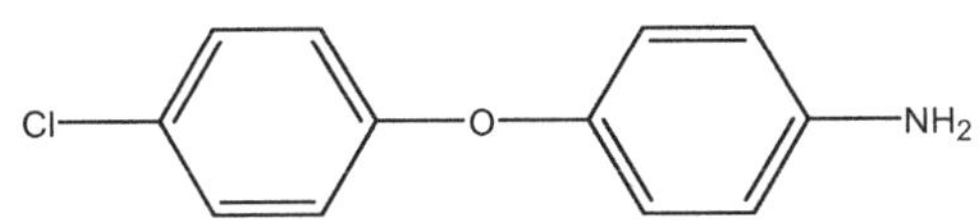	Gold	37.6 (no ubl+)	346 (ubl+)	

(continued)

Table B.1 (Continued)

2-Naphthylamine	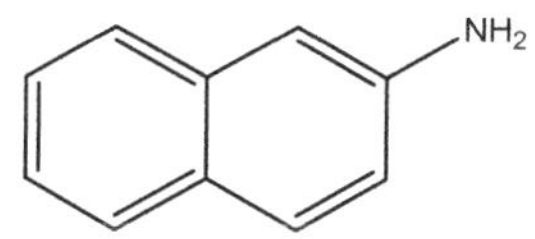	1	61.6 (ubl)	37.6 (no ubl+)	5.74 (ubl) Rhesus
2-Acetylaminofluorene	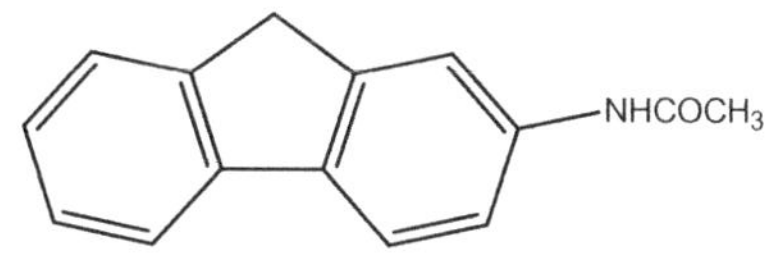	Gold	1.2 (no ubl+)	7.59 (ubl+)	17.4 (no ubl+) Hamster
N-hydroxy-2-acetylaminofluorene	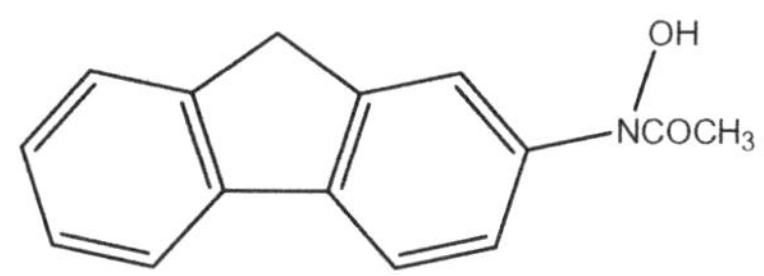	Gold	0.988 (no ubl+)	6.23 (ubl+)	2.1 (no ubl+) Hamster
2-Methoxy-3-aminodibenzofuran	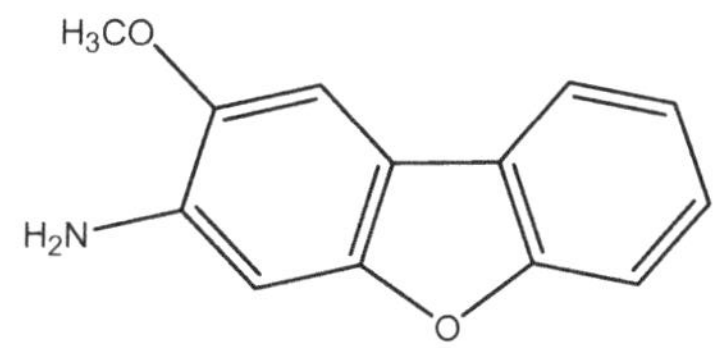	Gold	29 (ubl+)	Not tested	
1-Amino-2,4-dibromoanthraquinone	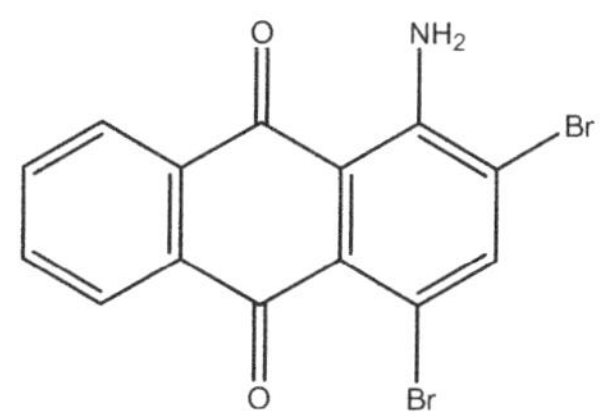	IARC? Gold	46 (ubl+)	477 (no ubl+)	
1,4,5,8-Tetraaminoanthraquinone (Disperse Blue 1)	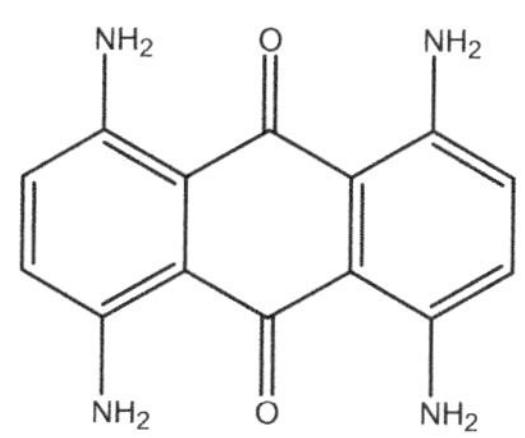	IARC	156 (ubl)	negative	
p-Quinone dioxime	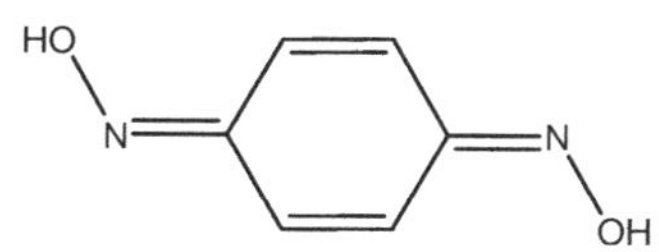	Gold	106 (ubl)	Negative (continued)	

Table B.1 (Continued)

Nitro Compounds					
2-Nitroanisole	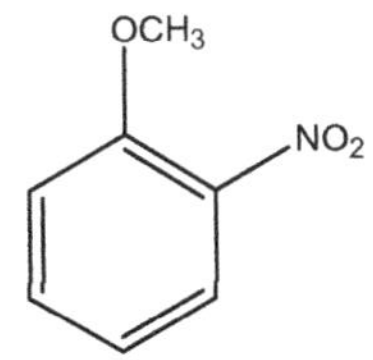	Gold	15.6 (ubl+)	178 (no ubl+)	
Aristolocholic Acid	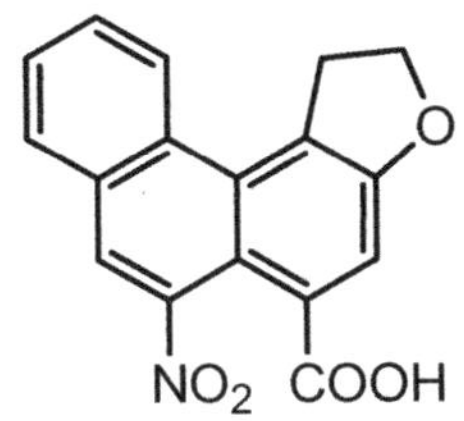	Gold no TD50			
Azo Compounds					
o-aminoazotoluene	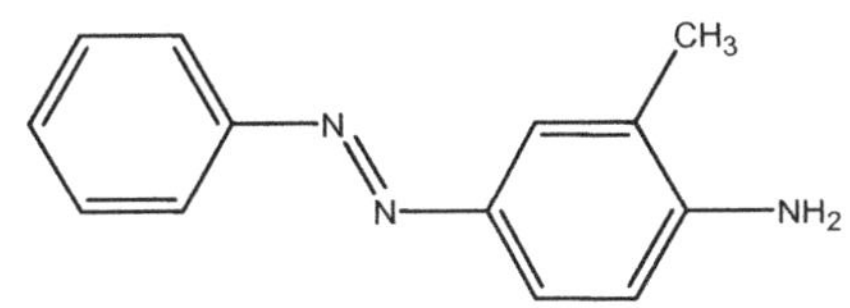	Gold	4.04 (no ubl+)	negative	Other ??
p-Dimethylaminoazotoluene	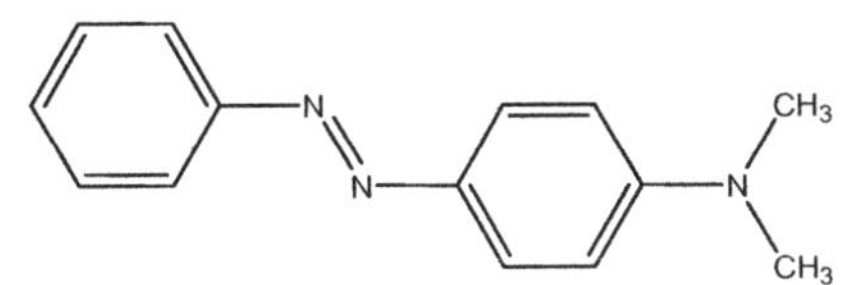				
Salicylazosulfapyridine	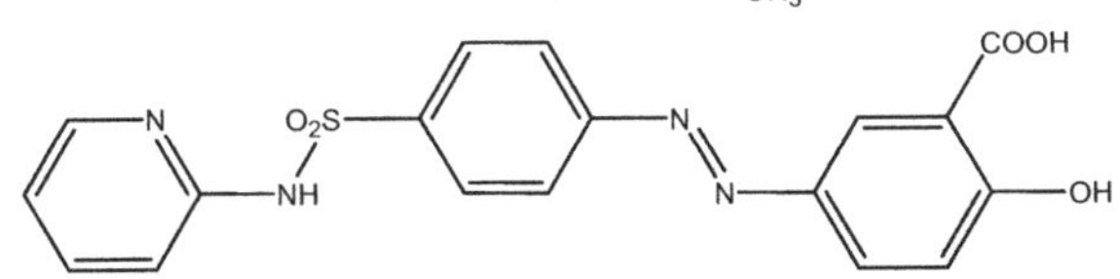	Gold	1590 (ubl+)	1030 (no ubl)	

(continued)

Table B.1 (Continued)

1-[(2,5-Dimethoxyphenyl)azo]-2-napthalenol	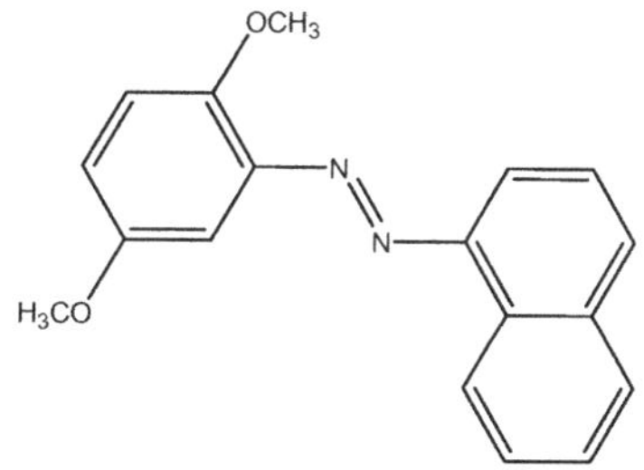					
1-(2,4,5-trimethylphenylazo)-2-napthol-3,6-disulfonic acid (Ponceau 3R)	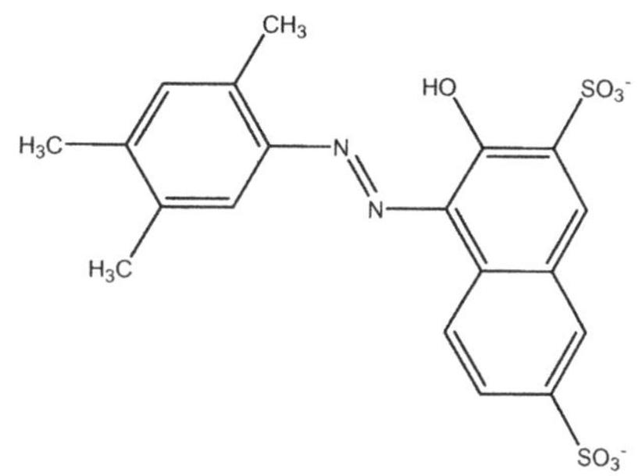	2,4,5-trimethyl aniline metabolite				
Direct Black 38	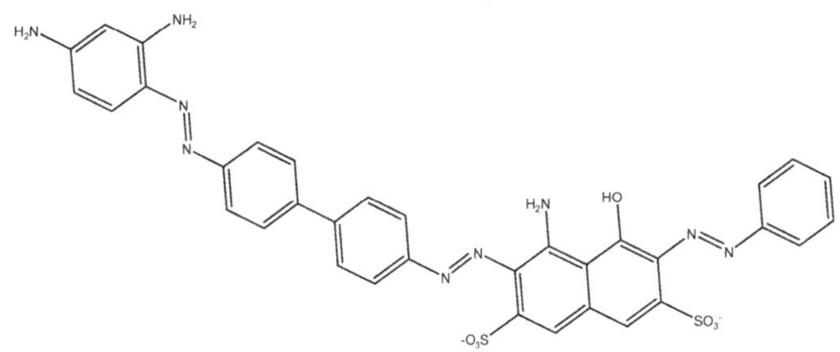		Gold	1.39 (no ubl+)	71.6 (no ubl+)	Other species? ?
Heterocyclic Amines						
Uracil	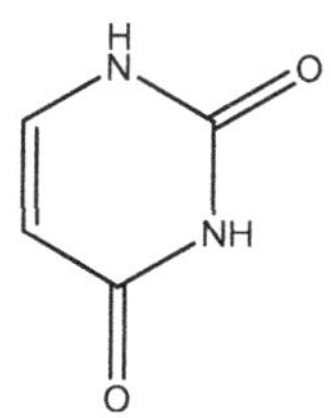		Gold	671 (ubl)	2750 (ubl)	

(continued)

Table B.1 (Continued)

Saccharin	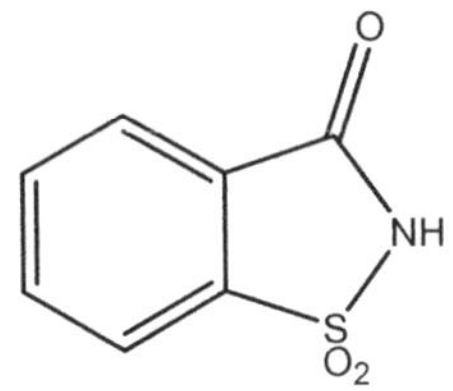	III	2140 (ubl)	negative	Monkey negative
2-Amino-3-methyl-9-H-pyrido-[2,3-*b*]indole	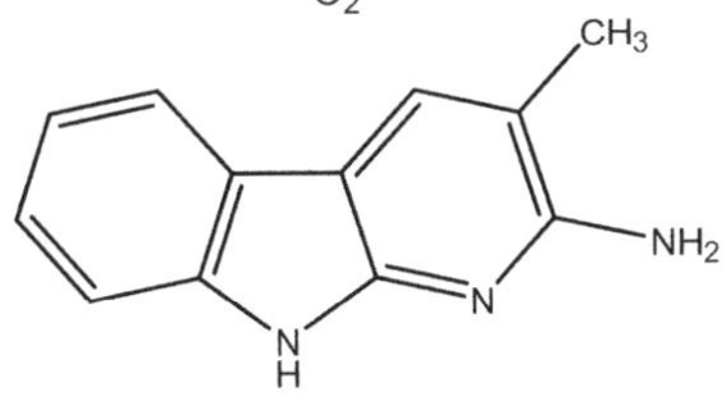				
2-Amino-3-methylimidazole[4,5-*f*]-quinoline (IQ)	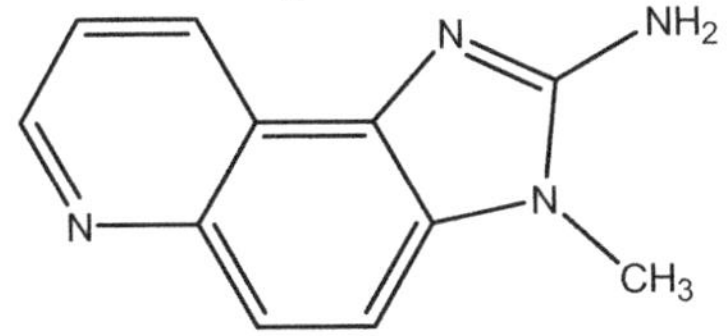	Gold			Cynom olgus 0.374 (no ubl+)
2-Amino-3,4-dimethylimidazole-[4,5-*f*]quinoline	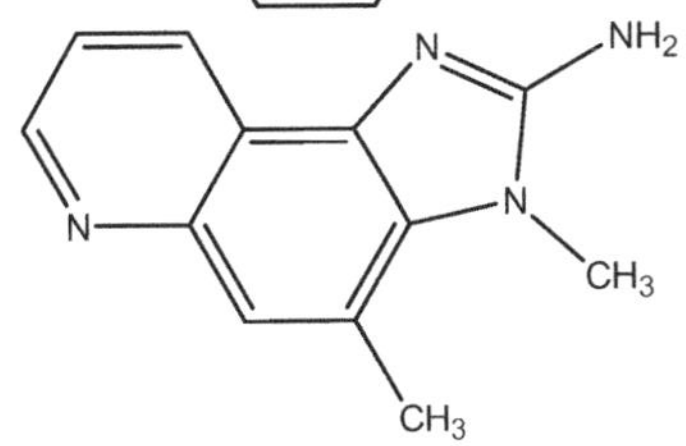				
3-Amino-1-methyl-5H-pyrido-[4,3-*b*]indole (Trp-P-2)	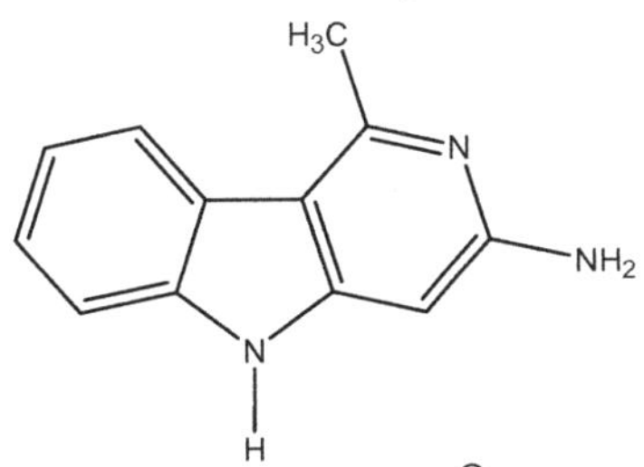	Gold	6.66 (ubl+)	12.6 (no ubl+)	
(5-nitro-2-thiazolyl)-1-imidazoline -2-thione(Niridazole)	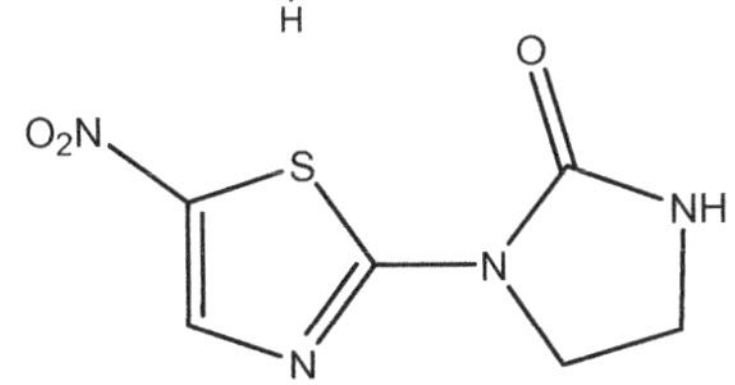		(ubl)	(ubl)	hamster

(continued)

Table B.1 (Continued)

2-Amino-4-(5-nitro-2-furyl)-thiazole	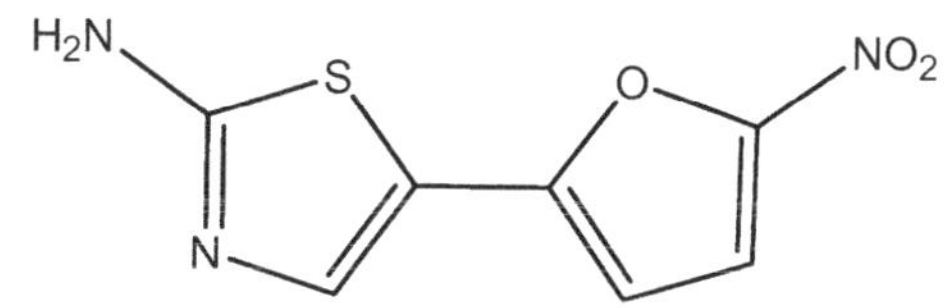	Gold	5.85 (ubl+)	7.87 (no ubl+)	
2-(2-Formylhydrazino)-4-(5-nitro-2-furyl)thiazole	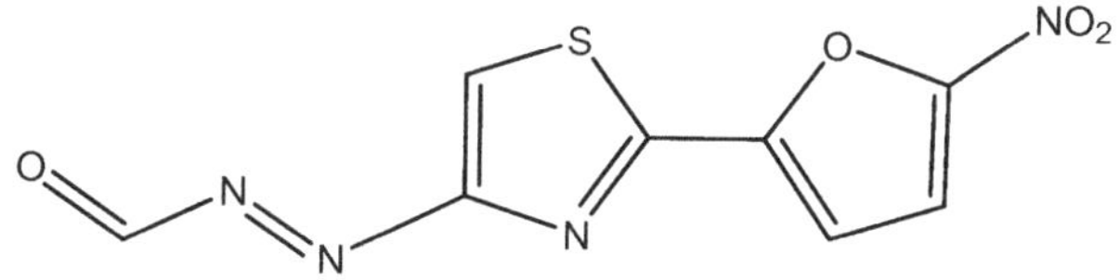	2B	5.06 (no ubl+)	10.8 (no ubl+)	16.6 (ubl) Hamster
N-[4-(5-nitro-2-furyl)-2-thiazolyl]-acetamide	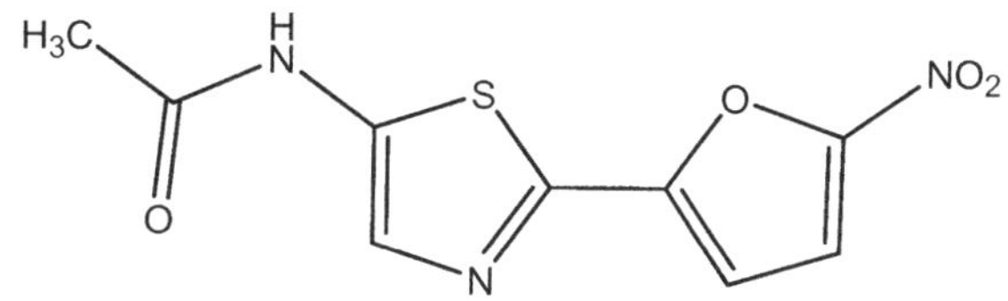	Gold	17.8 (no ubl+)	negative	8.91 (ubl+) Hamster
N-[4-(5-nitro-2-furyl)-2-thiazolyl]-formamide (FANFT)	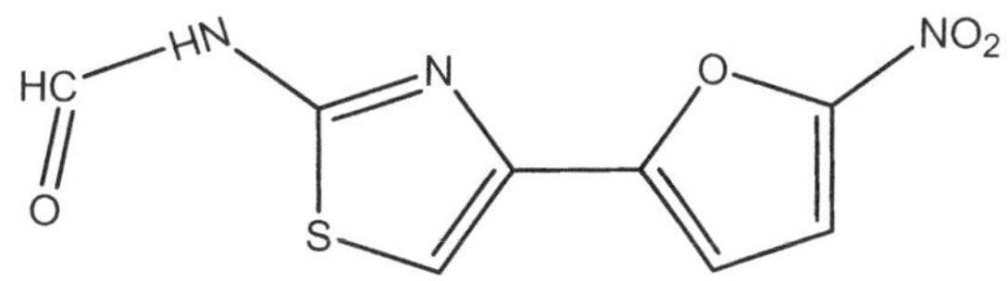	Gold	4.25 (ubl)	19.7 (ubl+)	9.17 (ubl+) Hamster
<u>**Nitrosamines and Nitrosamides**</u>					
N-Nitroso-di-n-butylamine	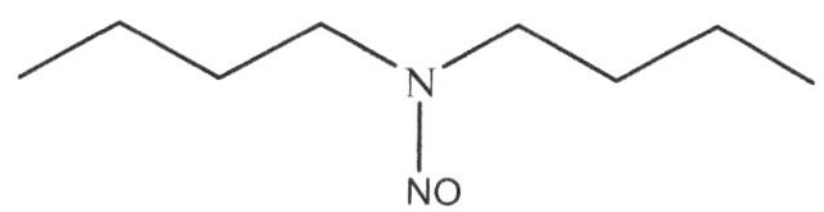	Gold	0.691 (ubl+)	1.09 (no ubl+)	
N-Butyl-N-(4-hydroxybutyl)-nitrosamine	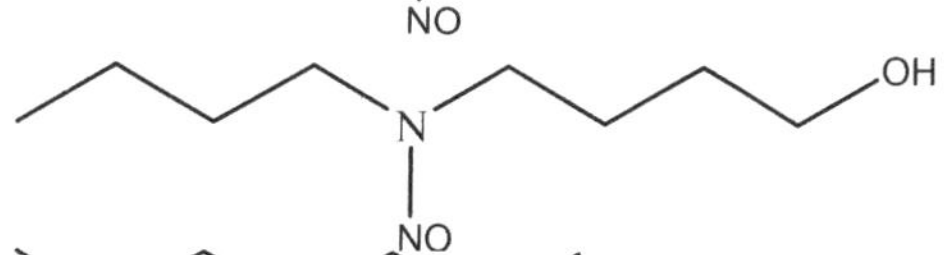	Gold	0.457 (ubl+)	Not tested	
N-Nitrosobis(2-oxopropyl)amine	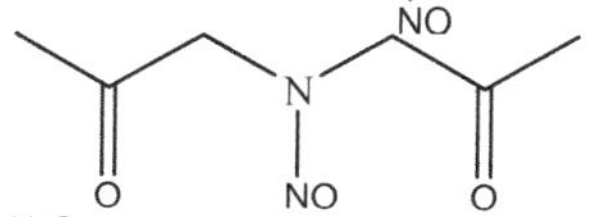	Gold	0.491 (ubl+)	Not tested	
N-Nitroso-N-methyldecylamine	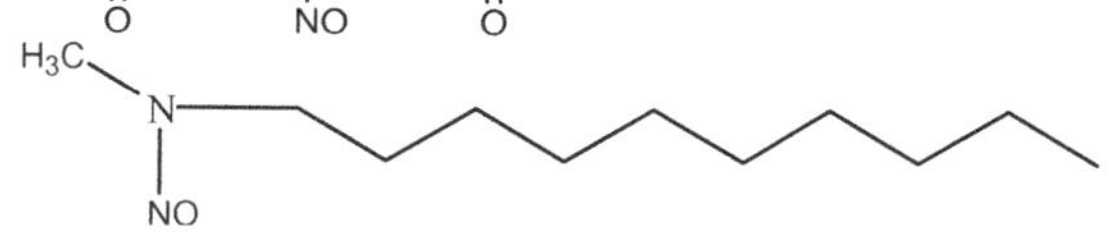	Gold	1.26 (ubl+)	Not tested	
N-Nitroso-N-methyl-N-dodecylamine	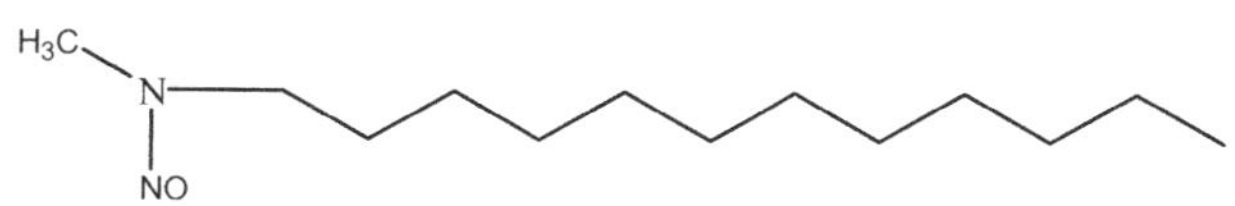	Gold	0.537 (ubl+)	Not tested	

(continued)

Table B.1 (Continued)

N-Nitroso-N-methyl-N-tetradecylamine	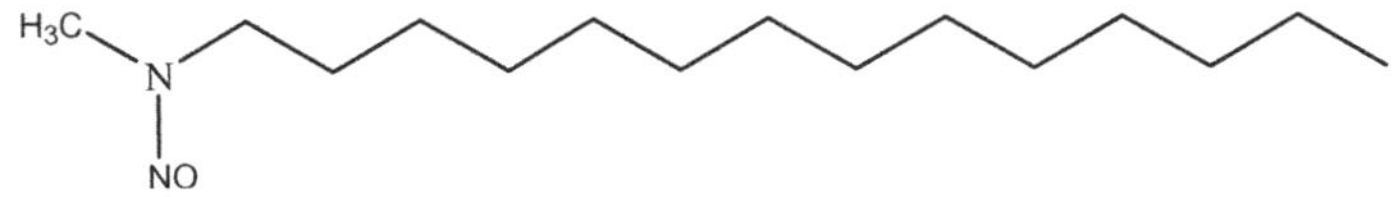	Gold	1.65 (ubl+)	Not tested	
N-Nitrosodiphenylamine	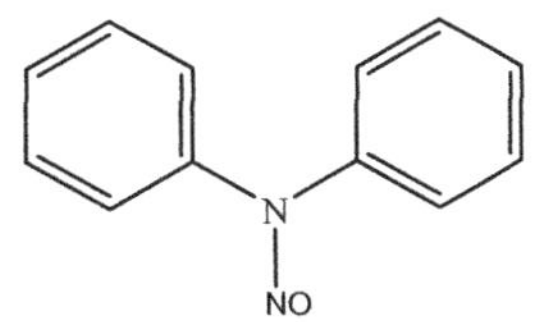	Gold	167 (ubl)	Negative	
N-Methyl-N-nitrosourea	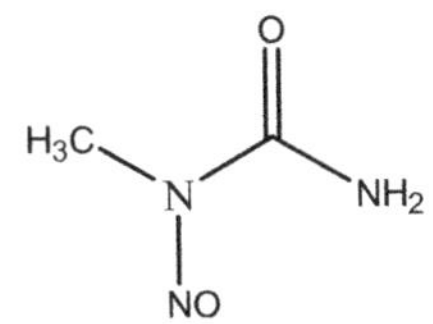	Gold	0.0927 (no ubl+)	1.23 (no ubl+)	Rhesus 7.08 (ubl??)
Halogenated HC					
1,3-dichloropropene, technical grade	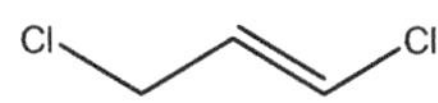		94 (no ubl+)	49.6 (ubl+)	
1,3-dichloropropene, reagent grade			Negati ve		
2,2-Bis(bromomethyl)-1,3-propanediol, technical grade	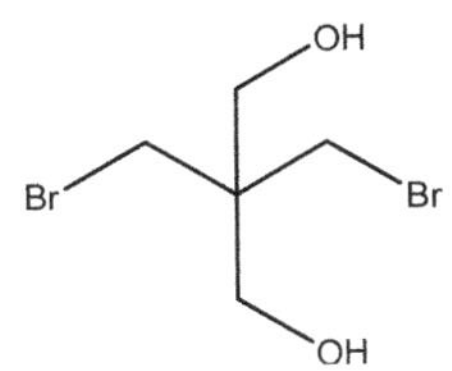	Gold	111 (ubl+)	137 (no ubl+)	
Miscellaneous					
Allyl isothiocyanate	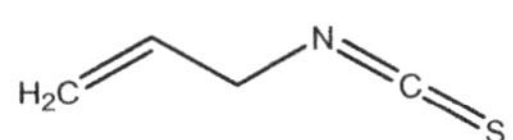	IARC ?	96 (ubl)	Negative	
α-Toluenesulfonamide	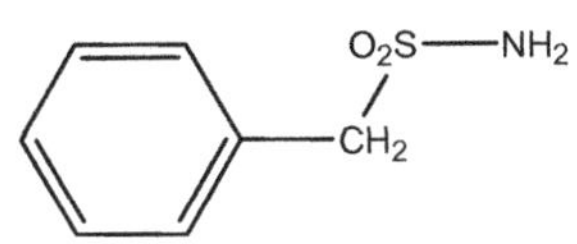	Gold	3960 (ubl)	Not tested	
Arsenic		1			

(continued)

Table B.1 (Continued)

Cyclophosphamide	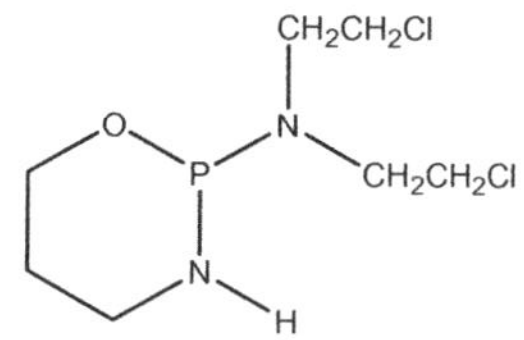	Gold	2.21 (ubl+)	5.96 (no ubl+)
Diethylene glycol	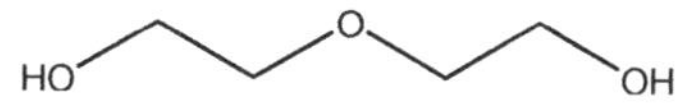	Gold	1660 (ubl)	Not tested
Fosetyl Al	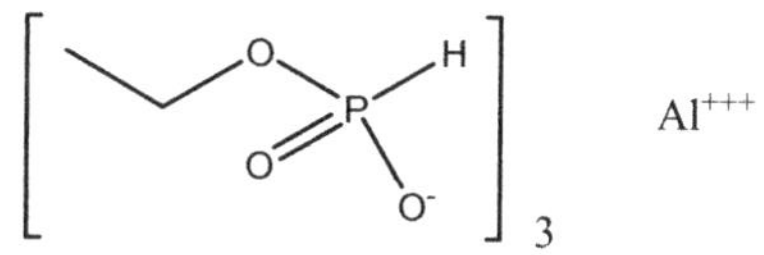	Gold	3660 (ubl)	Not tested
Nitrilotriacetic acid, sodium salt	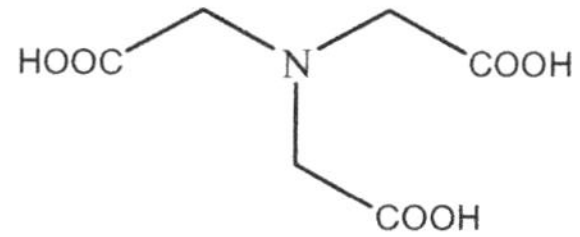	Gold	370 (ubl+)	2660 (ubl) acid, - salt
N-Oxydiethylene thiocarbamyl-N-oxydiethylene sulfenamide	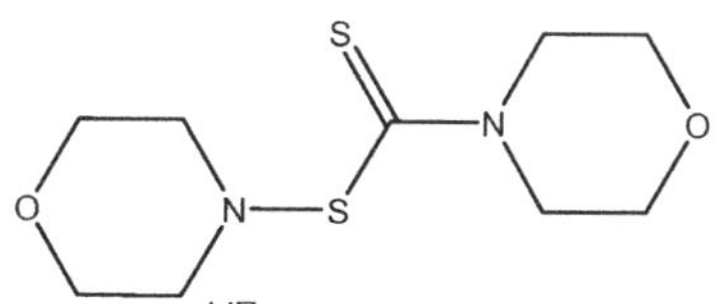	Gold	90.8 (ubl+)	Not tested
Phenylphenate - free acid	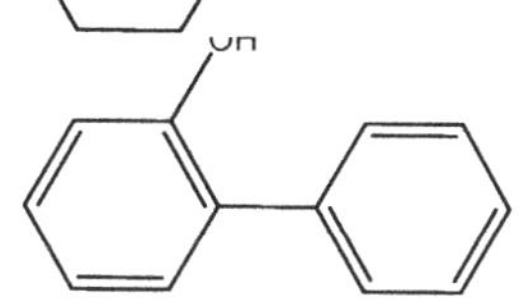	Gold	232 (ubl+)	Negative
1,2,4-trihydroxy-9,10-anthraquinone (Purpurin)	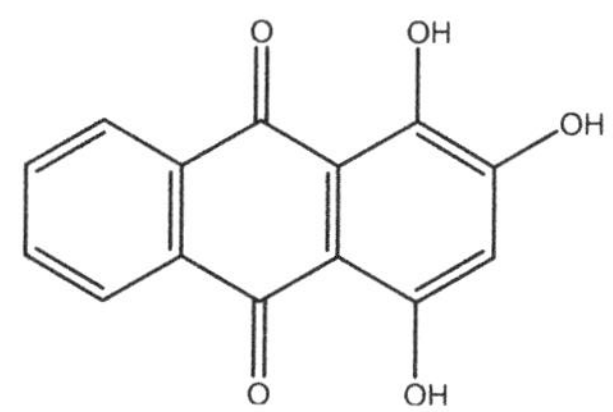	Gold	678 (ubl)	

(continued)

Table B.1 (Continued)

Quercetin	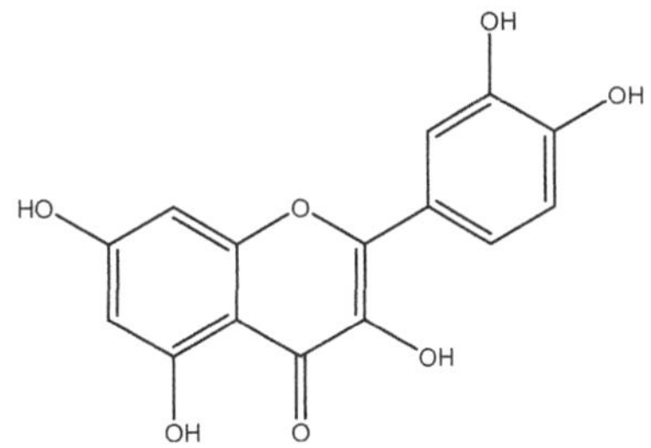	III	10.1 (ulbl+)	Negative	Hamster negative
<u>Mixtures</u>					
Bracken Fern					
Coal Tar Pitch					
Coal Tars					
Diesel engine exhaust					
Tobacco smoke					

ubl = observed urinary bladder cancer
ubl+ = observed urinary bladder cancer and other tumor sites
no ubl = bladder cancer not observed
no ubl+ = bladder cancer not observed but cancer in other sites

APPENDIX C

CHEMICAL CARCINOGENS WITH TD_{50}S BELOW 10 MG/KG DAY^{-1}

Table C.1
List of carcinogens that have been shown to have TD_{50}s of less than 10 mg/kg in rats and/or mice

Chemical	TD_{50} in rat mg/kg day^{-1}	TD_{50} in mice mg/kg day^{-1}
Acetaldehyde methylformylhydrazone		2.51
Acetone[4-(5-nitro-2-furyl)-2-thiazolyl]hydrazone	6.05000	.
4-Acetylaminobiphenyl	1.18000	.
2-Acetylaminofluorene s	1.22	7.59
Acronycine	0.505	I
Acrylamide	3.81	.
Actinomycin D	0.00111	.
Aflatoxicol	0.00247	.
Aflatoxin B1 s	0.0032	
Aflatoxin, crude	0.00299	0.343
Aldrin		1.27
1-Allyl-1-nitrosourea	0.341	.
2-Amino-5-(5-nitro-2-furyl)-1,3,4-oxadiazole	3.67	.
2-Amino-5-(5-nitro-2-furyl)-1,3,4-thiadiazole	0.662	.
2-Amino-4-(5-nitro-2-furyl)thiazole	5.85	7.87
2-Amino-4-(p-nitrophenyl)thiazole		9.95
o-Aminoazotoluene	4.04	
4-Aminodiphenyl		2.1
4-Aminodiphenyl.HCl	0.98	
2-Aminodiphenylene oxide		4.24
3-Aminotriazole s	9.94	25.3
1-Amyl-1-nitrosourea	0.555	
Aristolochic acid, sodium salt (77% AA I, 21% AA II)	0.0141	
Aroclor 1254		9.58
Aroclor 1260	1.74	.
5-Azacytidine	0.17	0.0774
Azaserine	0.793	.
Azathioprine [s]	–	8.92
Azoxymethane	0.0466	.
1-Azoxypropane	0.000241	.
2-Azoxypropane	0.00268	.
Benzidine	1.73	.
Benzidine.2HCl	.	19.7
Benzo(*a*)pyrene	0.956	11
1,4-Benzoquinone	.	5.07
Benzotrichloride	.	0.396
Benzoyl hydrazine	.	9.59
3-Benzylsydnone-4-acetamide	4.24	10.2
Bis-1,2-(chloromethoxy)ethane	.	4.62
Bis-1,4-(chloromethoxy)-*p*-xylene	.	3.11
Bis-(chloromethyl)ether	0.00357	0.182

(continued)

Table C.1 (Continued)

Chemical	TD_{50} in rat mg/kg day^{-1}	TD_{50} in mice mg/kg day^{-1}
4-Bis(2-hydroxyethyl)amino-2-(5-nitro-2-thienyl)quinazoline	3.14	.
C.I. direct black 38	1.39	71.6
C.I. direct blue 6	1.73	.
Bromate, potassium [s]	9.82	53.8
C.I. direct brown 95	2.07	.
Budesonide	0.291	.
N-Butyl-*N*-(4-hydroxybutyl)nitrosamine	0.457	.
N-*n*-Butyl-*N*-nitrosourea	0.517	.
Cadmium chloride [s]	0.0114	–
Cadmium sulphate (1:1) [s]	0.0217	–
Carbon tetrachloride	2.29	150
Carboxymethylnitrosourea	4.31	.
Chlorambucil	0.896	0.133
Chlordane, technical grade	–	1.37
[4-Chloro-6-(2,3-xylidino)-2-pyrimidinylthio]acetic acid [s]	4.36	10.8
4-Chloro-6-(2,3-xylidino)-2-pyrimidinylthio(*N*-b-hydroxyethyl)acetamide	6.49	44.6
p-Chloroaniline.HCl	7.62	89.5
1-Chloroethylnitroso-3-(2-hydroxypropyl)urea	0.124	.
Chloromethyl methyl ether [s]	5.5	.
1-(4-Chlorophenyl)-1-phenyl-2-propynyl carbamate	8.78	.
Chloroprene	12.5	8.17
1-Chloropropene	.	5.05
Chlorozotocin	0.0375	.
Ciprofibrate	2.48	6.2
Citrinin	7.48	.
Clivorine	0.5	.
Cobalt sulfate heptahydrate	0.137	0.756
Crotonaldehyde	4.2	.
Cupferron	8.35	585
Cyclophosphamide [s]	2.21	5.96
Cytembena	2.77	–
Dacarbazine	0.71	0.966
4,6-Diamino-2-(5-nitro-2-furyl)-*S*-triazine	1.71	.
2,4-Diaminotoluene	2.47	26.7
2,4-Diaminotoluene.2HCl	4.42	203
Dibenz(*a,h*)anthracene	.	5.88
3-Dibenzofuranamine	2.48	.
1,2-Dibromo-3-chloropropane	0.259	2.72
Dibromodulcitol	8.37	11
1,2-Dibromoethane	1.52	7.45
1,3-Dibutyl-1-nitrosourea	4.28	.
Dichloroacetylene	3.58	0.574

(continued)

Table C.1 (Continued)

Chemical	TD_{50} in rat mg/kg day^{-1}	TD_{50} in mice mg/kg day^{-1}
trans-1,4-Dichlorobutene-2	.	1.52
1,2-Dichloroethane	8.04	101
Dichlorvos	4.16	70.4
Dieldrin [s]	–	0.912
N,N-Diethyl-4-(4′-[pyridyl-1′-oxide]azo)aniline	1.63	.
Diethylacetamide	8.85	.
Diethylstilbestrol	0.223	0.0391
Diglycidyl resorcinol ether, technical grade	3.78	24.3
1,2-Dihydro-2-(5-nitro-2-thienyl)quinazolin-4(3*H*)-one	1.53	.
2,5-Dimethoxy-4′-aminostilbene	0.721	95.9
3,3′-Dimethoxybenzidine.2HCl	1.04	–
5,6-Dimethoxysterigmatocystin	0.364	.
N,N-Dimethyl-4-aminoazobenzene [s]	3.31	.
4,6-Dimethyl-2-(5-nitro-2-furyl)pyrimidine	1.39	.
Dimethylaminoethylnitrosoethylurea, nitrite salt	0.704	.
7,12-Dimethylbenz(*a*)anthracene	.	0.084
3,3′-Dimethylbenzidine.2HCl	0.629	28.6
Dimethylcarbamyl chloride [s]	.	5.37
1,1-Dimethylhydrazine [s]	–	3.96
1,2-Dimethylhydrazine.2HCl [s]	.	0.114
2-(2,2-Dimethylhydrazino)-4-(5-nitro-2-furyl)thiazole	0.41	.
Dimethylnitramine	0.547	.
Dinitrosohomopiperazine	0.0615	.
Dinitrosopiperazine	.	3.6
2,6-Dinitrotoluene	0.292	.
2,4-Dinitrotoluene (containing 1.0-1.5% 2,6-dinitrotoluene)	6.21	29.4
Dinitrotoluene, technical grade (2,4 (77%)- and 2,6 (19%)-)	8.02	.
Dipentylnitrosamine	4.03	.
a-Ecdysone	.	0.0358
Enovid	.	0.279
Epichlorohydrin	2.96	–
Estradiol mustard	–	1.45
Ethinyl estradiol	0.2	.
Ethionine	4.97	.
dl-Ethionine	9.11	71.4
Z-Ethyl-*O,N,N*-azoxyethane	0.022	.
Z-Ethyl-*O,N,N*-azoxymethane	0.0189	.
N-Ethyl-*N*-formylhydrazine	.	2.8
N-Ethyl-*N*′-nitro-*N*-nitrosoguanidine	.	2.84
1-Ethyl-1-nitrosourea	0.948	.
Ethylene imine	.	0.377
Ethylene thiourea	8.13	23.5
Ethylhydrazine.HCl	.	6.56
1-Ethylnitroso-3-(2-hydroxyethyl)-urea	0.522	.

(continued)

Table C.1 (Continued)

Chemical	TD_{50} in rat mg/kg day^{-1}	TD_{50} in mice mg/kg day^{-1}
1-Ethylnitroso-3-(2-oxopropyl)-urea	0.181	.
Ethylnitrosocyanamide	3.68	.
N-(2-Fluorenyl)-2,2,2-trifluoroacetamide	1.62	.
4′-Fluoro-4-aminodiphenyl	.	1.14
N-4-(4′-Fluorobiphenyl)acetamide	1.01	.
2-Fluoroethyl-nitrosourea	0.125	.
5-Fluorouracil	–	2.96
Formaldehyde [s]	1.35	43.9
Formic acid 2-[4-(5-nitro-2-furyl)-2-thiazolyl]hydrazide [s]	5.06	10.8
Fumonisin B_1	1.16	.
Furan	0.396	2.72
Glu-P-1	4.69	5.4
Glu-P-2	42.3	16
Glycidol [s]	4.28	34.7
HCDD mixture	0.000596	0.00143
Heptachlor	–	1.21
Hexachlorobenzene [s]	3.86	65.1
a-1,2,3,4,5,6-Hexachlorocyclohexane	11.2	6.62
Hexamethylphosphoramide	0.0387	.
Hexanal methylformylhydrazone	.	2.33
N-Hexylnitrosourea	0.513	.
Hydrazine [s]	0.613	2.93
Hydrazine sulfate [s]	40.8	7.59
2-Hydrazino-4-(*p*-aminophenyl)thiazole	1.03	11.3
2-Hydrazino-4-(5-nitro-2-furyl)thiazole	3.19	16.4
2-Hydrazino-4-(*p*-nitrophenyl)thiazole	3.21	10.6
Hydrazobenzene	5.59	26
N-Hydroxy-2-acetylaminofluorene [s]	0.988	6.23
1-(2-Hydroxyethyl)-nitroso-3-ethylurea	0.562	.
1-(2-Hydroxyethyl)-1-nitrosourea	0.244	0.818
4-(2-Hydroxyethylamino)-2-(5-nitro-2-thienyl)quinazoline	1.87	.
2-Hydroxyethylhydrazine [s]	.	0.397
1-(3-Hydroxypropyl)-1-nitrosourea	0.978	.
Indolidan	2.01	.
Indomethacin	1.15	.
IQ [s]	0.812	19.6
IQ.HCl	3.29	.
Isophosphamide	0.739	5.06
Kepone	2.96	0.982
Lasiocarpine	0.389	.
MeA-a-C acetate	6.7	22.2
MeIQ	.	12.3
MeIQx	1.66	24.3
Melphalan [s]	0.0938	0.15

(continued)

Table C.1 (Continued)

Chemical	TD_{50} in rat mg/kg day^{-1}	TD_{50} in mice mg/kg day^{-1}
Mercuric chloride	3.12	–
Mercurymethyl chloride	–	1.45
Metepa	4.46	.
Methapyrilene.HCl [s]	9.13	.
Methidathion	.	6.04
Methimazole	1.14	.
Methyl clofenapate	4.78	2.51
1-Methyl-1,4-dihydro-7-[2-(5-nitrofuryl)vinyl]-4-oxo-1,8-naphthyridine-3-carboxylate, potassium	.	8.03
3′-Methyl-4-dimethylaminoazobenzene [s]	3.28	.
N-Methyl-*N*,4-dinitrosoaniline	1.3	.
N-Methyl-*N*-formylhydrazine [s]	.	1.37
N-Methyl-*N′*-nitro-*N*-nitrosoguanidine [s]	0.803	2.03
2-Methyl-1-nitroanthraquinone	84.8	1.56
4-Methyl-1-[(5-nitrofurfurylidene)amino]-2-imidazolidinone	5.34	.
4-(4-*N*-Methyl-*N*-nitrosaminostyryl)quinoline	0.699	.
N-Methyl-*N*-nitrosobenzamide	3.23	.
N-(*N*-Methyl-*N*-nitrosocarbamoyl)-*l*-ornithine	0.787	.
3-Methylbutanal methylformylhydrazone	.	2.23
3-Methylcholanthrene [s]	0.491	.
4,4′-Methylene-bis(2-methylaniline)	7.38	.
Methylhydrazine [s]	.	7.55
Methylhydrazine sulfate	.	2.72
4-(Methylnitrosamino)-1-(3-pyridyl)-1-butanol	0.103	.
4-(Methylnitrosamino)-1-(3-pyridyl)-1-(butanone) [s]	0.0999	.
Methylnitrosocyanamide	0.48	.
Mirex	1.77	1.45
Mirex, photo-	1.46	.
Mitomycin-C	0.00102	.
Molybdenum trioxide	–	6.13
Monoacetyl hydrazine	.	9.85
Monocrotaline	0.94	.
4-Morpholino-2-(5-nitro-2-thienyl)quinazoline	5.03	.
l-5-Morpholinomethyl-3-[(5-nitrofurfurylidene)amino]-2-oxazolidinone.HCl	6.33	.
5-(5-Nitro-2-furyl)-1,3,4-oxadiazole-2-ol	8.61	.
N-[5-(5-Nitro-2-furyl)-1,3,4-thiadiazol-2-yl]acetamide	8.84	6.74
4-(5-Nitro-2-furyl)thiazole	7.68	.
N-[4-(5-Nitro-2-furyl)-2-thiazolyl]formamide [s]	4.25	19.7
3-Nitro-3-hexene	8.66	0.346
5-Nitroacenaphthene [s]	8.67	45.8
2-Nitrobutane	2.86	.
2-Nitrofluorene	0.285	.

(continued)

Table C.1 (Continued)

Chemical	TD_{50} in rat mg/kg day^{-1}	TD_{50} in mice mg/kg day^{-1}
1-[(5-Nitrofurfurylidene)amino]-2-imidazolidinone	5.26	.
Nitrogen mustard	0.0114	.
Nitrogen mustard *N*-oxide	0.764	.
3-Nitropentane	2.37	.
1-Nitropyrene	3.33	.
8-Nitroquinoline	9.82	.
Nitroso-Baygon	0.364	.
N-Nitroso-bis-(4,4,4-trifluoro-*N*-butyl)amine	0.748	.
1-Nitroso-5,6-dihydrothymine	–	.
1-Nitroso-5,6-dihydrouracil	0.0983	.
N-Nitroso-2,3-dihydroxypropyl-2-hydroxypropylamine [s]	0.0535	.
Nitroso-2,3-dihydroxypropyl-2-oxopropylamine [s]	0.0352	.
N-Nitroso-2,3-dihydroxypropylethanolamine [s]	5.98	.
1-Nitroso-3,5-dimethyl-4-benzoylpiperazine	9.66	.
1-Nitroso-1-hydroxyethyl-3-chloroethylurea	0.356	.
1-Nitroso-1-(2-hydroxypropyl)-3-chloroethylurea	0.873	.
N-Nitroso-(2-hydroxypropyl)-(2-hydroxyethyl)amine	1.02	.
N-Nitroso-3-hydroxypyrrolidine	7.65	.
N-Nitroso-*N*-isobutylurea	4.73	.
N-Nitroso-*N*-methyl-*N*-dodecylamine	0.537	.
N-Nitroso-*N*-methyl-4-fluoroaniline	0.255	.
Nitroso-*N*-methyl-*N*-(2-phenyl)ethylamine	0.00998	.
N-Nitroso-*N*-methyl-*N*-tetradecylamine	1.65	.
N-Nitroso-*N*-methyldecylamine	1.26	.
N-Nitroso-*N*-methylurea [s]	0.0927	1.23
3-Nitroso-2-oxazolidinone	0.385	.
Nitroso-2-oxopropylethanolamine [s]	1.8	.
di(*N*-Nitroso)-perhydropyrimidine	0.166	.
Nitroso-1,2,3,6-tetrahydropyridine	0.0601	.
N-Nitroso(2,2,2-trifluoroethyl)ethylamine	2.52	.
N-Nitroso-2,2,4-trimethyl-1,2-dihydroquinoline polymer	3.31	.
1-Nitroso-3,4,5-trimethylpiperazine [s]	0.151	.
N-Nitrosoallyl-2,3-dihydroxypropylamine	0.825	.
N-Nitrosoallyl-2-hydroxypropylamine	0.877	.
N-Nitrosoallyl-2-oxopropylamine [s]	0.335	.
N-Nitrosoallylethanolamine	0.491	.
Nitrosoamylurethan	1.01	.
N-Nitrosobenzthiazuron	1.13	.
N-Nitrosobis(2-hydroxypropyl)amine	0.846	.
N-Nitrosobis(2-oxopropyl)amine	0.491	.
Nitrosodibutylamine	0.691	1.09
N-Nitrosodiethanolamine	3.17	.
N-Nitrosodiethylamine [s]	0.0265	.
N-Nitrosodimethylamine [s]	0.0959	0.189

(continued)

Table C.1 (Continued)

Chemical	TD$_{50}$ in rat mg/kg day^{-1}	TD$_{50}$ in mice mg/kg day^{-1}
N-Nitrosodipropylamine [s]	0.186	.
Nitrosoethylmethylamine	0.0503	.
Nitrosoethylurethan	0.0904	.
Nitrosoheptamethyleneimine	0.0378	.
N-Nitrosohexamethyleneimine	.	0.528
N-Nitrosomethyl-2,3-dihydroxypropylamine [s]	0.646	.
N-Nitrosomethyl-(2-hydroxyethyl)amine	1.29	.
N-Nitrosomethyl-(3-hydroxypropyl)amine	1.66	.
N-Nitrosomethyl-2-hydroxypropylamine	0.0463	.
N-Nitrosomethyl(2-oxopropyl)amine	0.0172	.
N-Nitrosomethyl-(2-tosyloxyethyl)amine	4.8	.
2-Nitrosomethylaminopyridine	0.214	.
Nitrosomethylaniline	0.142	.
Nitrosomethylundecylamine	2.37	.
N-Nitrosomorpholine [s]	0.109	.
N′-Nitrosonornicotine-1-*N*-oxide [s]	0.876	.
N-Nitrosopiperazine	8.78	.
N-Nitrosopiperidine [s]	1.43	1.3
N-Nitrosopyrrolidine [s]	0.799	0.679
N-Nitrosothialdine	0.483	.
N-Nitrosothiomorpholine	5.39	.
Nonabromobiphenyl	.	3.61
Norlestrin [s]	1.94	1.34
Ochratoxin A	0.103	6.41
N-(9-Oxo-2-fluorenyl)acetamide	6.17	.
4,4′-Oxydianiline	9.51	33.6
Ozone	–	1.88
Pentanal methylformylhydrazone	.	3.42
n-Pentylhydrazine.HCl	.	5.87
Petasitenine	0.922	.
Phenesterin	0.523	0.616
Phenobarbital [s]	–	7.37
Phenoxybenzamine.HCl	1.09	5.36
1-Phenyl-3,3-dimethyltriazene	2.31	.
PhIP.HCl	1.91	33.2
Phorbol	.	2.21
Polybrominated biphenyl mixture	0.322	0.332
Prednisolone	1.53	.
Procarbazine	4.01	.
Procarbazine.HCl [s]	0.351	0.558
Propane sultone	3.84	.
b-Propiolactone	1.46	1.24
N-*N*′-Propyl-*N*-formylhydrazine	.	8.79
N-Propyl-*N*′-nitro-*N*-nitrosoguanidine	1.31	.

(continued)

Table C.1 (Continued)

Chemical	TD_{50} in rat mg/kg day^{-1}	TD_{50} in mice mg/kg day^{-1}
N-Propyl-*N*-nitrosourea	3.77	.
C.I. acid red 114	3.89	.
Reserpine	0.306	5.02
Selenium diethyldithiocarbamate	.	1.49
Selenium sulfide	8.01	69.3
Senkirkine	1.7	.
Sodium dichromate	4.64	.
Sterigmatocystin [s]	0.152	0.908
Streptozotocin	0.963	0.272
Strobane	.	0.884
Symphytine	1.91	.
T-2 toxin	.	0.883
Tamoxifen citrate	3.71	4.39
2,3,7,8-Tetrachlorodibenzo-*p*-dioxin	0.0000457	0.000156
Tetranitromethane	0.447	1.19
Thio-tepa	0.164	0.223
Thioacetamide	11.5	8.81
4,4′-Thiodianiline	3.71	33.2
b-Thioguanine deoxyriboside	2.1	I
Toxaphene	–	9.09
Trenimon	0.00504	.
Triamcinolone acetonide	0.053	.
1,2,3-Trichloropropane	1.35	0.875
2,2,2-Trifluoro-*N*-[4-(5-nitro-2-furyl)-2-thiazolyl]acetamide	6.79	9.98
2,4,5-Trimethylaniline	33.6	6.13
Tris-1,2,3-(chloromethoxy)propane	.	3.44
Tris(2,3-dibromopropyl)phosphate	3.83	128
Trp-P-1 acetate	0.575	40.7
Trp-P-2 acetate	6.66	12.6
Vinyl carbamate	.	0.124
Vinyl fluoride	20	8.11

APPENDIX D

ADDITIONAL REACTIONS OF DISINFECTANTS WITH SUB-STRUCTURES OF NATURAL ORGANIC MATTER

ADDITIONAL REACTIONS OF DISINFECTANTS WITH SUB-STRUCTURES OF NATURAL ORGANIC MATTER

This appendix contains a summary of all model compound data collected as part of this project, but not presented in Chapter 5. While all of these helped to focus the authors' mechanistic projections, those that appear below are of lesser importance to the central thesis, and as a result are presented in an appendix. A large fraction of the model compounds used in this study were nitrogenous compounds and non-nitrogenous biopolymer constituents. The compounds presented below generally fall into one of the following categories:

- Aniline derivatives
- Amino sugars
- Pyridine and derivatives
- Syringyl units and derivatives

Much of the model compound data are shown in a multi-graphic format. These typically show chlorine demand, TOX formation, "unknown TOX" formation in a simple bar chart. Below this chart are pie charts showing the breakdown of TOX species for the individual compounds. This is helpful in comparing the spectrum of byproducts formed from the particular model compounds to those seen in real waters. Finally below these are the structures for the compounds. These data are presented in molar yields which are easier to use in interpreting the reaction mechanisms, than the %age of TOX as shown in the pie charts.

AMINO BENZENE COMPOUNDS

Figure D.1 shows these data for two simple aromatic amines: 2-aminopheno and 1,4-phenylenediamine. Both show moderate levels of reactivity somewhat lower than is typical of phenolic compounds. The aminophenol produces a substantial amount of small halogenated products, whereas the diamine produces very little of any TOX.

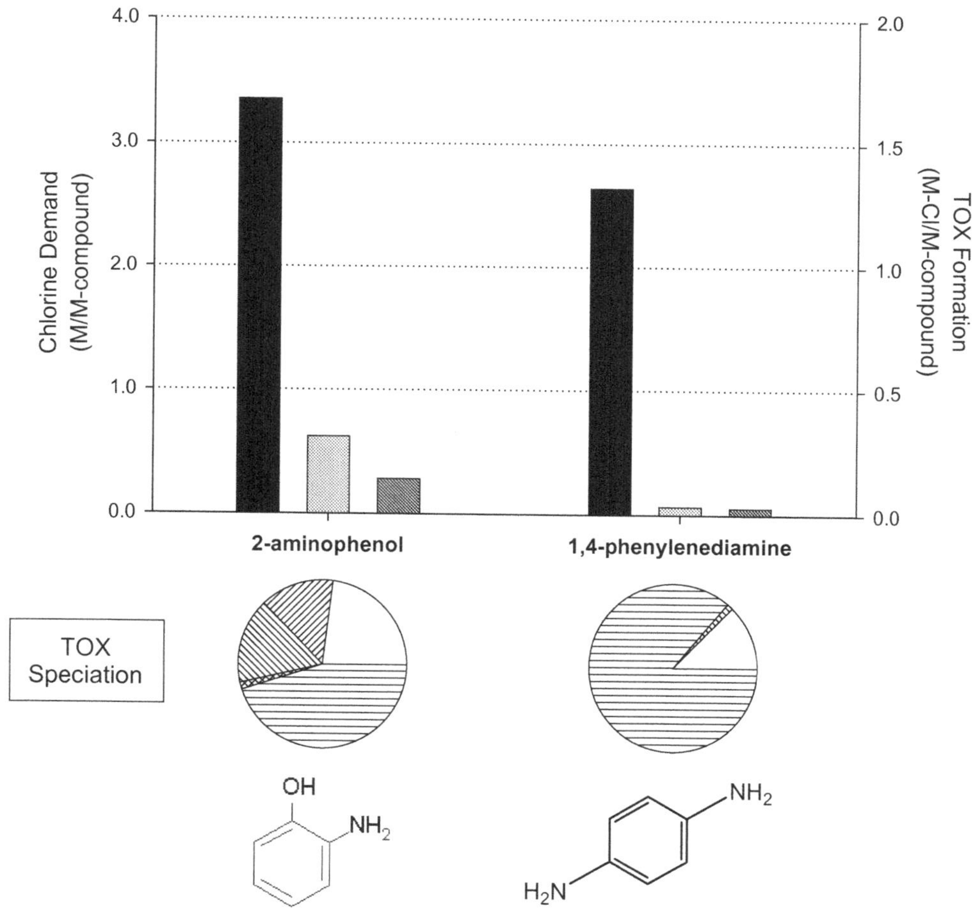

Figure D.1 Reactivity and DBP yield for 2-aminophenol and 1,4-phenlenediamine

Amino Sugars and Derivatives

Figure D.2 shows chlorine reactivity and byproduct data for a set of amino sugars and a structurally related compound [i.e., (+)-N-acetylmuramic acid, (-)-N-acetylneuraminic acid and coniferyl alcohol]. These data show a broad range of reactivity, with little occurring with the muramic acid, more with the neuraminic acid, and extensive degradation of the coniferyl alcohol. Very little identifiable TOX is formed from any of these, although the unidentified TOX can be substantial. Figure D.3 presents the concentrations of particular DBP groups and species formed from chlorination of these model compounds. Molar yields are all quite low, indicating that know DBP formation is result of a major degradation pathway.

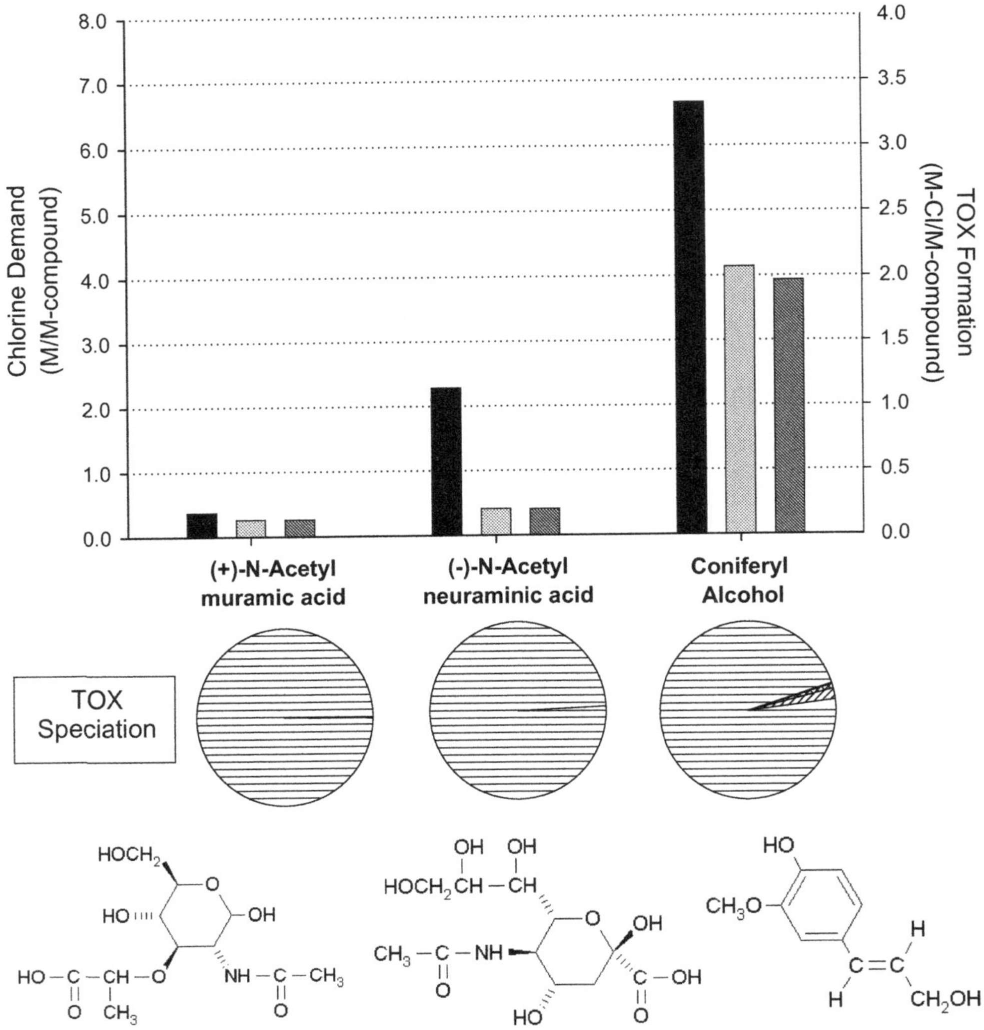

Figure D.2 Reactivity and DBP yield for (+)-N-acetylmuramic acid, (-)-N-acetylneuraminic acid and coniferyl alcohol

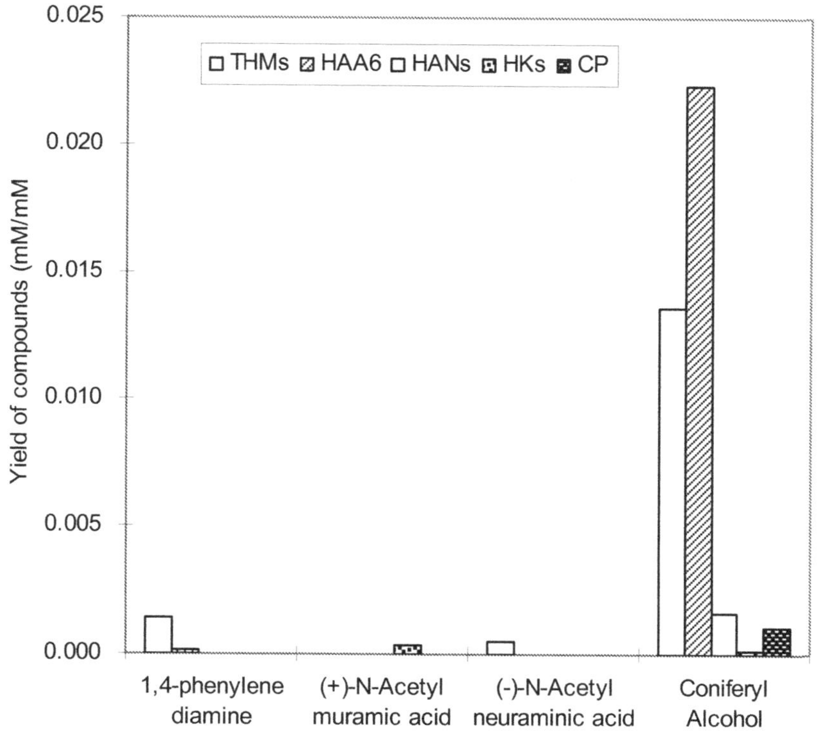

Figure D.3 Selected DBP yield for 1,4-phenylene diamime, (+)-N-acetylmuramic acid, (-)-N-acetylneuraminic acid and coniferyl alcohol

NITROGEN HETEROCYCLES

Figure D.4 shows chlorine reactivity and byproduct data for pyridine and a set of pyridine derivatives [i.e., pyridine, 4-hydroxypyridine, 4-pyridinealdoxime and 3-pyridinepropanol]. These data show a broad range of reactivity, with little reaction occurring except with the aldoxime. Very little identifiable TOX is formed from any of these, although the unidentified TOX can be substantial. Figure D.5 presents the concentrations of particular DBP groups and species formed from chlorination of these model compounds. Molar yields are all quite low, indicating that know DBP formation is result of a major degradation pathway.

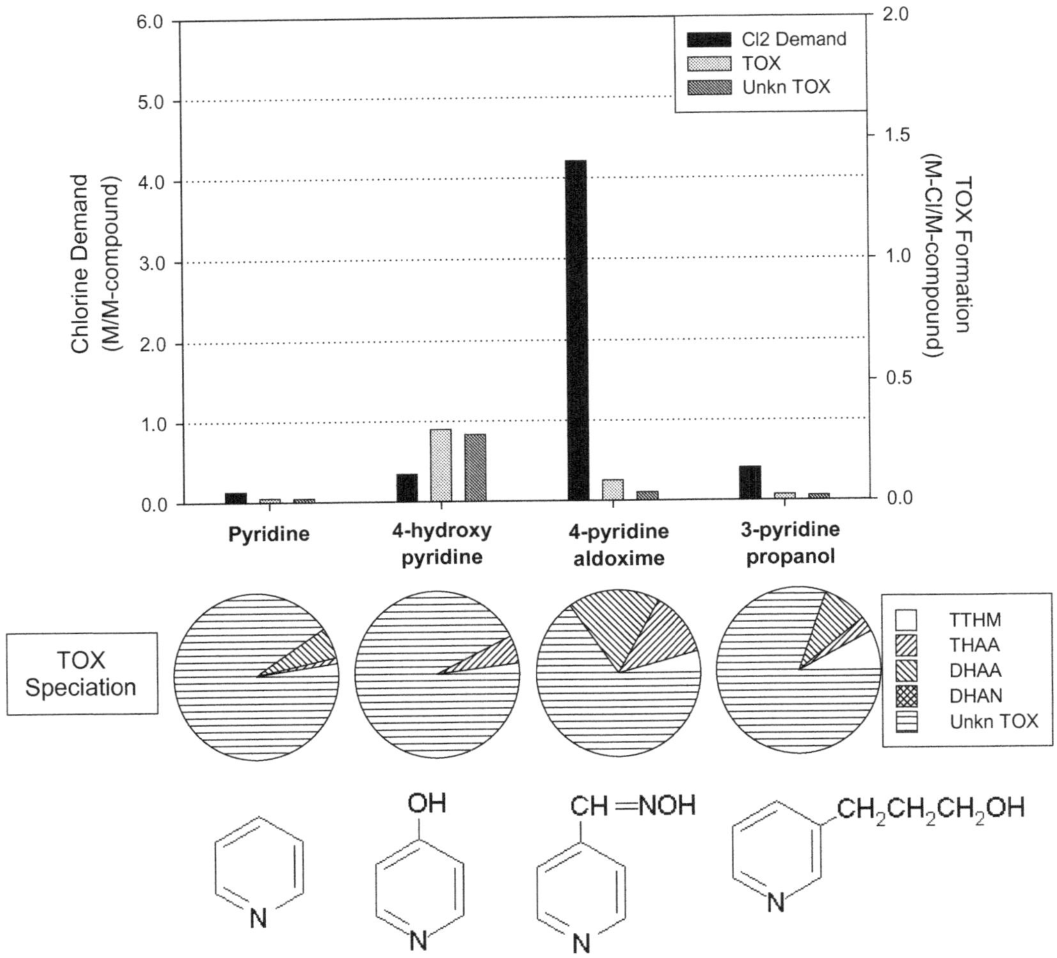

Figure D.4 Reactivity and DBP yield for pyridine, 4-hydroxypyridine, 4-pyridinealdoxime and 3-pyridinepropanol

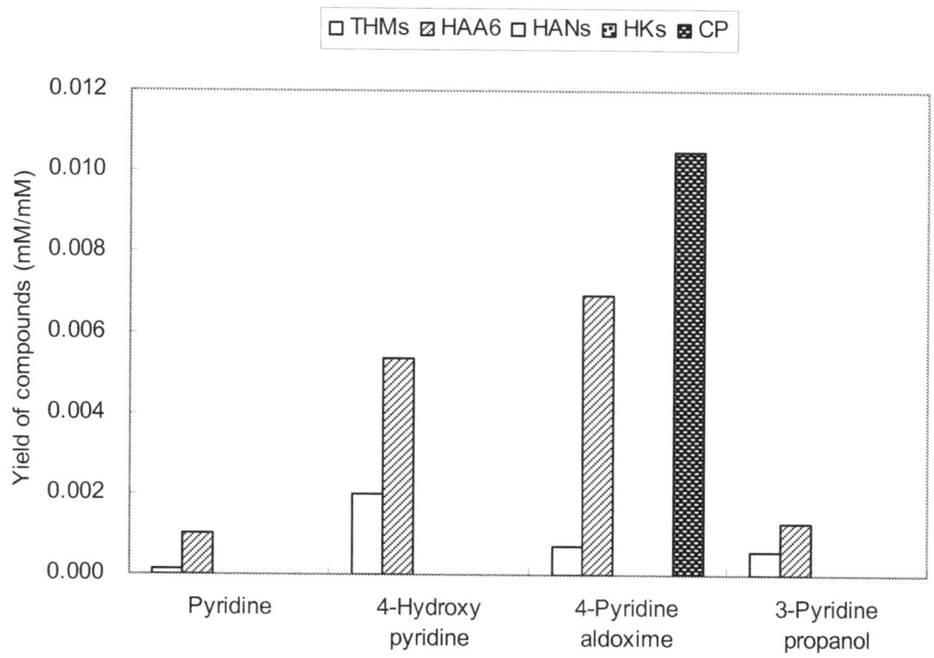

Figure D.5 Selected DBP yield for pyridine, 4-hydroxypyridine, 4-pyridinealdoxime and 3-pyridinepropanol

Figure D.6 shows chlorine reactivity and byproduct data for Nicotinic Acid and Derivatives (i.e., 2-hydroxy, 6-hydroxy, and nicotinamide). These data show little reactivity, except for the amide, which has a low-moderate level. Very little identifiable TOX is formed from any of these, and the overall TOX is rather small. Figure D.7 presents the concentrations of particular DBP groups and species formed from chlorination of these model compounds. Molar yields are all quite low, indicating that know DBP formation is result of a major degradation pathway.

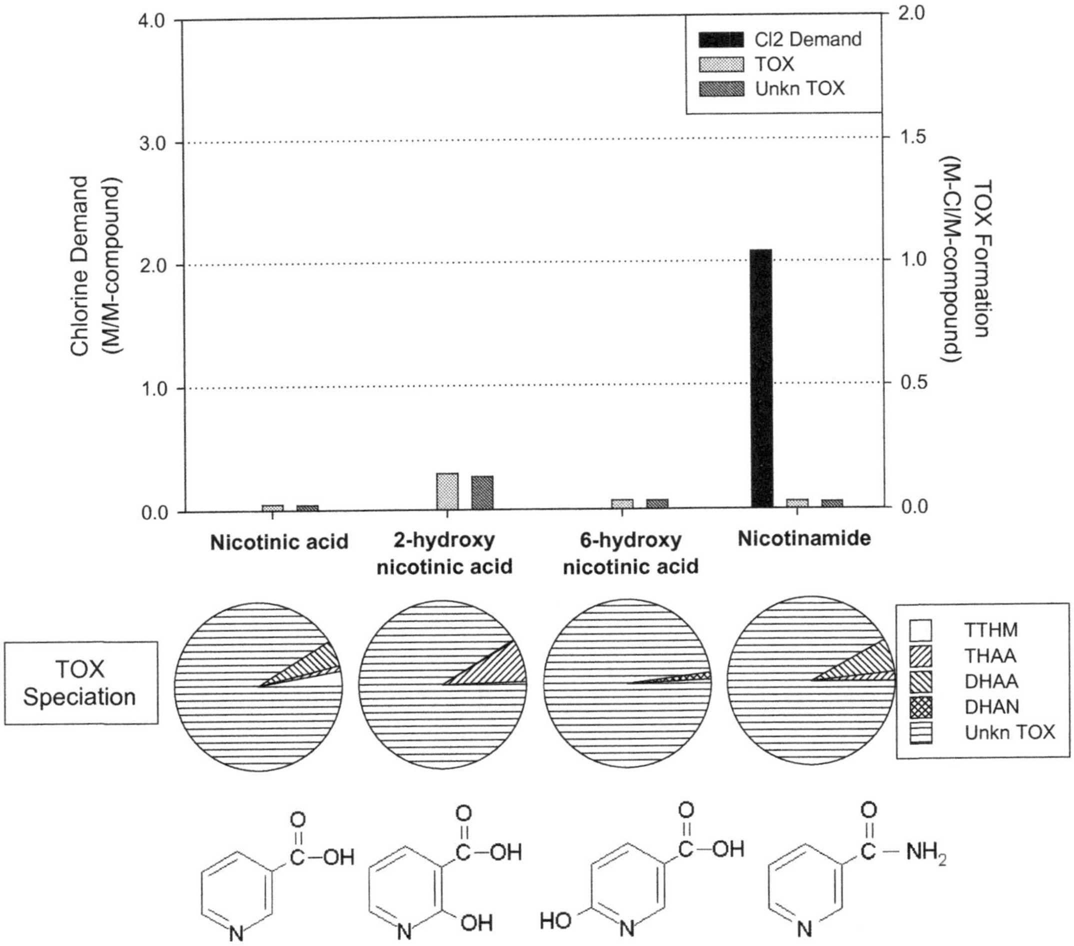

Figure D.6 Reactivity and DBP yield for nicotinic acid and derivatives

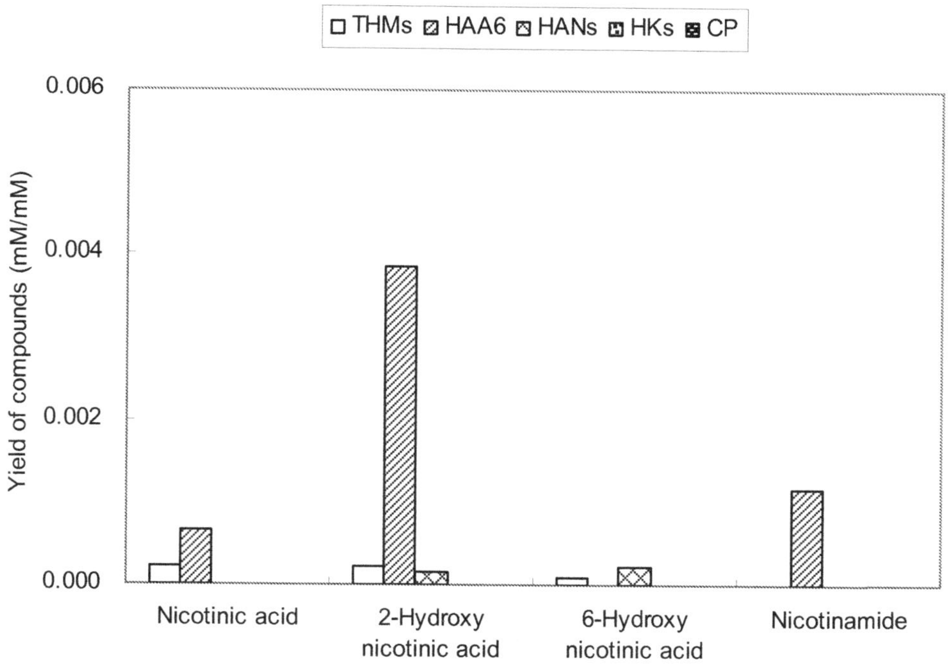

Figure D.7 Selected DBP yield for nicotinic acid and derivatives

TRIMETHOXY BENZOIC ACID DERIVATIVES

Figure D.8 shows chlorine reactivity and byproduct data for the first set of syringyl derivatives (i.e., the aldehyde, acid and ketone forms). The trimethoxyderivatives were expected to be far less reactive due to their lack of a ring hydroxyl group. This was, in fact, the case as all had chlorine demands in the range of 0-1 M/M. As expected the acetophenone produced about 12 % yield of chloroform and some trichloropropanone due to reaction with the side chain. These compounds show some TOX formation which is probably attributed to some ring chlorination, as there is very little general oxidation of the molecules.

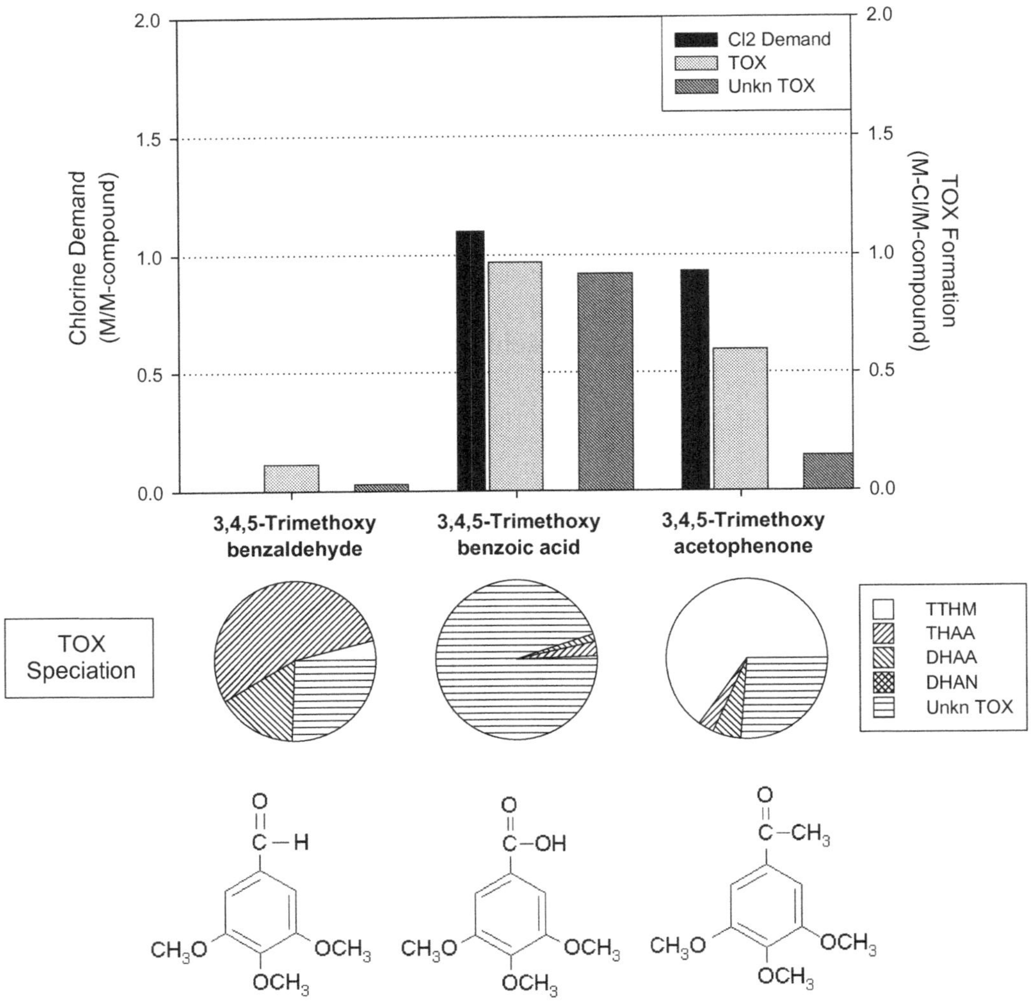

Figure D.8 Reactivity and chlorine incorporation for trimethoxy aromatic compounds with syringyl substitution patterns I

Figure D.9 presents the concentrations of particular DBP groups and species formed from chlorination of these model compounds. Molar yields are all quite low, indicating that know DBP formation is result of a major degradation pathway.

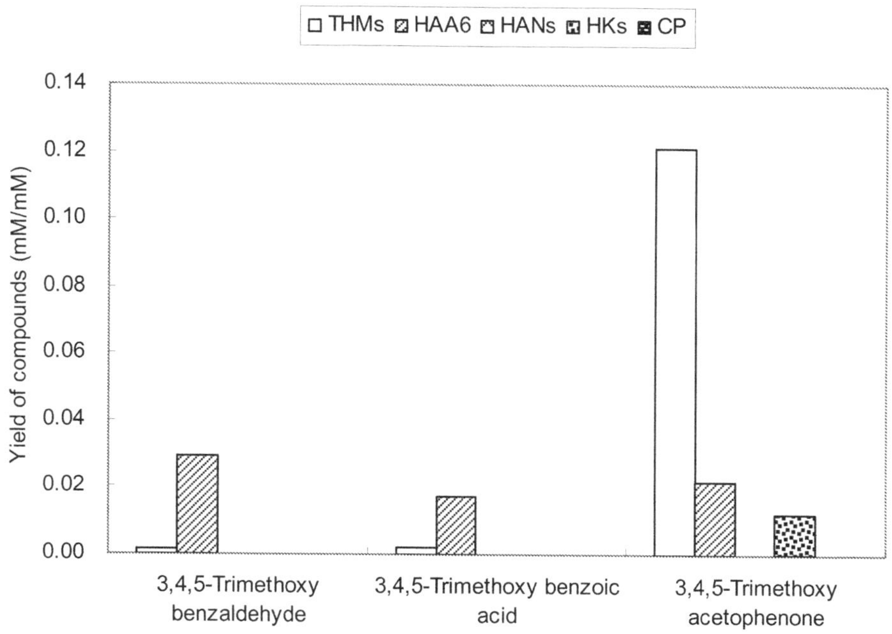

Figure D.9 Yields of selected DBPs for trimethoxy aromatic compounds with syringyl substitution patterns I

Figure D.10 shows chlorine reactivity and byproduct data for the second set of syringyl derivatives (i.e., the phenylacetic, benzamide, and propionate forms). All showed low to moderate reactivity which is typical of this group, except for the benzamide, which was highly reactive. This compound must have undergone some ring cleavage. Figure D.11 presents the concentrations of particular DBP groups and species formed from chlorination of these model compounds. Molar yields are all quite low, indicating that know DBP formation is result of a major degradation pathway. However, the propionate showed TOX levels that were nearly identical to the chlorine demand. This suggests almost no oxidation, and a major product or set of products that are among the unknown TOX.

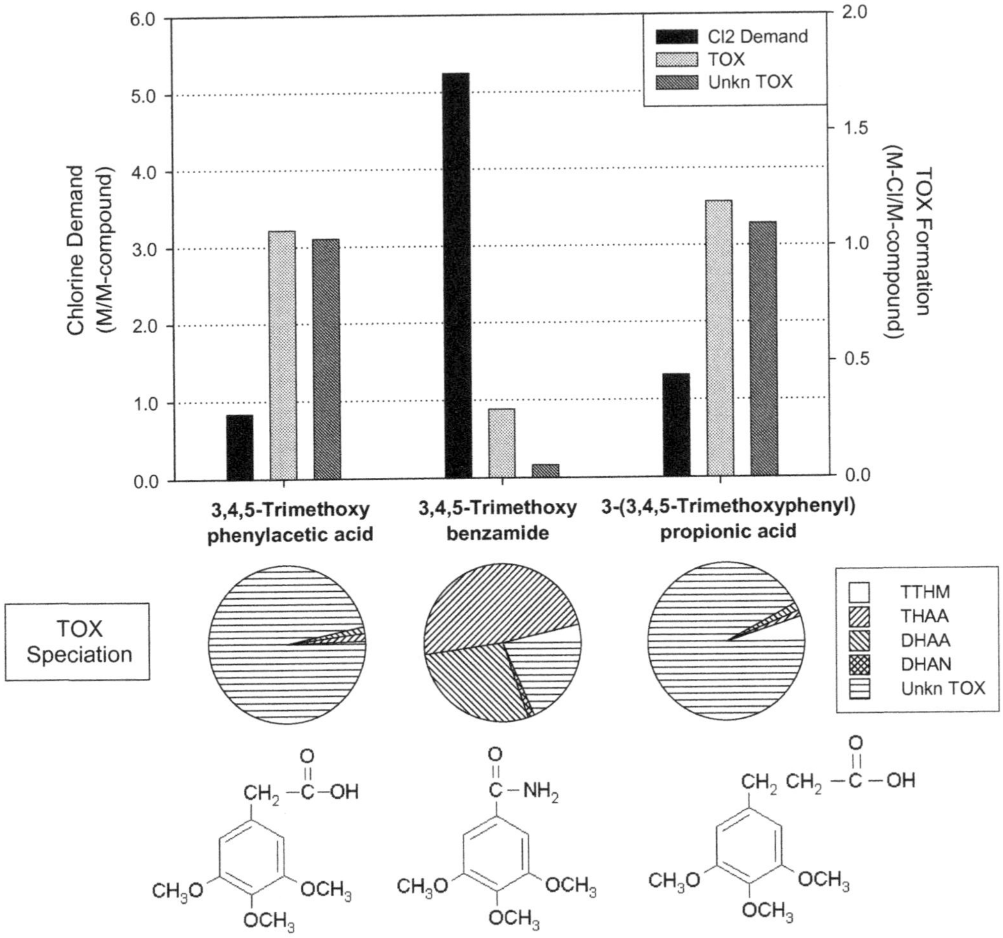

Figure D.10 Reactivity and chlorine incorporation for trimethoxy aromatic compounds with syringyl substitution patterns II

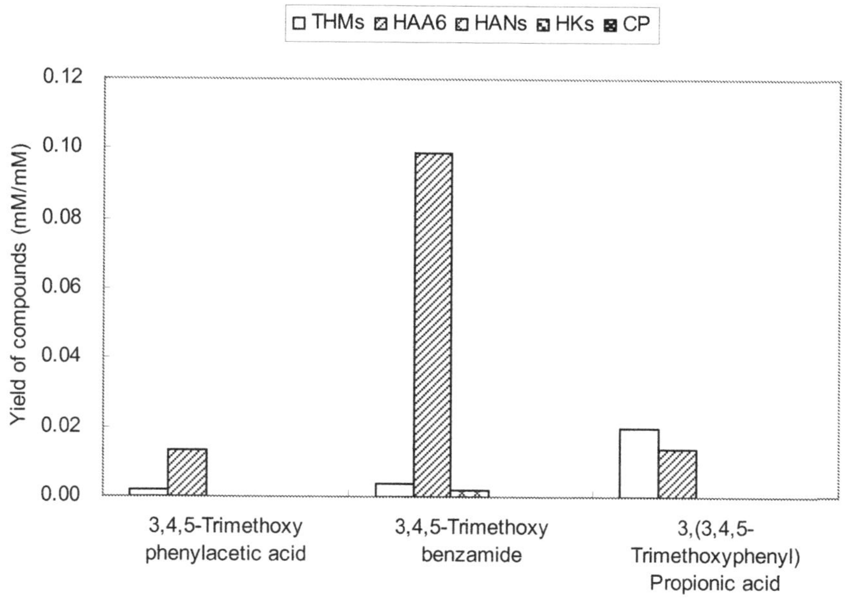

Figure D.11 Yields of selected DBPs for trimethoxy aromatic compounds with syringyl substitution patterns II

Figure D.12 shows chlorine reactivity and byproduct data for the third set of syringyl derivatives (i.e., the alcohol, cyanide, and two with multifunctional alkyl chains). All showed moderate reactivity probably reflecting at least some reaction with the substituent groups. Figure D.13 presents the concentrations of particular DBP groups and species formed from chlorination of these model compounds. Molar yields are all quite low, except for one compound. The trimethoxy benzoylacetate ester exhibited about a 40% molar yield of haloacetic acids.

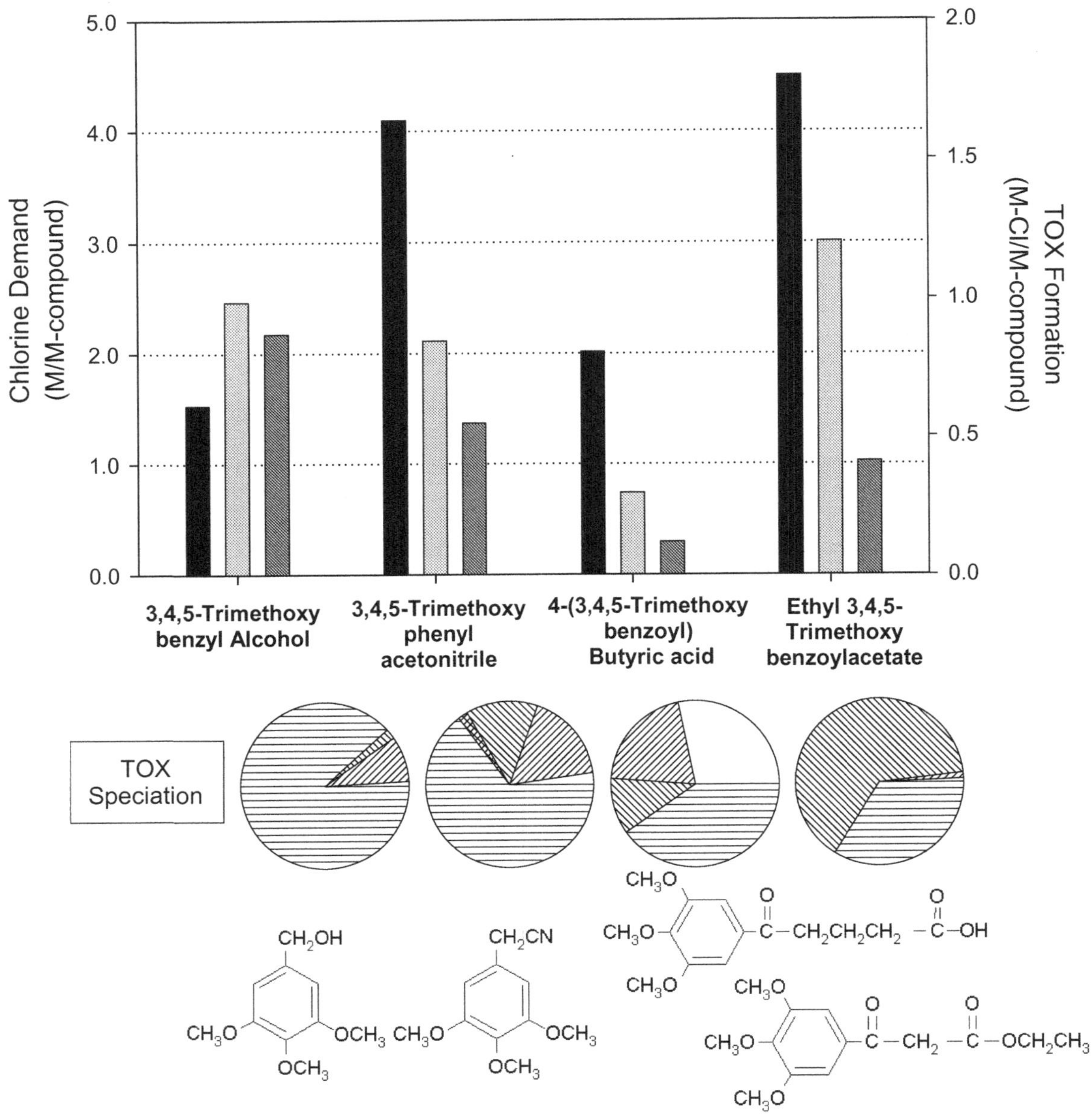

Figure D.12 Reactivity and chlorine incorporation for trimethoxy aromatic compounds with syringyl substitution patterns III

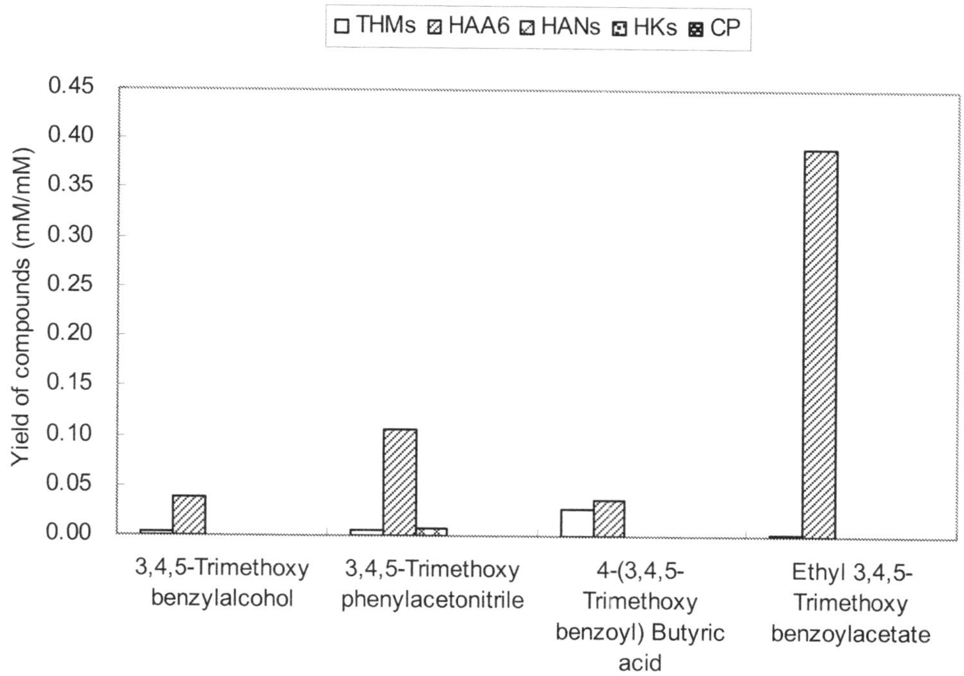

Figure D.13 Yields of selected DBPs for trimethoxy aromatic compounds with syringyl substitution patterns III

Figure D.14 shows chlorine reactivity and byproduct data for the fourth set of syringyl derivatives. These all have ethoxy rather than methoxy substuents, and they include the acid and alcohol. Figure D.15 presents the concentrations of particular DBP groups and species formed from chlorination of these model compounds. The data in these two figures are quite similar to those for the methoxy forms (i.e., syringic acid and 3,4,5- trimethoxy benzyl alcohol). This is a clear indication that the type of alkoxy subsituent isn't important in determining reactivity or pathway.

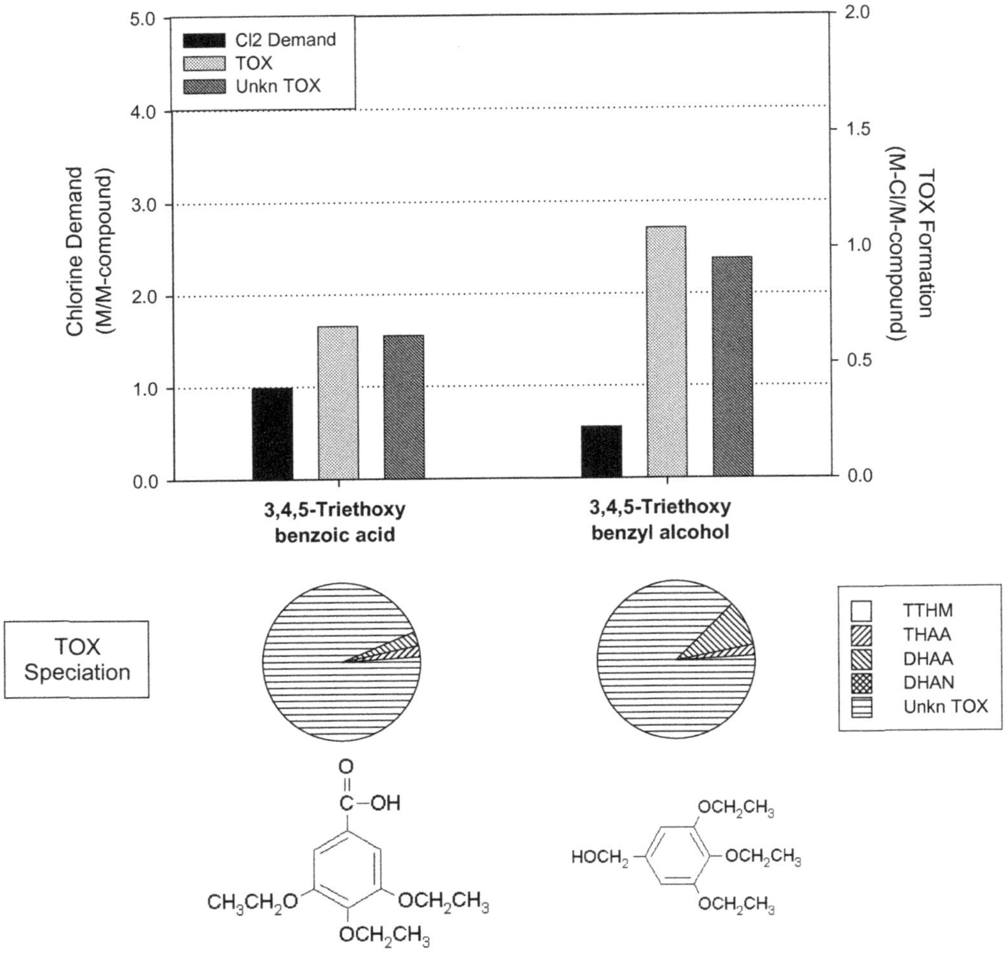

Figure D.14 Reactivity and chlorine incorporation for triethoxy aromatic compounds with syringyl substitution patterns

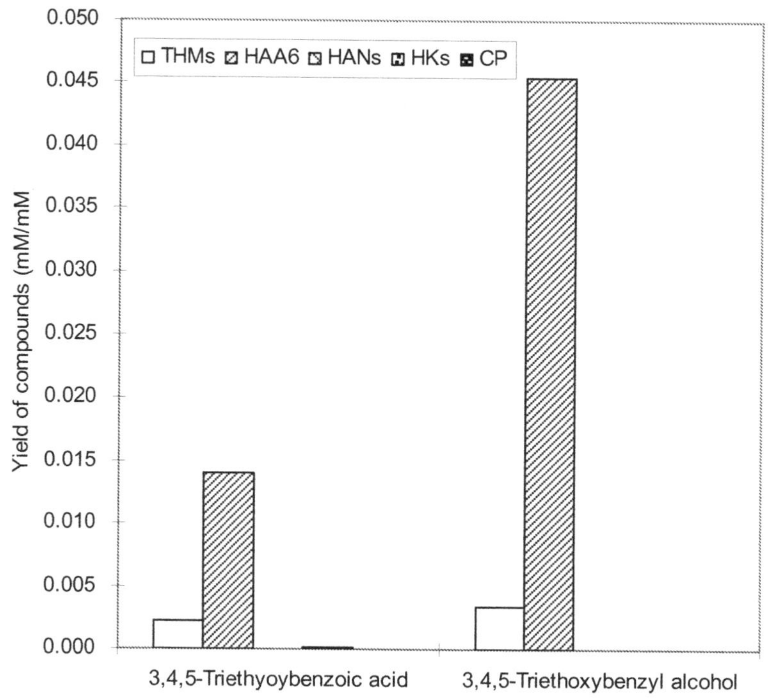

Figure D.15 Yields of selected DBPs for triethoxy aromatic compounds with syringyl substitution patterns

Figure D.16 shows chlorine reactivity and byproduct data for the fifth set of syringyl derivatives. These all have a central OH group instead of the methoxy. They also have a variety of alkyl side chains. As expected, the presence of the phenolic-OH imparts a high level of reactivity. Most of this is directed at oxidation as the TOX formation is still rather low. Figure D.17 presents the concentrations of particular DBP groups and species formed from chlorination of these model compounds. Again these levels are all low. The low level of THMs is especially striking.

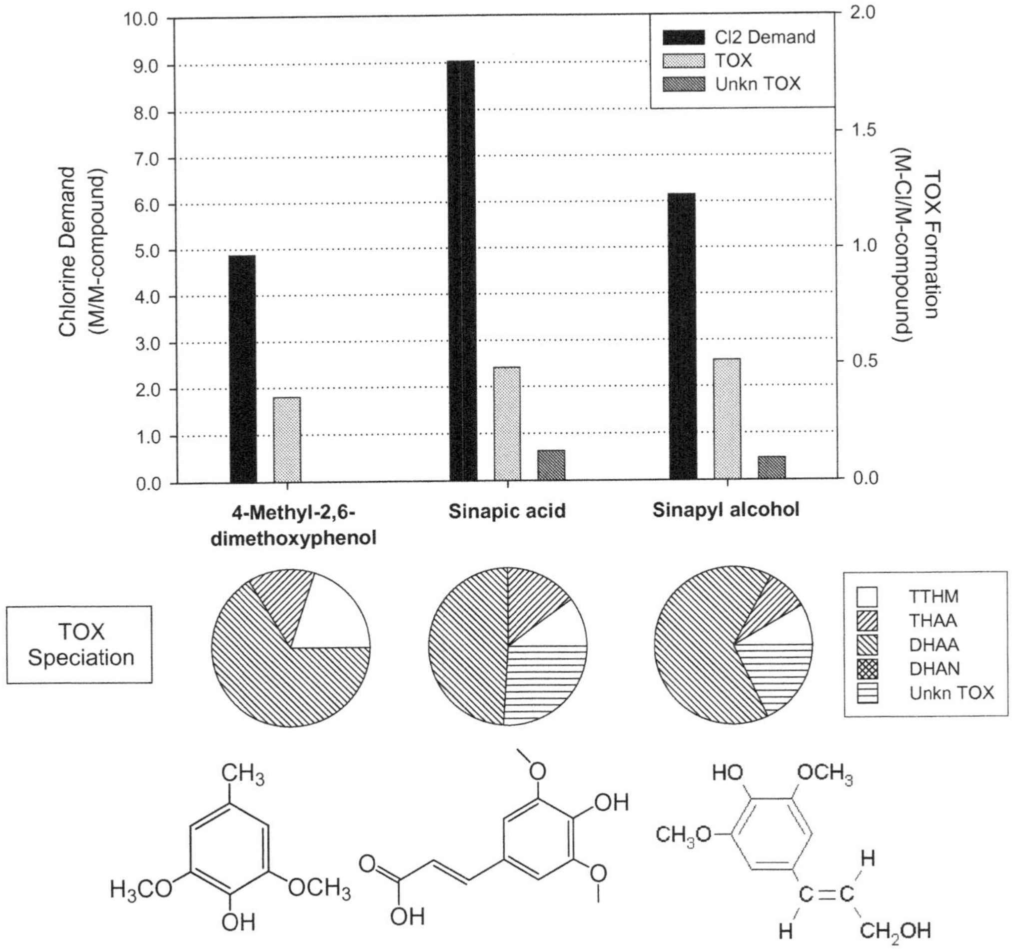

Figure D.16 Reactivity and chlorine incorporation for 2,6-dimethoxy-4-alkylphenols I

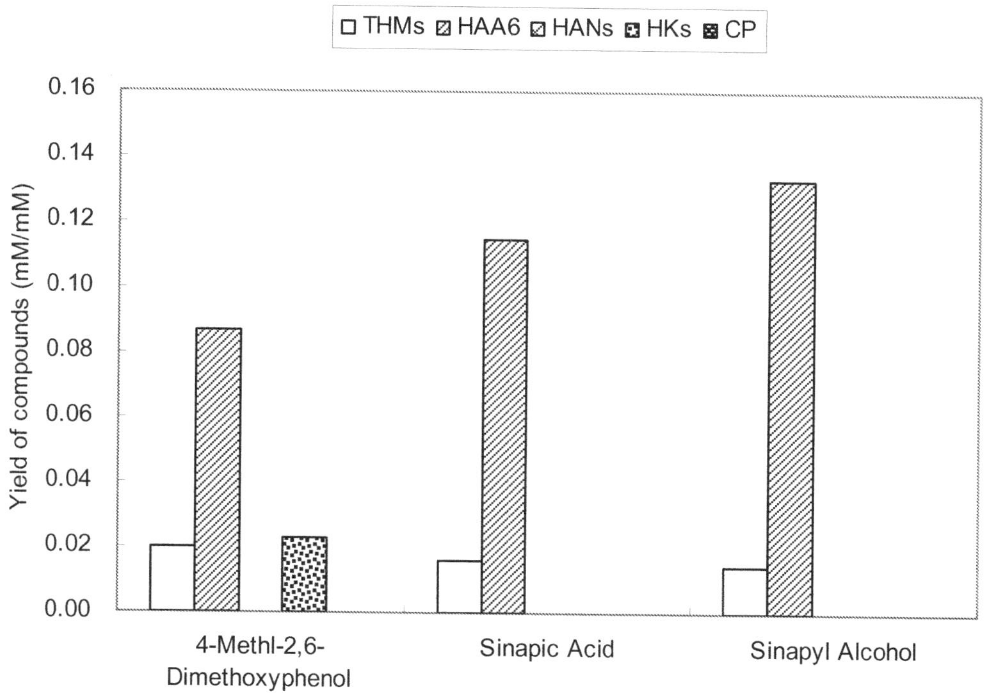

Figure D.17 Yields of selected DBPs for 2,6-dimethoxy-4-alkylphenols I

Figure D.18 shows chlorine reactivity and byproduct data for the sixth set of syringyl derivatives. Like the previous set, these all have a central OH group instead of the methoxy. This particular group has a different set of alkyl side chains. Again, the presence of the phenolic-OH imparted a high level of reactivity, and most of this is directed at oxidation reactions. Figure D.19 presents the concentrations of particular DBP groups and species formed from chlorination of these model compounds. These levels are rather low. The methyl syringate produced a TCAA molar yield of about 40%. This was a bit higher than had been seen with free syringic acid. In addition, much less TOX had been measured with the methyl ester. The net result was a TCAA yield that exceeded the TOX yield, producing a negative value for the "unknown TOX" calculation. On possible explanation is that the TCAA was hydrolyzed from a larger TOX precursor during extraction and analysis, and that this larger precursor was very hydrophilic and not well captured by the TOX method.

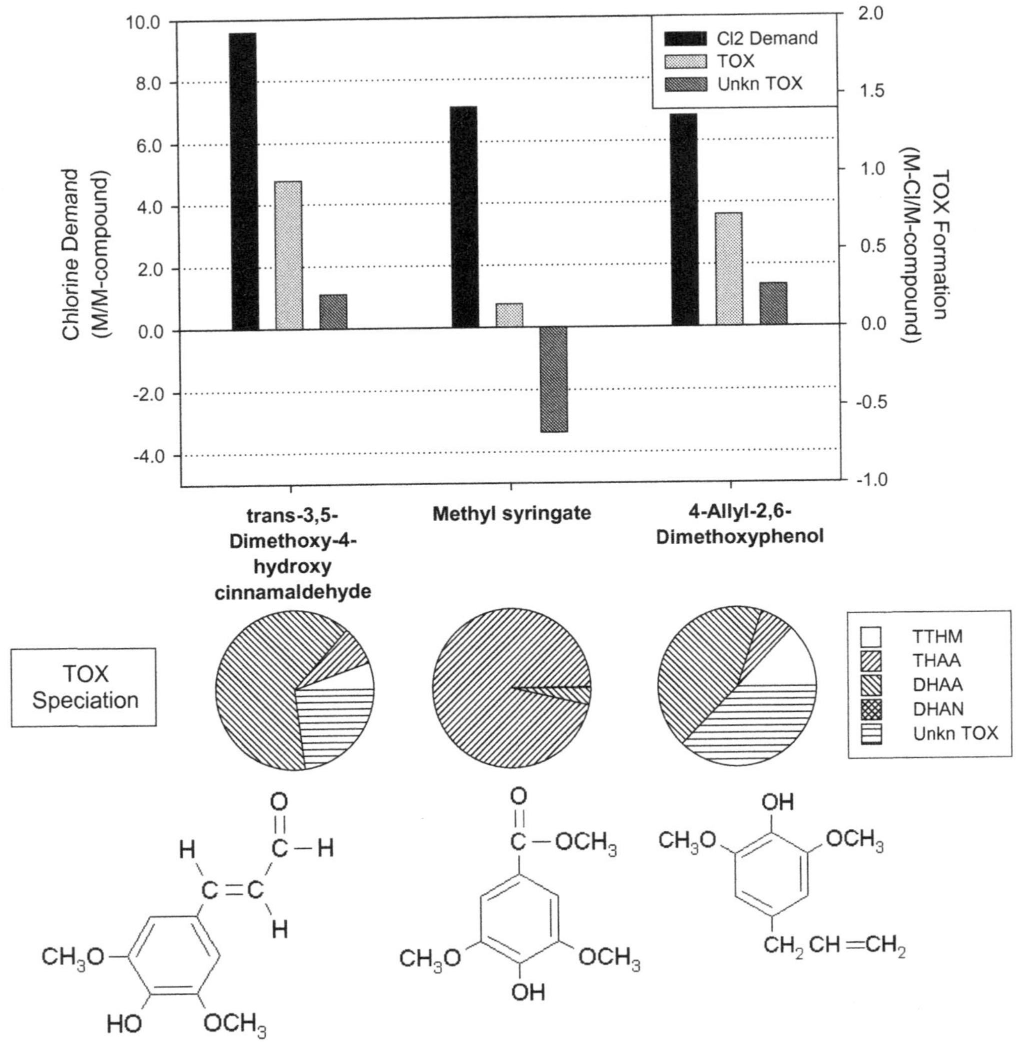

Figure D.18 Reactivity and chlorine incorporation for 2,6-dimethoxy-4-alkylphenols II

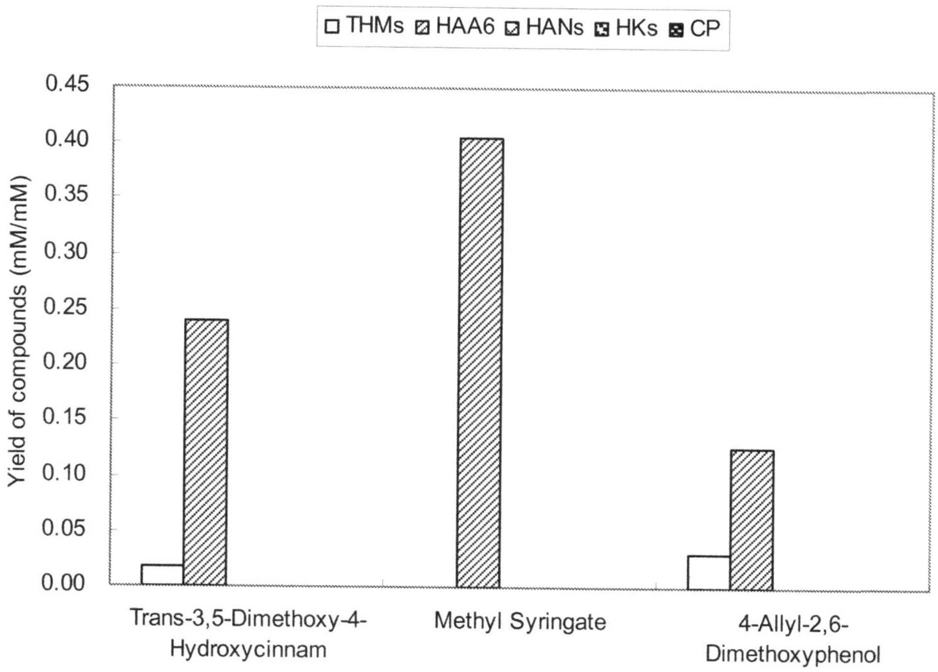

Figure D.19 Yields of Selected DBPs for 2,6-dimethoxy-4-alkylphenols II

APPENDIX E

TABLES AND FIGURES OF QSTR-RELATED DATA

Table E.1
Putative disinfection by-products subjected to QTSR analysis

Category	No.	Chemical
K01		**Saturated haloacids**
K01	C01	2,2,6,6,6-Pentachloro-4-hydroxy-hexanoic acid
K01	C02	2,2,6,6-Tetrachloro-4,5-dihydroxy-hexanoic acid
K01	C03	2,2-Dibromo-4-hydroxy-pentanoic acid
K01	C04	2,2-Dibromobutanoic acid
K01	C05	2,2-Dibromopentanoic acid
K01	C06	2,2-Dichloro-3-hydroxypropanoic acid
K01	C07	2,2-Dichloro-4-hydroxy-pentanoic acid
K01	C08	2,2-Dichlorobutanoic acid
K01	C09	2,2-Dichlorohexanoic acid
K01	C10	2,2-Dichloropentanoic acid
K01	C11	2,2-Dichloropropanoic acid
K01	C12	2,3,3,5,5,5-Hexachloro-4-hydroxy-pentanoic acid
K01	C13	2,3,4,4,4-Pentachloro-3-hydroxy-butanoic acid
K01	C14	2,4-Dichloro-3-hydroxybutanoic acid
K01	C15	2,5,5,5-Tetrachloro-4-hydroxy-pentanoic acid
K01	C16	2,5,5-Trichloro-4-oxopentanoic acid
K01	C17	2,5,5-Trichloropentanoic acid
K01	C18	2,5-Dichloropentanoic acid
K01	C19	2,6,6,6-Tetrachloro-4,5-dihydroxy-hexanoic acid
K01	C20	2,6,6,6-Tetrachloro-4-hydroxy-hexanoic acid
K01	C21	2-Chlorohexanoic acid
K01	C22	2-Bromo-2-chloro-4-hydroxy-pentanoic acid
K01	C23	2-Bromo-2-chlorobutanoic acid
K01	C24	2-Bromo-5-chloropentanoic acid
K01	C25	2-Bromobutanoic acid
K01	C26	2-Chlorobutanoic acid
K01	C27	2-Chloroheptanoic acid
K01	C28	2-Chlorohexanoic acid
K01	C29	2-Chloropentanoic acid
K01	C30	2-Chloropropanoic acid
K01	C31	3,3,3-Trichloro-2-hydroxy-2-methylpropanoic acid
K01	C32	3,3,3-Trichloro-2-hydroxypropanoic acid
K01	C33	3,3,3-Trichloropropanoic acid
K01	C34	3,3,4,4,4-Pentachloro-2-hydroxy-butanoic acid
K01	C35	3,3,5,5,5-Pentachloro-4-hydroxy-pentanoic acid
K01	C36	3,3-Dibromo-3-chloropropanoic acid
K01	C37	5,5-Dibromo-4-oxopentanoic acid
K01	C38	3-(Dichloromethyl)-2,5,5,5-tetrahloro-4-oxopentanoic acid
K01	C39	3-Bromo-3,3-dichloropropanoic acid
K01	C40	4,4,4-Trichloro-2-hydroxybutanoic acid
K01	C41	4,4,4-Trichloro-3-hydroxybutanoic acid
K01	C42	5,5,5-Trichloro-4-hydroxy-pentanoic acid

(continued)

Table E.1 (Continued)

K01	C43	5,5,5-Trichloropentanoic acid
K01	C44	2-Bromo-4-chlorobutanoic acid
K01	C45	2-Bromo-3,3-dichloropropanoic acid
K01	C46	2,4-Dichlorobutanoic acid
K01	C47	2,3,3-Trichloropropanoic acid
K01	C48	2,2,3-Trichloropropanoic acid
K01	C49	2,2-Dibromo-4-hydroxybutanoic acid
K01	C50	2,4,4-Trichlorobutanoic acid
K01	C51	4,4,4-Trichlorobutanoic acid
K02		**Unsaturated haloacid**
K02	C01	2,3,3-Tribromopropenoic acid
K02	C02	2,3,3-Trichloropropenoic acid
K02	C03	2,3,5,5-Tetrachloro-4-oxopentenoic acid
K02	C04	2,3,5-Trichloro-4-oxopentenoic acid
K02	C05	2,3-Dichloro-4-oxopentenoic acid
K02	C06	2,3-Dichloro-2-butenoic acid
K02	C07	2,5,5,5-Tetrachloro-4-oxopentenoic acid
K02	C08	2,5,5-Trichloro-4-oxopentenoic acid
K02	C09	2-Chloro-3-(dichloromethyl)-4-oxo-3-pentenoic acid
K02	C10	2-Chloro-3-(dichloromethyl)-4-oxo-4-pentenoic acid
K02	C11	2-Chloro-3-(dichloromethyl)-4-oxo-2-pentenoic acid
K02	C12	2-Chloro-2-butenoic acid
K02	C13	3,3-Dibromopropenoic acid
K02	C14	3,3-Dichloropropenoic acid
K02	C15	3,5,5,5-Tetrachloro-4-oxo-2-methyl-3-pentenoic acid
K02	C16	3,5,5,5-Tetrachloro-4-oxo-2-methylpentenoic acid
K02	C17	3,5,5,5-Tetrachloro-4-oxo-2-pentenoic acid
K02	C18	3,5,5-Trichloro-4-oxo-2-pentenoic acid
K02	C19	3-Bromo-3-chloropropenoic acid
K02	C20	3-Chloro-2-butenoic acid
K02	C21	4-Chloro-2-pentenoic acid
K02	C22	5,5,5-Trichloro-4-oxo-2-pentenoic acid
K02	C23	cis-2,3-Dibromopropenoic acid
K02	C24	cis-2,3-Dichloropropenoic acid
K02	C25	cis-4-Bromo-2-butenoic acid
K02	C26	2,3,4,5,6,6,6-Heptachloro-5-oxo-2-hexenoic acid
K02	C27	2,3,5,5,5-Pentachloro-4-hydroxy--3-pentenoic acid
K02	C28	2,3,5,5,5-Pentachloro-2-pentenoic acid
K02	C29	2,5,5,5-Tetrachloro-4-hydroxy-2-pentenoic acid
K02	C30	2,3,4,4-Tetrachloro-2-butenoic acid
K02	C31	trans-2,3-Dibromo-2-butenoic acid
K02	C32	trans-2,3-Dibromopropenoic acid
K02	C33	trans-2,3-Dichloropropenoic acid
K02	C34	trans-4-Bromo-2-butenoic acid
K02	C35	2,3,4-Tribromo-2-butenoic acid

(continued)

Table E.1 (Continued)

K02	C36	3,4,4-Trichloro-2-methyl-4-oxo-2-butenoic acid
K02	C37	2,3,4-Trichloro-2-butenoic acid
K02	C38	2,3,4-Trichlorocyclohexadien-5-one carboxylic acid
K02	C39	2,3,4,5,6,6,6-Heptachloro-5-hydroxy-2-hexenoic acid
K02	C40	2-Bromo-3-iodopropenoic acid
K02	C41	2-Iodo-3-bromopropenoic acid
K02	C42	2-Bromo-3-iodopropenoic acid
K03		**Haloalcohol**
K03	C01	1,4-Dibromo-2,3-butanediol
K03	C02	2-Bromoethanol
K03	C03	3-Chloro-2-pentanol
K03	C04	3-Chloro-2-methylbutanol
K03	C05	Bromochloroethanol
K03	C06	Bromodichloroethanol
K03	C07	2-Chloro-4-dimethylbutanol
K03	C08	Chlorodinaphthol
K03	C09	Chloroisobutanol
K03	C10	Dibromochloroethanol
K03	C11	Dibromoethanol
K03	C12	Dichloroethanol
K03	C13	1,1,2,2,3-Pentachloroisobutanol
K03	C14	3,4,4,4-Tetrachloro-2-butanol
K03	C15	Tribromoethanol
K03	C16	Trichloroethanol
K03	C17	2,5,5-Trichloropentanol
K03	C18	5,5,5-Trichloropentanol
K03	C19	3-Chloro-2-butanol
K03	C20	2,3,3,4-Tetrachloroisobutanol
K04		**Dihaloacetaldehydes**
K04	C01	2-Phenyl-2,2-dichloroacetaldehyde
K04	C02	Bromochloroacetaldehyde
K04	C03	Dibromoacetaldehyde
K04	C04	Dichloroacetaldehdye
K05		**Unsaturated haloketones**
K05	C01	1,1,1,3,-Tetrachloro-3-buten-2-one
K05	C02	1,1,1,3,4-Pentachloro-3-buten-2-one
K05	C03	1-Chloro-3-buten-2-one
K05	C04	3-Bromo-4-methyl-3-penten-2-one
K05	C05	3-Chloro-3-buten-2-one
K05	C06	Trichlorocyclopentenedione
K06		**Haloalkenes**
K06	C01	Bromocyclohexane (misplaced in database)
K06	C02	2,4-Dibromo-1-butene
K06	C03	Hexachlorocyclopentadiene
K06	C04	Pentachloropropene

(continued)

Table E.1 (Continued)

K06	C05	Tetrachlorocyclopropene
K07		**Haloesters**
K07	C01	1-Chloroethanolacetate
K07	C02	2,3-Dibromopropylacetate
K07	C03	2-Bromoethanolacetate
K07	C04	2-Chloroethanolacetate
K07	C05	Methyl bromochloroacetate
K07	C06	Methyl dibromoacetate
K08		**Halophenols**
K08	C01	2,4,6-Tribromophenol
K08	C02	2,4,6-Trichlorophenol
K08	C03	2,4-Dibromophenol
K08	C04	2,4-Dichlorophenol
K08	C05	2,6-Dibromo-1,4-dihydroxybenzene
K08	C06	2,6-Dibromo-4-methyl phenol
K08	C07	2,6-Dibromophenol
K08	C08	2,6-Dichloro-1,4-dihydroxybenzene
K08	C09	2-Bromo-4,6-dichlorophenol
K08	C10	2-Bromophenol
K08	C11	2-Chlorophenol
K08	C12	3-Bromophenol
K08	C13	4-Bromo-2,6-di-tert-butyl phenol
K08	C14	4-Chloro-2,6-di-tert-butyl phenol
K08	C15	4-Chloro-2-methoxy-5-methylphenol
K08	C16	Pentachlorophenol
K08	C17	Trichloro-trihydroxybenzene
K09		**Other haloalkanes**
K09	C01	1,1-Dibromopropane
K09	C02	1,2,3-Trichloropropane
K09	C03	1-Bromohexane
K09	C04	2-Bromobutane
K09	C05	Bromochloromethane
K09	C06	Carbon tetrachloride
K09	C07	Chloroform
K09	C08	Chloromethane
K09	C09	Dichloroiodomethane
K09	C10	Dibromomethane
K09	C11	Dichloromethane
K09	C12	Hexachloroethane
K09	C13	Tribromochloromethane
K10		**Saturated haloaldehydes**
K10	C01	2,2,3-Trichlorobutanal
K10	C02	2,2,3-Trichloropropanal
K10	C03	2,2,4-Trichlorobutanal
K10	C04	2,2-Dichloropropanal

(continued)

Table E.1 (Continued)

K10	C05	2,3,3-Trichlorobutanal
K10	C06	2,3,3-Trichloropropanal
K10	C07	2,3,4-Trichlorobutanal
K10	C08	2,3-Dichloropropanal
K10	C09	2,4,4-Trichlorobutanal
K10	C10	2-Bromo-2-methylpropanal
K10	C11	3,3,3-Trichloropropanal
K10	C12	3,3,4-Trichlorobutanal
K10	C13	3,3-Dichloropropanal
K10	C14	3,4,4-Trichlorobutanal
K11		**Saturated halodicarboxylic acids**
K11	C01	2,2,3,3-Hexachlorohexanedioic acid
K11	C02	2,2,3,3-Tetrachloro-4-hydroxypentanedioic acid
K11	C03	2,2,3-Trichloro-4-hydroxypentanedioic acid
K11	C04	2,2,4,4-Tetrachlorohexanedioic acid
K11	C05	2,2,5,5-Tetrachlorohexandioic acid (misnamed in database)
K11	C06	2,2-Dichloro-4-hydroxypentanedioic acid
K11	C07	2,2-Dichlorobutanedioic acid
K11	C08	2,2-Dichlorohexanedioic acid
K11	C09	2,2-Dichlorooctanedioic acid
K11	C10	2,2-Dichloropentanedioic acid
K11	C11	2,2-Dichloropropanedioic acid
K11	C12	2,3-Dichlorobutanedioic acid
K11	C13	2,3-Dichloropentanedioic acid
K11	C14	2,4-Dichloropentanedioic acid
K11	C15	2-(Dichloromethyl)-2,3-dichloro-3-hydroxy-butanedioic acid
K11	C16	2-Chloro-2-methylpropanedioic acid
K11	C17	2-Chloroheptanedioic acid
K11	C18	2-Chlorohexanedioic acid
K11	C19	2-Chloromethylpropanedioic acid
K11	C20	2-Chlorononanedioic acid
K11	C21	2-Chlorooctanedioic acid
K11	C22	3-Chlorohexanedioic acid
K11	C23	Chlorobutanedioic acid
K11	C24	Trichlorobutanedioic acid
K11	C25	2,2,5,5-Tetrachlorohexanedioic acid
K12		**Saturated haloketones**
K12	C01	1,1,1,3,3-Pentabromo-3-chloropropanone
K12	C02	1,1,1,3,3-Pentachloropropanone
K12	C03	1,1,1,3-Tetrabromo-3-chloropropanone
K12	C04	1,1,1,3-Tetrachloropropanone
K12	C05	1,1,1-Tribromopropanone
K12	C06	1,1,1-Trichloro-2-butanone
K12	C07	1,1,1-Trichloro-2-pentanone
K12	C08	1,1,1-Trichloropropanone

(continued)

Table E.1 (Continued)

K12	C09	1,1,3,3-Tetrabromopropanone
K12	C10	1,1,3,3-Tetrachloropropanone
K12	C11	1,1,3-Tribromo-3-chloropropanone
K12	C12	1,1,3-Tribromopropanone
K12	C13	1,1,3-Trichloro-2-butanone
K12	C14	1,1,3-Trichloropropanone
K12	C15	1,1-Dibromo-3,3-Dichloropropanone
K12	C16	1,1-Dibromo-3-chloropropanone
K12	C17	1,1-Dibromopropanone
K12	C18	1,1-Dichloro-2-butanone
K12	C19	1,1-Dichloropropanone
K12	C20	1,3-Dibromo-1,3-dichloropropanone
K12	C21	1,3-Dibromopropanone
K12	C22	1,3-Dichloro-2-butanone
K12	C23	1-Bromo-3-chloropropanone
K12	C24	3,3-Dichloro-2-butanone
K12	C25	3-Chloro-2-butanone
K12	C26	5-Bromo-2-pentanone
K12	C27	Bromopropanone
K12	C28	Chloropropanone
K12	C29	Hexachloropropanone
K12	C30	1-Bromo-1,3,3-trichloropropanone
K13		**Trihaloacetaldehydes**
K13	C01	Bromodichloroacetaldehyde
K13	C02	Dibromochloroacetaldehyde
K13	C03	Tribromoacetaldehyde
K13	C04	Trichloroacetaldehyde
K14		**Trihaloacetates**
K14	C01	Bromodichloroacetate
K14	C02	Dibromochloroacetate
K14	C03	Tribromoacetate
K14	C04	Trichloroacetate
K15		**Unsaturated haloaldehydes**
K15	C01	2,3,3-Trichloropropenal
K15	C02	2,3-Dichlorobutenal
K15	C03	2,3-Dichloropropenal
K15	C04	2,4-Dichlorobutenal
K15	C05	2-Bromo-3-chlorobutenal
K15	C06	2-Bromo-4-chlorobutenal
K15	C07	2-Bromobutenal
K15	C08	2-Bromopentadienal
K15	C09	2-Chloropropenal
K15	C10	3,3-Dichloropropenal
K16		**Unsaturated halodicarboxylic acids**
K16	C01	2-Carboxy-3,5,5,5-tetrachloro-4-oxopentenedioc acid

(continued)

Table E.1 (Continued)

K16	C02	2-Chloro-2-pentenedioc acid
K16	C03	2-Chloro-3-(trichloroacetyl)butenedioc acid
K16	C04	2-Chloro-3-pentenedioc acid
K16	C05	2-Chloro-3,4-dicarboxy-2-pentendioc acid
K16	C06	3-Chloro-2-pentenedioc acid
K16	C07	3-Chloro-3-pentenedioc acid
K16	C08	cis-2-Bromo-3-methylbutenedioc acid
K16	C09	trans-2-bromobutenoic acid
K16	C10	trans-2-chlorobutenedioc acid
K16	C11	cis-Dichlorobutenedioc acid
K16	C12	Dichloropentenedioc acid
K16	C13	Tetrachloropentenedioc acid
K16	C14	trans-2,3-Dibromobutenedioc acid
K16	C15	cis-2-Bromobutenedioc acid
K16	C16	cis-2-Chlorobutenedioc acid
K16	C17	trans-Dichlorobutenedioc acid
K16	C18	Trichloropentenedioc acid
K16	C19	2-Iodo-3-methylbutenedioc acid
K17		**Haloacetonitriles**
K17	C01	Bromochloroacetonitrile
K17	C02	Bromodichloroacetonitrile
K17	C03	Dibromoacetonitrile
K17	C04	Dibromochloroacetonitrile
K17	C05	Dichloroacetonitrile
K17	C06	Tribromoacetonitrile
K17	C07	Trichloroacetonitrile
K18		Halotricarboxylic acids
K18	C01	2-Chloro-3-Dicarboxy-2-butenenoic acid
K18	C02	Chloroethenetricarboxylic acid
K19		**Monohaloacetates**
K19	C01	Bromoacetate
K19	C02	chloroacetate
K19	C03	Iodoacetate
K20		**Dihaloacetates**
K20	C01	Dichloroacetate
K20	C02	Bromochloroacetate
K20	C03	Dibromoacetate
K20	C04	Bromoiodoacetate
K20	C05	Diiodoacetate
K21		**MX and analogs**
K21	C01	4-Bromochloromethyl-MX
K21	C02	4-Dibromomethyl-MX
K21	C03	5-Hydroxy-4-trichloromethyl-2-furanone
K21	C04	5-Hydroxy-5-trichloromethyl-2-furanone
K21	C05	CMCF

(continued)

Table E.1 (Continued)

K21	C06	E-MX
K21	C07	MCA
K21	C08	MX
K21	C09	Open-MX (same as E-MX)
K21	C10	Ox-EMX
K21	C11	Red-MX
K21	C12	Tribromomethyl-MX
K22		**Aromatic haloacetaldehydes**
K22	C01	(2-Chloro-4-hydroxyphenyl)-ethanal
K22	C02	(3-Chloro-4-hydroxyphenyl)-ethanal
K22	C03	2-Chloro-4-hydroxybenzaldehyde
K22	C04	3-Bromo-4-hydroxy-5-methoxy-benzaldehyde
K22	C05	3-Chloro-4-hydroxybenzaldehyde
K23		**Haloamides**
K23	C01	3-Bromo-3-Chloropropamide
K23	C02	3-Bromo-3,3-dichloropropamide
K23	C03	3,3-Dibromopropamide
K23	C04	3,3,-Dibromo-3-chloropropamide
K23	C05	3,3-Dichloropropamide
K23	C06	3,3,3-Tribromopropamide
K23	C07	3,3,3-Trichloropropamide
K23	C08	Bromochloroacetamide
K23	C09	Bromodichloroacetamide
K23	C10	Dibromoacetamide
K23	C11	Dibromochloroacetamide
K23	C12	Dichloroacetamide
K23	C13	Tribromoacetamide
K23	C14	Trichloroacetamide
K23	C15	2-Bromopropamide
K23	C16	2,2-Dibromopropamide
K23	C17	2-Chloropropamide
K23	C18	2,2-Dichloropropamide
K23	C19	2,3-Dibromopropamide
K23	C20	2-3-Dichloropropamide
K23	C21	2-Bromo-3-Chloropropamide
K23	C22	2-Bromobutamide
K23	C23	2,2-Dibromobutamide
K23	C24	2-Chlorobutamide
K23	C25	2,2-Dichlorobutamide
K24		**Haloaromatic acids**
K24	C01	2,6-Dichloro-3-carboxy-5-formylphenol
K24	C02	2,4-Dichlorophenylacetic acid
K24	C03	2,5-Dibromobenzoic acid
K24	C04	2,5-Dichloro-4-hydroxybenzoic acid
K24	C05	2,6-Dibromo-4-hydroxybenzoic acid

(continued)

Table E.1 (Continued)

K24	C06	2,6-Dibromobenzoic acid
K24	C07	2,4-Dichloro-3-carboxy-6-formylphenol
K24	C08	2,4-Dichloro-3-carboxy-4-formylphenol
K24	C09	2,6-Dichloro-4-hydroxybenzoic acid
K24	C10	2,6-Dichlorophenylacetic acid
K24	C11	2-Chloro-3-carboxy-6-formylphenol
K24	C12	2-Chloro-4-hydroxy-1,3,5-tricarboxylic acid
K24	C13	2-Formyl-3-chloro-4-carboxyphenol
K24	C14	3-Formyl-4-chloro-5-carboxyphenol
K24	C15	2-Formyl-4-chloro-5-carboxyphenol
K24	C16	2-Chlorobenzoic acid
K24	C17	2-Chlorophenylacetic acid
K24	C18	3,5-Dibromobenzoic acid
K24	C19	3,5-Dichloro-4-hydroxybenzoic acid
K24	C20	3,5-Dichlorobenzoic acid
K24	C21	3-Chloro-6-hydroxy-1,2,5-tricarboxylic acid
K24	C22	4-Chloro-6-hydroxy-1,2,5-tricarboxylic acid
K24	C23	4-Chlorobenzoic acid
K24	C24	4-Chlorophenylacetic acid
K24	C25	5-Chloro-2-hydroxybenzoic acid
K24	C26	2,6-Dichloro-3,5-dicarboxyphenol
K24	C27	2,4-Dichloro-3,6-dicarboxyphenol
K24	C28	2-Chloro-3,6-dicarboxyphenol
K24	C29	2,4-Dicarboxy-3-chlorophenol
K24	C30	3,5-Dicarboxy-4-chlorophenol
K24	C31	2,5-Dicarboxy-4-chlorophenol
K25		**Halogenated aromatic amines**
K25	C01	2,4,6-Tribromobenzenamine
K25	C02	2,4,6-Trichlorobenzenamine
K25	C03	3,5-Dibromo-2,4,6-Triaminobenzoic acid
K25	C04	3,5-Dibromo-2-aminobenzoic acid
K25	C05	3,5-Dichloro-2-aminobenzoic acid
K25	C06	4,6-Dibromo-2-aminobenzoic acid
K25	C07	4,6-Dichloro-2-aminobenzoic acid
K26		**Halonitromethanes**
K26	C01	Bromodichloronitromethane
K26	C02	Bromonitromethane
K26	C03	Dibromochloronitromethane
K26	C04	Dibromonitromethane
K26	C05	Tribromonitromethane
K26	C06	Trichloronitromethane
K27		**Halopyrroles**
K27	C01	2,3,5-Tribromopyrrole
K27	C02	2,3,5-Trichloropyrrole
K27	C03	5-Bromo-2,3-dichloropyrrole

(continued)

Table E.1 (Continued)

K28		**Halothiophenes**
K28	C01	2,3,4-Tribromothiophene
K28	C02	2,3,5-Tribromothiophene
K28	C03	2,3-Dibromo-4,5-dichlorothiophene
K28	C04	2,4-Dibromo-3,5-dichlorothiophene
K28	C05	Bromotrichlorothiophene
K28	C06	Tetrabromothiophene
K28	C07	Tetrachlorothiophene
K29		**Other haloaromatics**
K29	C01	(2-Chloroethenyl)-benzene
K29	C02	1,2-Dichlorobenzene
K29	C03	1,3-Dichlorobenzene
K29	C04	1,4-Dichlorobenzene
K29	C05	2,4-Dichloroethylbenzene
K29	C06	2-Chlorotoluene
K29	C07	3,5-Dichloro-2-methoxytoluene
K29	C08	3,5-Dichloro-4-methoxytoluene
K29	C09	3,5-Dichloroethylbenzene
K29	C10	3,5-Dichlorotoluene
K29	C11	3-Chlorotoluene
K29	C12	4-Chlorotoluene
K29	C13	Chlorobenzene
K30		**Other halonitriles**
K30	C01	2,3,3-Trichloropropenenitrile
K30	C02	2,3-Dibromopropionitrile
K30	C03	2,3-Dichloropropionitrile
K30	C04	2,3-Dichloro-3-bromopropionitrile
K30	C05	3,3,3-Trichloropropionitrile
K30	C06	3,3-Dibromopropionitrile
K30	C07	3,3-Dichloropropionitrile
K30	C08	3,3-Dichloropropenenitrile
K30	C09	3,4-Dichlorobutanenitrile
K30	C10	3-Bromopropionitrile
K30	C11	cis-2,3,4-Trichloro-2-butenenitrile
K30	C12	Cyanogen bromide
K30	C13	Cyanogen chloride
K30	C14	trans-2,3,4-Trichloro-2-butenenitrile
K30	C15	3-Bromopropionitrile
K30	C16	2-Bromoproprionitrile
K30	C17	3-Chloroproprionitrile
K30	C18	2-Chloropropionitrile
K30	C19	2,2-Dichloropropionitrile
K30	C20	2,2-Dibromoproprionitrile

Table E.2
Selected carcinogens from the Carcinogenic Potency Database used to compare LOAELs predicted by TOPKAT, actual experimentally determined LOAELs, and TD_{50}s

Carcinogen	Predicted LOAEL	Rat TD_{50}	Actual LOAEL	Predicted LOAEL OPS or other error*	Rat TD_{50}
Hexachlorobenzene	2.9	3.86	0.25		
Cyclophosphamide	1.9	2.21	1		
Furan	8.9	0.396	1.43		
Acrylamide	4.4	3.81	2		
Hexachlorobutadiene	165.3	65.8	2		
1,2,3-Trichloropropane	4.7	1.35	2.14		
Tris(2,3-dibromopropyl)phosphate	3.1	3.83	2.5		
Dichlorovos	10.4	4.16	2.9		
2,4-Diaminotoluene	39.2	2.47	3.95		
Hexachloroethane	492.1	55.4	7.14		
Azobenzene	52.7	24.1	10		
1,2-Dibromo-3-chloropropane	6.7	0.259	10.72		
N,N-Diethylthiourea	11.5	24	12.5		
Coumarin	59.6	39.2	17.86		
Benzene	24.5	169	17.9		
Benzofuran	26.1	424	21.43		
Captan	41.5	2080	25		
1,2-Dibromoethane	60.3	1.52	27.14		
1,2-Dichloroethane	84.6	8.04	33.6		
Methylene chloride	107.8	724	50		
m-Cresidine	78.3	470	57.14		
Tribromomethane	99.4	648	71.43		
p-Cresidine	71.1	98	75		
Chlorobenzene	59.4	247	85.72		
Tris-(2-Chloromethyl)phosphate	55.2	86.7	88		
4-Chloro-m-phenylenediamine	100.2	315	100		
1,4-Dichlorobenzene	48.9	644	107		
o-Anisidine	184.5	29.1	250		
4-Chloro-o-phenylenediamine	118.9	214	250		
2,4,6-Trichlorophenol	116.3	405	250		
Tetrachloroethylene	193.3	101	336.4		
Trichloroethylene	28	668	357		
11-Aminoundecanoic acid	211.5	1100	375		
Butyl benzylphthalate	650.7	1040	600		
Acetamide	75.9	160			
Acetominophen	226.7	495			

(continued)

Table E.2 (Continued)

2-Acetoaminofluorene	84.2	1.22		
Acrylonitrile	1.8	16.9		
Aflatoxin B1	37	0.0032		
2-Amino-4-nitrophenol	102.6	839		
2-Amino-5-nitrophenol	103.7	111		
4-Amino-2-nitrophenol	94.3	309		
4-Aminodiphenyl.HCl	45.4	0.98		
3-Aminotriazole	106	9.94		
1-Amyl-1-nitrosourea	9.1	0.555		
Aniline.HCl	42.2	269		
Atrazine	24.1	31.7		
Azaserine			9.1	0.793
Azoxymethane			149.7	0.0466
1-Azoxypropane			196.2	0.000241
2-Azoxypropane			9.8	0.00268
Barbital sodium	13.8	45.7		
Benzidine	40.6	1.73		
Benzo(a)pyrene	712.9	0.856		
Bromodichloromethane	35.7	72.5		
2.2-Bis(bromomethyl)-1,3-propanediol	347.1	111		
Bis-(chloromethyl)ethere	56.5	0.000357		
Bis-(2,3-dibromopropyl)phosphate, Mg			347.1	32
Bromomethane	64	149		
1,3-Butadiene	201.1	261		
t-Butyl alcohol	1300	64.6		
N-Butyl-N-(4-hydroxybutyl)nitrosamine	8.2	0.457		
Carbon tetrachloride	820.7	2.29		
Chlorambucil	4.7	0.896		
2-Chloro-1,1,1-trifluoroethane	111	87.3		
p-Chloroaniline,HCl			111	7.62
Chlorofluoromethane	20.5	27.5		
Chloroform	64.3	262		
Chloromethyl methyl ether	36.6	5.5		
3-(Chloromethyl)pyridine.HCl			36.6	433
Chloroprene	23.2	12.5		
Chlorozotocin			1.6	0.375
Ciprofibrate			528	2.48
Clofibrate	473.6	169		
Crotonaldehyde	85.4	4.2		

(continued)

Table E.2 (Continued)

Cupferron	19	8.35		
DDT			65.8	84.7
Dehydroepiandrosterone			2.1	68.1
Dehydroepiandrosterone acetate	10.8	31.4		
Dibromodulcitol	19	8.37		
Dibromomannitol	19	27.6		
1,3-Dibutyl-1-nitrosourea	0.626	4.28		
Dichloroacetic acid	5.9	161		
Dichloroacetylene	34.8	3.58		
3,3-Dichlorobenzidine	38.9	28.1		
Diethylacetamide			15.9	8.85
Diethylene glycol	1500	1660		
Diethylstilbesterol			14.9	0.223
Epichlorhydrin	124.1	2.96		
Ethinyl estradiol			10000	0.2
Formaldehyde	59.9	1.35		
N-Hydroxy-2-acetylaminofluorene	63.3	0.988		
Iodinated glycerol (70/30 mixture)	14.3	101		
Kepone	56.5	2.96		
Methyl carbamate			44.3	56.6
Methyl clofenapate	54.1	4.78		
N-Methyl-N,4-dinitrosoaniline	5.6	1.3		
4,4'-Methylene-bis(2-chloroaniline)	31.3	19.3		
Mirex	10.4	1.77		
Photo-Mirex	10.4	1.46		
Nafenopin	3.4	11		
2-Nitroanisole	107.1	15.6		
1,1,1.2-Tetrafluoroethane	95.6	29900		
Tetrafluoroethylene	116.6	107		
Urethane	57.1	41.3		
Vinyl acetate	77.4	341		
Vinyl bromide	14	18.5		
Vinyl chloride	13.7	19.1		
Vinyl fluoride	10.1	20		
NDMA			2.7	0.096
NDEA			7.14	0.265

* Predicted LOAELs that were outside the optimal predictive space of the model or which have fragments not represented in the training set of the model.

Table E.3
Data on halogenated compounds abstracted from the Carcinogenicity Potency Database that was used to compare the LOAEL predicted by TOPKAT with the actual experimentally LOAEL and TD_{50}

Carcinogen	Predicted LOAEL	Rat TD_{50}	Actual LOAEL	Predicted LOAEL OPS or other error*	Rat TD_{50}
Bis-(chloromethyl)ether	56.5	0.000357			
1,2-Dibromo-3-chloropropane	6.7	0.259	10.72		
Chlorambucil	4.7	0.896			
1,2,3-Trichloropropane	4.7	1.35	2.14		
Photo-Mirex	10.4	1.46			
1,2-Dibromoethane	60.3	1.52	27.14		
Mirex	10.4	1.77			
Carbon tetrachloride	820.7	2.29			
Epichlorhydrin	124.1	2.96			
Kepone	56.5	2.96			
Dichloroacetylene	34.8	3.58			
Tris(2,3-dibromopropyl)phosphate	3.1	3.83	2.5		
Hexachlorobenzene	2.9	3.86	0.25		
Dichlorovos	10.4	4.16	2.9		
Chloromethyl methyl ether	36.6	5.5			
1,2-Dichloroethane	84.6	8.04	33.6		
Dibromodulcitol	19	8.37			
Chloroprene	23.2	12.5			
Vinyl bromide	14	18.5			
Vinyl chloride	13.7	19.1			
4,4'-Methylene-bis(2-chloroaniline)	31.3	19.3			
Vinyl fluoride	10.1	20			
Chlorofluoromethane	20.5	27.5			
Dibromomannitol	19	27.6			
3,3-Dichlorobenzidine	38.9	28.1			
Hexachloroethane	492.1	55.4	7.14		
Hexachlorobutadiene	165.3	65.8	2		
Bromodichloromethane	35.7	72.5			
Tris-(2-Chloromethyl)phosphate	55.2	86.7	88		
2-Chloro-1,1,1-trifluoroethane	111	87.3			
Iodinated glycerol (70/30 mixture)	14.3	101			
Tetrachloroethylene	193.3	101	336.4		
Tetrafluoroethylene	116.6	107			
2.2-Bis(bromomethyl)-1,3-propanediol	347.1	111			
Bromomethane	64	149			

(continued)

Table E.3 (Continued)

Dichloroacetic acid	5.9	161			
4-Chloro-o-phenylenediamine	118.9	214	250		
Chlorobenzene	59.4	247	85.72		
Chloroform	64.3	262			
4-Chloro-m-phenylenediamine	100.2	315	100		
2,4,6-Trichlorophenol	116.3	405	250		
1,4-Dichlorobenzene	48.9	644	107		
Tribromomethane	99.4	648	71.43		
Trichloroethylene	28	668	357		
Methylene chloride	107.8	724	50		
1,1,1.2-Tetrafluoroethane	95.6	29900			
Bis-(2,3-dibromopropyl)phosphate, Mg				347.1	32
p-Chloroaniline,HCl				111	7.62
3-(Chloromethyl)pyridine.HCl				36.6	433
Chlorozotocin				1.6	0.375

* Predicted LOAELs that were outside the optimal predictive space of the model or which have fragments not represented in the training set of the model.

Table E.4
Unsaturated haloacids predicted to be carcinogens and have chronic LOAELs below 10 mg/kg day^{-1}

Index Name	Chemical	No. Pos. Cancer Pred. (8 poss.)	No. outside OPS*	LOAEL mg/kg	LOAEL Lower Limit mg/kg	LOAEL Upper Limit mg/kg
K02C38	2,3,4-Trichlorocyclohexadien-5-one carboxylic acid	3	3	0.22	0.05	0.90
K02C05	2,3-Dichloro-4-oxopentenoic acid	6	2	0.61	0.09	4.1
K02C18	3,5,5-Trichloro-4-oxo-2-pentenoic acid	4	1	0.64	0.10	4.0
K02C19	3-Bromo-3-chloropropenoic acid	2	0	1.1	0.17	6.3
K02C25	cis-4-Bromo-2-butenoic acid	3	0	1.5	0.27	8.4
K02C34	Trans-4-Bromo-2-butenoic acid	3	0	1.5	0.27	8.4
K02C21	4-Chloro-2-pentenoic acid	2	0	1.7	0.30	9.3
K02C13	3,3-Dibromopropenoic acid	3	0	1.9	0.33	11
K02C14	3,3-Dichloropropenoic acid	4	0	2.5	0.45	14
K02C20	3-Chloro-2-butenoic acid	3	1	2.8	0.53	15
K02C22	5,5,5-Trichloro-4-oxo-2-pentenoic acid	4	2	3.8	0.70	21
K02C04	2,3,5-Trichloro-4-oxopentenoic acid	4	3	4.3	0.68	27
K02C17	3,5,5,5-Tetrachloro-4-oxo-2-pentenoic acid	6	4	4.3	0.79	24
K02C08	2,5,5-Trichloro-4-oxopentenoic acid	4	2	4.7	0.85	26
K02C03	2,3,5,5-Tetrachloro-4-oxopentenoic acid	2	2	5.7	1.0	31
K02C10	2-Chloro-3-(dichloromethyl)-4-oxo-4-pentenoic acid	2	2	5.7	1.0	31
K02C23	cis-2,3-Dibromopropenoic acid	4	1	5.9	1.0	34
K02C32	Trans-2,3-Dibromopropenoic acid	4	1	5.9	1.0	34
K02C30	2,3,4,4-Tetrachloro-2-butenoic acid	4	4	6.7	1.2	35
K02C24	cis-2,3-Dichloropropenoic acid	4	1	8.1	1.5	45
K02C33	Trans-2,3-Dichloropropenoic acid	4	0	8.1	1.5	45
K02C09	2-Chloro-3-(dichloromethyl)-4-oxo-3-pentenoic acid	1	1	8.4	1.6	44
K02C27	2,3,5,5,5-Pentachloro-4-hydroxy--3-pentenoic acid	4	3	9.2	1.7	49
K02C35	2,3,4-Tribromo-2-butenoic acid	4	4	9.6	1.7	53

* OPS = optimal predictive space of the model

Table E.5
Reduction of the list of unsaturated haloacids predicted to be carcinogenic to those for which there was a positive prediction in two species

Index Number	Chemical	Pred. Ames' Mut.	Male rat	Female rat	Male mice	Female mice	LOAEL mg/kg
K02C05	2,3-Dichloro-4-oxopentenoic acid	+	pos*	pos	pos	pos	0.61
K02C18	3,5,5-Trichloro-4-oxo-2-pentenoic acid	+	pos	pos*	pos	pos	0.64
K02C19	3-Bromo-3-chloropropenoic acid	+		pos		pos	1.1
K02C25	cis-4-Bromo-2-butenoic acid	+	pos	pos		pos	1.5
K02C34	trans-4-Bromo-2-butenoic acid	+	pos	pos		pos	1.5
K02C13	3,3-Dibromopropenoic acid	+		pos	pos	pos	1.9
K02C14	3,3-Dichloropropenoic acid	-	pos	pos		pos	2.5
K02C20	3-Chloro-2-butenoic acid	+		pos		pos	2.8
K02C23	cis-2,3-Dibromopropenoic acid	+	pos*	pos	pos	pos	5.9
K02C32	trans-2,3-Dibromopropenoic acid	+	pos*	pos	pos	pos	5.9
K02C24	cis-2,3-Dichloropropenoic acid	-	pos*	pos		pos	8.1
K02C33	trans-2,3-Dichloropropenoic acid	-	pos	pos		pos	8.1

* Outside the optimal predictive space of the model, but within permissible limits
† Outside the optimal predictive space of the model and exceeds permissible limits

Table E.6

Similarity indices for predictions of carcinogenicity in the unsaturated haloacid group of disinfection by-products. The strongest criteria to be satisfied are a probability of carcinogenic activity of greater than 0.7 accompanied by a similarity index that is < 0.2. Chemicals with similarity indices < 0.3 are included so that those approaching the criteria can also be viewed

Index Number	Chemical	NTP Male rat	Simil. index	FDA Male rat	Simil. index	NTP Female rat	Simil. index	FDA Female rat	Simil. index	NTP Male mice	Simil. index	FDA Male mice	Simil. index	NTP Female mice	Simil. index	FDA Female mice	Simil. index
K02	Unsaturated haloacid																
K02C05	2,3-Dichloro-4-oxopentenoic acid	0.018	0.175	0.999	0.398	1.0*	0.533	0.968	0.437	0.959	0.214	1.0*	0.254	0.984	0.344	0.272	0.271
K02C18	3,5,5-Trichloro-4-oxo-2-pentenoic acid	0.052	0.141	1.0	0.337	0.0*	0.624	1.0*	0.334	0.0*	0.416	1.0	0.376	1.0	0.132	0.104	0.336
K02C19	3-Bromo-3-chloropropenoic acid	0.001	0.140	0.154	0.325	1.0	0.417	0.539	0.340	0.409	0.207	0.0†	0.420	0.994	0.266	0.0	0.312
K02C25	cis-4-Bromo-2-butenoic acid	0.849	0.393	0.0	0.213	1.0	0.395	0.163	0.293	0.002	0.152	0.093	0.094	0.0	0.378	1.0	0.177
K02C34	Trans-4-Bromo-2-butenoic acid	0.849	0.393	0.0	0.213	1.0	0.395	0.163	0.293	0.002	0.152	0.093	0.094	0.0	0.378	1.0	0.177
K02C13	3,3-Dibromopropenoic acid	0.139	0.054	0.066	0.327	0.032	0.402	0.999	0.266	1.0	0.187	0.0*	0.500	0.974	0.290	0.004	0.170
K02C14	3,3-Dichloropropenoic acid	0.003	0.129	0.968	0.217	0.999	0.338	0.885	0.236	0.410	0.223	0.0	0.108	0.999	0.251	0.026	0.230
K02C20	3-Chloro-2-butenoic acid	0.002	0.156	0.022	0.365	1.0†	0.492	0.739	0.264	0.163	0.303	0.0	0.387	0.995	0.182	0.0	0.304
K02C23	cis-2,3-Dibromopropenoic acid	0.011	0.034	1.0*	0.313	0.087	0.313	1.0	0.383	0.983	0.263	0.0†	0.361	0.957	0.302	0.0	0.203
K02C32	Trans-2,3-Dibromopropenoic acid	0.011	0.034	1.0*	0.313	0.087	0.313	1.0	0.383	0.983	0.263	0.0*	0.361	0.957	0.302	0.0	0.203
K02C24	cis-2,3-Dichloropropenoic acid	0.001	0.140	1.0†	0.348	1.0	0.324	0.998	0.333	0.001	0.220	0.125	0.375	0.998	0.310	0.001	0.209
K02C33	Trans-2,3-Dichloropropenoic acid	0.001	0.140	1.0†	0.348	1.0	0.324	0.998	0.333	0.001	0.220	0.125	0.375	0.998	0.310	0.001	0.209

* Outside the optimal prediction space of the model, but within permissible limits

† Outside the optimal prediction space of the model and exceeds permissible limits

Table E.7

Unsaturated haloacids predicted to be developmental toxicants

Index number	Chemical	Prob. Develop. Toxicity	Pred. outside OPS?*	LOAEL mg/kg	LOAEL Lower Limit mg/kg	LOAEL Upper Limit mg/kg
K02C38	2,3,4-Trichlorocyclohexadien-5-one carboxylic acid	1	yes	0.22	0.055	0.90
K02C20	3-Chloro-2-butenoic acid	0.854		2.8	0.53	15
K02C12	2-Chloro-2-butenoic acid	0.851		21	4.5	97
K02C36	3,4,4-Trichloro-2-methyl-4-oxo-2-butenoic acid	0.728		43	9.4	195

* OPS = optimal predictive space of the model

Table E.8

Haloalcohols predicted to be carcinogenic by at least one model within a TOPKAT® model in which the chronic LOAEL was predicted to be less than 10 mg/kg day^{-1}

Index Number	Chemical	Pred. Ames' Mut.	No. Pos. Cancer Pred. (8 poss.)	No. outside OPS*	LOAEL mg/kg	LOAEL Lower Limit mg/kg	LOAEL Upper Limit mg/kg
K03C07	2-Chloro-4-dimethylbutanol	+	1	1	1.1	0.1	10.7
K03C05	Bromochloroethanol	+	3	2	2.4	0.4	13.8
K03C12	Dichloroethanol	+	2	0	5.2	0.9	29.8
K03C03	3-Chloro-2-pentanol	+	1	1	8.5	1.6	46.4
K03C17	2,5,5-Trichloropentanol	-	2	1	10.0	1.9	54.0

* OPS = optimal predictive space of the model

Table E.9

Haloalcohols predicted to be developmental toxicants by TOPKAT®.

Index number	Chemical	Prob. Develop. Toxicity	Pred. outside OPS?*	LOAEL mg/kg	LOAEL Lower Limit mg/kg	LOAEL Upper Limit mg/kg
K03C23	1-Bromo-2-hydroxy-2-methylbutane	100%		62.2	14.3	270.8
K03C13	1,1,2,2,3-Pentachloroisobutanol	100%		147.8	34.2	638.5
K03C08	Chlorodinaphthol	100%		375	72.2	1900
K03C21	1-Bromo-2-hydroxy-2-methylpropane	100%		453.2	81.0	2500.0
K03C15	Tribromoethanol	94%		158.5	34.1	736.7
K03C16	Trichloroethanol	94%		193.3	43.9	850.8
K03C04	3-Chloro-2-methylbutanol	92%		79	10.9	573.5
K03C10	Dibromochloroethanol	79%		88.6	19.6	401
K03C06	Bromodichloroethanol	79%		95.2	21.2	426.8
K03C02	2-Bromoethanol	79%		225.2	41.2	1200

* OPS = optimal predictive space of the model

Table E.10
Haloaldehydes predicted to be carcinogenic by at least one TOPKAT model and predicted to have a chronic LOAEL less than 10 mg/kg day^{-1}

Index Number	Chemical	Pred. Ames' Mut.	No. Pos. Cancer Pred.	No. outside OPS*	LOAEL mg/kg	LOAEL Lower Limit mg/kg	LOAEL Upper Limit mg/kg
K04	**Dihaloacetaldehydes**						
K04C02	Bromochloroacetaldehyde	+	2	1	2.2	0.4	13.2
K04C04	Dichloroacetaldehdye	+	3	0	5.8	1.1	29.3
K10	Saturated haloaldehydes						
K10C07	2,3,4-Trichlorobutanal	+	5	4	1.0	0.2	5.9
K10C08	2,3-Dichloropropanal	+	2	0	1.1	0.2	6.3
K10C09	2,4,4-Trichlorobutanal	-	3	2	2.1	0.4	12.1
K10C14	3,4,4-Trichlorobutanal	-	4	4	2.5	0.5	13.8
K10C13	3,3-Dichloropropanal	+	2	0	2.7	0.5	14.6
K10C06	2,3,3-Trichloropropanal	-	4	3	5.3	1.0	28.1
K15	**Unsaturated haloaldehydes**						
K15C03	2,3-Dichloropropenal	+	4	1	1.0	0.2	5.9
K15C08	2-Bromopentadienal	+	2	1	1.4	0.2	7.7
K15C09	2-Chloropropenal	+	3	0	1.7	0.3	9.4
K15C06	2-Bromo-4-chlorobutenal	+	4	1	1.7	0.3	9.7
K15C04	2,4-Dichlorobutenal	+	5	1	2.0	0.4	11.0
K15C07	2-Bromobutenal	+	2	0	2.6	0.5	13.9
K15C01	2,3,3-Trichloropropenal	+	3	1	3.1	0.6	16.6
K15C05	2-Bromo-3-chlorobutenal	+	5	2	3.3	0.6	17.5
K15C02	2,3-Dichlorobutenal	+	3	0	3.7	0.7	19.5

* OPS = optimal predictive space of the model

Table E.11

Summary of haloaldehydes that were predicted to be carcinogenic in two species by TOPKAT® carinogenesis modules. Only those chemicals with predicted chronic LOAELs below 10 mg/kg are included

Index Number	Chemical	Pred. Ames' Mut.	Male rat	Female rat	Male mice	Female mice	LOAEL mg/kg
K04	Dihaloacetaldehydes						
K04C04	Dichloroacetaldehdye	+	pos		pos	pos	5.8
K10	Saturated haloaldehydes						
K10C13	3,3-Dichloropropanal	+	pos		pos		2.7
K15	Unsaturated haloaldehydes						
K15C03	2,3-Dichloropropenal	+	pos	pos	pos	pos*	1.0
K15C06	2-Bromo-4-chlorobutenal	+	pos	pos	pos		1.7
K15C09	2-Chloropropenal	+		pos		pos	1.7
K15C04	2,4-Dichlorobutenal	+	pos	pos	pos		2.0
K15C01	2,3,3-Trichloropropenal	+		pos	pos*	pos	3.1
K15C05	2-Bromo-3-chlorobutenal	+	pos*	pos	pos	pos	3.3
K15C02	2,3-Dichlorobutenal	+	pos*	pos	pos	pos	3.7

* Outside the optimal predictive space of the model, but within permissible limits

† Outside the optimal predictive space of the model and exceeds permissible limits

Table E.12

Similarity indices for haloaldehydes predicted to be carcinogenic by TOPKAT®. The strongest criteria to be satisfied are a probability of carcinogenic activity of greater than 0.7 accompanied by a similarity index that < 0.2. Chemicals with similarity indices < 0.3 are included so that those approaching the criteria can also be viewed

Rats

Index number	Chemical	NTP Male rat	Simil. index	FDA Male rat	Simil. index	NTP Female rat	Simil. index	FDA Female rat	Simil. index
K04C04	Dichloroacetaldehdye	0.422	0.1	0.822	0.18	0.0*	0.735	0.075*	0.353
K15C03	2,3-Dichloropropenal	0.001	0.151	0.807	0.236	0.005	0.314	0.894	0.285
K15C09	2-Chloropropenal	0.002	0.143	0.0*	0.613	0.999	0.357	0.957	0.273
K15C04	2,4-Dichlorobutenal	0.993	0.418	0.999*	0.373	1.0	0.376	0.267	0.369
K15C05	2-Bromo-3-chlorobutenal	0.001	0.144	0.988*	0.630	1.0†	0.523	0.978	0.557
K10C13	3,3-Dichloropropanal	0.999	0.370	0.430	0.190	0.0	0.256	0.0	0.409
K15C01	2,3,3-Trichloropropenal	0.0*	0.163	0.006†	0.380	1.0	0.274	0.0	0.345
K15C02	2,3-Dichlorobutenal	0.0*	0.163	0.006†	0.380	1.0	0.274	0.0	0.345

Mice

Index number	Chemical	NTP Male mice	Simil. index	FDA Male mice	Simil. index	NTP Female mice	Simil. index	FDA Female mice	Simil. index
K04C04	Dichloroacetaldehdye	0.000	0.130	0.997	0.251	1.000	0.13	0.0†	0.39
K15C03	2,3-Dichloropropenal	0.448	0.180	0.942	0.153	1.0*	0.153	0.0*	0.230
K15C09	2-Chloropropenal	0.0	0.062	0.024	0.404	1.0	0.084	0.0†	0.298
K15C04	2,4-Dichlorobutenal	1.0	0.354	0.968	0.197	0.015	0.351	0.0†	0.277
K15C05	2-Bromo-3-chlorobutenal	0.000	0.107	0.833	0.450	1.000	0.193	0.0†	0.326
K10C13	3,3-Dichloropropanal	0.0	0.152	0.998	0.211	0.0	0.198	0.0†	0.413
K15C01	2,3,3-Trichloropropenal	0.020	0.157	1.0†	0.228	1.000	0.331	0.0†	0.261
K15C02	2,3-Dichlorobutenal	0.020	0.157	1.0†	0.228	1.000	0.331	0.0†	0.261

* Outside the optimal prediction space of the model, but within permissible limits

† Outside the optimal prediction space of the model and exceeds permissible limits

Table E.13
Halogenated haloaldehydes predicted to be developmental toxicants, no restriction on chronic LOAEL

Index Number	Chemical	Prob. of develop. toxicity	Outside OPS?*	LOAEL mg/kg	LOAEL Lower Limit mg/kg	LOAEL Upper Limit mg/kg
K10	Saturated haloaldehydes					
K10C10	2-Bromo-2-methylpropanal	100%		60	14	257
K10C02	2,2,3-Trichloropropanal	98%		39	8.6	173
K10C04	2,2-Dichloropropanal	93%		62	14	268
K13	Trihaloacetaldehydes					
K13C03	Tribromoacetaldehyde	78%		63	14	286
K13C04	Trichloroacetaldehyde	78%		72	17	312
K13C02	Dibromochloroacetaldehyde	39%		28	5.9	129
K13C01	Bromodichloroacetaldehyde	39%		29	6.2	135

* OPS = optimal predictive space of the model

Table E.14
Haloketones predicted to be carcinogenic by one or more TOPKAT® models and for which the chronic NOAEL was predicted to below 10 mg/kg day^{-1}

Index number	Chemical	Pred. Ames' Mut.	# Pos. Cancer Pred. (8 poss.)	No. outside OPS*	LOAEL mg/kg	LOAEL Lower Limit mg/kg	LOAEL Upper Limit mg/kg
K05	**Unsaturated haloketones**						
K05C06	Trichlorocyclopentenedione	+	3	3	0.5	0.1	1.7
K05C03	1-Chloro-3-buten-2-one	+	2	1	3.0	0.6	16
K12	**Saturated haloketones**						
K12C11	1,1,3-Tribromo-3-chloropropanone	+	5	2	5.1	0.9	28

* OPS = optimal predictive space of the model

Table E.15

Probability scores and similarity indices for halogenated ketones predicted to be carcinogenic in eight carcinogenesis models

Index number	Chemical	NTP Male rat	Simil. index	FDA Male rat	Simil. index	NTP Female rat	Simil. index	FDA Female rat	Simil. index	NTP Male mice	Simil. index	FDA Male mice	Simil. index	NTP Female mice	Simil. index	FDA Female mice	Simil. index
K12C11	1,1,3-Tribromo-3-chloropropanone	0.143	0.079	0.914	0.286	0.0†	0.608	0.997†	0.395	0.999	0.384	0.882	0.452	0.501	0.215	0.907†	0.449

* Outside the optimal prediction space of the model, but within permissible limite

† Outside the optimal prediction space of the model and exceeds permissible limits

Table E.16
Haloketones predicted to be developmental toxicants

Index Name	Chemical	DTP Pred	# OPS*	LOAEL mg/kg	LOAEL Lower Limit mg/kg	LOAEL Upper Limit mg/kg
K05	**Unsaturated haloketones**					
K05C06	Trichlorocyclopentenedione	94%		0.5	0.1	1.7
K05C03	1-Chloro-3-buten-2-one	93%		3.0	0.6	15.6
K05C04	3-Bromo-4-methyl-3-penten-2-one	89%		84.9	19.5	369.6
K12	**Saturated haloketones**					
K12C21	1,3-Dibromopropanone	100%		59.3	13.2	265.6
K12C29	Hexachloropropanone	100%		327.7	74.6	1400.0
K12C23	1-Bromo-3-chloropropanone	100%		29.0	6.5	130.6
K12C04	1,1,1,3-Tetrachloropropanone	99%		99.3	23.0	428.7
K12C27	Bromopropanone	98%		47.3	10.9	205.3
K12C28	Chloropropanone	98%		47.5	11.0	205.1
K12C24	3,3-Dichloro-2-butanone	98%		118.9	27.2	519.4
K12C05	1,1,1-Tribromopropanone	95%		138.4	30.1	636.3
K12C08	1,1,1-Trichloropropanone	95%		169.2	38.6	741.4
K12C06	1,1,1-Trichloro-2-butanone	94%		157.5	36.1	687.2
K12C01	1,1,1,3,3-Pentabromo-3-chloropropanone	92%		82.6	17.4	392.9
K12C26	5-Bromo-2-pentanone	79%		53.6	12.3	233.8

* OPS = Optimal predictive space of the model

Table E.17
Members of the halonitriles, halonitromethane, and halopyrroles groups of disinfection by-products that were predicted to be carcinogenic by at least one TOPKAT® model and to have a chronic LOAEL of less than 10 mg/kg day^{-1}

Index number	Chemical	Pred. Ames' Mut.	No. Pos. Cancer Pred. (8 poss.)	No. outside OPS*	LOAEL mg/kg	LOAEL Lower Limit mg/kg	LOAEL Upper Limit mg/kg
K17	**Haloacetonitriles**						
K17C01	Bromochloroacetonitrile	+	2	1	0.1	0.0	0.7
K17C05	Dichloroacetonitrile	+	4	1	0.3	0.0	2.0
K17C03	Dibromoacetonitrile	+	2	1	0.9	0.2	5.9
K17C02	Bromodichloroacetonitrile	+	2	0	1.5	0.2	9.3
K17C04	Dibromochloroacetonitrile	+	2	1	1.5	0.2	8.9
K17C06	Tribromoacetonitrile	+	3	1	4.0	0.6	26
K17C07	Trichloroacetonitrile	+	4	1	4.5	0.7	29
K30	**Other halonitriles**						
K30C03	2,3-Dichloropropionitrile	+	4	1	0.0	0.0	0.4
K30C09	3,4-Dichlorobutanenitrile	+	4	2	0.1	0.0	0.6
K30C04	2,3-Dichloro-3-bromopropionitrile	+	3	1	0.1	0.0	0.7
K30C07	3,3-Dichloropropionitrile	+	1	0	0.1	0.0	0.9
K30C02	2,3-Dibromopropionitrile	+	4	2	0.1	0.0	1.1
K30C13	Cyanogen chloride	+	1	0	0.2	0.0	1.5
K30C12	Cyanogen bromide	+	3	0	0.3	0.0	1.6
K30C06	3,3-Dibromopropionitrile	+	2	0	0.3	0.0	2.5
K30C11	cis-2,3,4-Trichloro-2-butenenitrile	+	5	5	0.7	0.1	4.5
K30C14	trans-2,3,4-Trichloro-2-butenenitrile	+	5	5	0.7	0.1	4.5
K30C08	3,3-Dichloropropenenitrile	+	5	0	0.8	0.1	5.0
K30C01	2,3,3-Trichloropropenenitrile	+	3	2	1.0	0.2	6.4
K30C21	2,2,3-Trichloropropionitrile	+	3	2	1.7	0.3	11
K30C10	3-Bromobutanenitrile	+	4	2	2.5	0.3	18
K30C20	2,2-Dibromopropionitrile	+	1	0	2.8	0.4	19
K30C19	2,2-Dichloropropionitrile	+	3	0	2.9	0.4	20
K30C18	2-Chloroproprionitrile	+	1	0	3.8	0.4	36
K30C05	3,3,3-Trichloropropionitrile	-	4	1	3.9	0.6	26
K26	**Halonitromethanes**						
K23C07	Dichloronitromethane	-	2	0	1.9	0.34	10
K26C02	Bromonitromethane	+	1	0	3.0	0.6	15
K26C04	Dibromonitromethane	+	2	0	6.4	1.2	33

* Outside the optimal prediction space of the model, but within permissible limits

Table E.18
Members of the halonitrile and halonitromethane groups that are predicted to be carcinogenic in at least one sex of two species

Index Number	Chemical	Male rat	Female rat	Male mice	Female mice	LOAEL mg/kg
K17	**Haloacetonitriles**					
K17C05	Dichloroacetonitrile	pos		pos	pos	0.3
K17C02	Bromodichloroacetonitrile	pos			pos	1.5
K17C07	Trichloroacetonitrile	pos		pos*	pos	4.5
K30	**Other halonitriles**					
K30C03	2,3-Dichloropropionitrile		pos	pos	pos	0.04
K30C02	2,3-Dibromopropionitrile		pos	pos	pos	0.15
K30C12	Cyanogen bromide		pos	pos		0.25
K30C06	3,3-Dibromopropionitrile		pos		pos	0.35
K30C08	3,3-Dichloropropenenitrile	pos		pos	pos	0.77
K30C15	3-Bromopropionitrile		pos	pos	pos	0.89
K30C21	2,2,3-Trichloropropionitrile	pos		pos†	pos*	
K30C10	3-Bromobutanenitrile		pos	pos	pos*	2.5
K30C19	2,2-Dichloropropionitrile	pos		pos	pos	2.9
K30C05	3,3,3-Trichloropropionitrile	pos		pos*	pos	3.9
K26	**Halonitromethanes**					
K26C07	Dichloronitromethane		pos	pos		1.9
K26C04	Dibromonitromethane		pos	pos		6.4

* Outside the optimum predictive space of the model, but is within permissible limits.
† Outside the optimum predictive space of the model and exceeds permissible limits

Table E.19

Predicted probabilities that the listed halonitriles are carcinogenic and the associated similarity indices of the closet compound in the data base for each of the eight models used

Index Name	Chemical	NTP Male rat	Simil. index	FDA Male rat	Simil. index	NTP Female rat	Simil. index	FDA Female rat	Simil. Index	NTP Male mice	Simil. index	FDA Male mice	Simil. index	NTP Female mice	Simil. index	FDA Female mice	Simil. index
K17C05	Dichloroacetonitrile	0.413	0.101	0.719	0.2	0.000	0.795	0.000	0.374	0.000	0.208	0.999	0.280	0.994	0.02	0.777*	0.443
K17C07	Trichloroacetonitrile	0.89	0.03	1.000	0.1	0.000	0.356	0.050	0.135	0.000	0.302	0.938*	0.161	0.609	0.533	1.000	0.29
K30C15	3-Bromopropionitrile	0.059	0.088	0.001	0.326	1.000	0.140	0.959	0.220	0.0	0.235	0.903	0.885	0.999	0.147	1.000	0.238
K30C02	2,3-Dibromopropionitrile	0.029	0.058	0.000	0.375	1.000	0.151	0.0*	0.498	0.985*	0.224	0.540	0.873	1.000	0.122	1.0†	0.361
K30C08	3,3-Dichloropropenenitrile	0.004	0.167	0.794	0.214	0.241	0.258	0.162	0.306	0.997	0.224	0.980	0.182	0.998	0.146	0.991	0.287
K30C19	2,2-Dichloropropionitrile	0.190	0.073	0.984	0.242	0.0	0.559	0.287	0.286	0.0	0.233	0.725	0.277	0.144	0.392	0.763	0.263
K30C05	3,3,3-Trichloropropionitrile	0.894	0.101	0.875	0.165	0.000	0.424	0.029†	0.195	0.000	0.303	0.010	0.195	0.229	0.458	1.000	0.307
K17C02	Bromodichloroacetonitrile	0.273	0.094	0.990	0.2	0.002	0.571	0.000	0.266	0.000	0.350	0.144*	0.294	0.609	0.533	1.000	0.326
K30C03	2,3-Dichloropropionitrile	0.029	0.058	0.08	0.266	0.998	0.130	0.0*	0.486	0.001	0.212	1.0	0.222	1.0	0.029	1.0*	0.029
K30C21	2,2,3-Trichloropropionitrile	0.487	0.068	0.984	0.201	0.0	0.459	0.0*	0.318	0.0	0.338	0.999†	0.090	0.593	0.538	1.0*	0.320
K30C06	3,3-Dibromopropionitrile	0.290	0.080	0.001	0.295	1.000	0.138	0.001	0.456	0.001	0.315	0.543	0.873	0.007	0.286	1.000	0.301
K30C12	Cyanogen bromide	0.089	0.056	0.017	0.288	0.999	0.239	0.959	0.220	0.0	0.128	0.753	0.844	0.011	0.000	0.055	0.252

* Outside the optimal prediction space of the model, but within permissible limite

† Outside the optimal prediction space of the model and exceeds permissible limits

Table E.20
Halonitriles, halonitromethanes, and halopyrroles predicted by TOPKAT® to be developmental toxicants

Index number	Chemical	Prob. Develop. Toxicity	Pred. outside OPS?*	LOAEL mg/kg	LOAEL Lower Limit mg/kg	LOAEL Upper Limit mg/kg
K17	**Haloacetonitriles**					
K17C06	Tribromoacetonitrile	1		4	0.62	26
K17C07	Trichloroacetonitrile	1		4.5	0.70	29
K17C02	Bromodichloroacetonitrile	0.998		1.5	0.25	9.3
K17C04	Dibromochloroacetonitrile	0.998		1.5	0.25	8.9
K17C05	Dichloroacetonitrile	0.726		0.29	0.04	2
K17C03	Dibromoacetonitrile	0.726		0.94	0.15	5.9
K30	**Other halonitriles**					
K30C10	3-Bromobutanenitrile	1		2.5	0.34	18
K30C05	3,3,3-Trichloropropionitrile	1		3.9	0.58	26
K30C15	3-Bromopropionitrile	1		0.89	0.14	5.7
K30C17	3-Chloropropionitrile	1		0.93	0.14	6
K30C20	2,2-Dibromopropionitrile	1		2.8	0.42	19
K30C19	2,2-Dichloropropionitrile	1		2.9	0.44	20
K30C21	2,2,3-Trichloropropionitrile	1		1.7	0.27	11
K30C16	2-Bromopropionitrile	0.995		20	2.2	179
K30C18	2-Chloropropionitrile	0.995		3.8	0.41	36
K30C07	3,3-Dichloropropionitrile	0.988		0.11	0.013	0.93
K30C06	3,3-Dibromopropionitrile	0.988		0.35	0.05	2.5
K30C08	3,3-Dichloropropenenitrile	0.975		0.77	0.12	5
K30C03	2,3-Dichloropropionitrile	0.974		0.04	0.005	0.36
K30C02	2,3-Dibromopropionitrile	0.974		0.15	0.021	1.1
K30C01	2,3,3-Trichloropropenenitrile	0.923		1	0.16	6.4
K30C13	Cyanogen chloride	0.869		0.23	0.04	1.5
K30C12	Cyanogen bromide	0.869		0.25	0.04	1.6
K26	**Halonitromethanes**					
K26C02	Bromonitromethane	0.979		3	0.59	15
K26C06	Trichloronitromethane	0.863		34	7.2	157
K26C05	Tribromonitromethane	0.863		38	8.1	182
K27	**Halopyrroles**					
K27C01	2,3,5-Tribromopyrrole	0.837		564	132	2400

* OPS = optimal predictive space of the model

Table E.21

Haloalkanes and haloalkenes predicted to have chronic LOAELs below 10 mg/kg day^{-1} and to be carcinogenic by at least one TOPKAT® model

Index Name	Chemical	Pred. Ames' Mut.	No.Pos. Cancer Pred.	No. outside OPS*	LOAEL mg/kg	LOAEL Lower Limit mg/kg	LOAEL Upper Limit mg/kg
K09	**Other haloalkanes**						
K09C02	1,2,3-Trichloropropane	+	6	2	4.7	0.8	27
K06	**Haloalkenes**						
K06C03	Hexachlorocyclopentadiene	-	4	3	0.01	0.003	0.04
K06C05	Tetrachlorocyclopropene	-	4	3	1.5	0.4	4.7

* OPS = optimal predictive space of the model

Table E.22
Members of the haloalkane and haloalkene groups that are predicted to be carcinogenic in more than one species and have a similarity index of < 0.3

Index number	Chemical	NTP Male rat	Simil. index	FDA Male rat	Simil. index	NTP Female rat	Simil. index	FDA Female rat	Simil. index	NTP Male mice	Simil. index	FDA Male mice	Simil. index	NTP Female mice	Simil. index	FDA Female mice	Simil. index
K09C02	1,2,3-Trichloropropane	0.683	0.0	0.958	0.130	0.905	0.0	0.0*	0.411	1.0	0.0	1.0†	0.138	1.0	0.0	1.0†	0.380

* Outside the optimal prediction space of the model, but within permissible limite
† Outside the optimal prediction space of the model and exceeds permissible limits

Table E.23

Other haloalkanes (K09) and haloalkenes (K06) predicted to be developmental toxicants

Index Number	Chemical	DTP Pred	Outside OPS?*	LOAEL mg/kg	LOAEL Lower Limit mg/kg	LOAEL Upper Limit mg/kg
K09	**Other haloalkanes**					
K09C12	Hexachloroethane	98%		492.1	107.1	2300.0
K09C10	Dibromomethane	95%		89.1	19.0	418.1
K09C11	Dichloromethane	95%		107.8	24.7	471.1
K06	**Haloalkenes**					
K06C01	Bromocyclohexene	97%	yes	664.4	205.8	2100.0

* OPS = optimal predictive space of the model

Table E.24
Haloesters, haloamides, and halothiophenes predicted to be carcinogenic in at least one TOPKAT® model and to have a chronic LOAEL < 10 mg/kg day^{-1}

Index Name	Chemical	Pred. Ames' Mut.	No. Pos. Cancer Pred. (8 poss.)	No. outside OPS*	LOAEL mg/kg	LOAEL Lower Limit mg/kg	LOAEL Upper Limit mg/kg
K07	**Haloesters**						
K07C05	Methyl bromochloroacetate	-	3	2	3.9	0.8	20
K23	**Haloamides**						
K23C20	2,3-Dichloropropamide	+	6	1	0.50	0.08	3.2
K23C01	3-Bromo-3-Chloropropamide	+	3	2	0.6	0.1	3.7
K23C05	3,3-Dichloropropamide	+	2	0	1.2	0.2	7.3
K23C08	Bromochloroacetamide	+	5	2	1.5	0.3	8.5
K23C19	2,3-Dibromopropamide	+	4	3	1.8	0.32	9.7
K23C21	2-Bromo-3-chloropropamide	+	3	2	2.2	0.38	12.7
K23C12	Dichloroacetamide	+	3	1	3.2	0.6	17
K23C03	3,3-Dibromopropamide	+	3	1	3.6	0.6	20
K28	**Halothiophenes**						
K28C07	Tetrachlorothiophene	-	3	1	7.7	1.3	41

* OPS = optimal predictive space of the model

Table E.25
Haloesters, haloamides, and halothiophenes predicted by TOPKAT® models to be carcinogenic in both mice and rats

Index Number	Chemical	Pred. Ames' Mut.	Male rat	Female rat	Male mice	Female mice	LOAEL mg/kg
K23C20	2,3-Dichloropropamide	+		pos*	pos*	pos	0.50
K23C08	Bromochloroacetamide	+	pos		pos	pos	1.5
K23C19	2,3-Dibromopropamide	+	pos	pos*	pos	pos	1.8
K23C21	2-Bromo-3-chloropropamide	+	pos	pos*	pos*		2.2
K23C12	Dichloroacetamide	+	pos		Pos	pos*	3.2

* Outside the optimal predictive space of the model, but within permissible limits
† Outside the optimal predictive space of the model and exceeds permissible limits

Table E.26

Probabilities and similarity indices for each TOPKAT® carcinogenesis model for haloamides

Rats

Index number	Chemical	NTP male rat	Simil. index	FDA Male rat	Simil. index	NTP Female rat	Simil. index	FDA Female Rat	Simil. Index
K23C20	2,3-Dichloropropamide	0.028	0.073	0.988	0.251	1.0*	0.344	0.0*	0.355
K23C08	Bromochloroacetamide	0.012	0.065	0.925	0.299	0.0†	0.613	0.011*	0.389
K23C21	2-Bromo-3-chloropropamide	0.028	0.073	0.829	0.290	1.0*	0.318	0.002*	0.379
K23C05	3,3-Dichloropropamide	0.081	0.091	0.996	0.172	0.117†	0.439	0.0*	0.384
K23C12	Dichloroacetamide	0.219	0.080	0.999	0.191	0.0†	0.729	0.10*	0.373
K23C19	2,3-Dibromopropamide	0.028	0.073	0.136	0.355	1.0*	0.314	0.057*	0.427

Mice

Index number	Chemical	NTP male mice	Simil. index	FDA male mice	Simil. index	NTP female mice	Simil. index	FDA female mice	Simil. index
K23C20	2,3-Dichloropropamide	0.990	0.246	1.0	0.244	0.961	0.129	1.0	0.244
K23C08	Bromochloroacetamide	1.0*	0.332	0.798	0.485	0.995	0.178	0.925*	0.433
K23C21	2-Bromo-3-chloropropamide	1.0*	0.198	0.233	0.383	0.0	0.193	0.0	0.077
K23C05	3,3-Dichloropropamide	0.006	0.248	0.003	0.090	0.009	0.089	0.005*	0.424
K23C12	Dichloroacetamide	0.122	0.312	1.000	0.337	0.271	0.081	0.997*	0.403
K23C19	2,3-Dibromopropamide	1.0*	0.226	0.628	0.222	0.975	0.587	1.0	0.321

* Outside the optimal prediction space of the model, but within permissible limite

† Outside the optimal prediction space of the model and exceeds permissible limits

Table E.27

Haloesters, haloamides, and halothiophenes predicted to be developmental toxicants

Index number	Chemical	Prob. Develop. Toxicity	Pred. outside OPS?*	LOAEL mg/kg	LOAEL Lower Limit mg/kg	LOAEL Upper Limit mg/kg
K07	**Haloesters**					
K07C03	2-Bromoethanolacetate	0.979		33	7.4	144
K07C04	2-Chloroethanolacetate	0.979		37	8.3	160
K23	**Haloamides**					
K23C06	3,3,3-Tribromopropamide	1.0		24	4.8	116
K23C13	Tribromoacetamide	1.0		26	5.5	124
K23C14	Trichloroacetamide	1.0		28	6.1	133
K23C07	3,3,3-Trichloropropamide	1.0		33	7	156
K23C16	2,2-Dibromopropamide	1.0		23	4.9	110
K23C18	2,2-Dichloropropamide	1.0		24	5.0	114
K23C21	2,2,3-Trichloropropionamide	1.0		16	3.3	74
K23C04	3,3-Dibromo-3-chloropropamide	0.999		17	3.5	79
K23C02	3-Bromo-3-dichloropropamide	0.999		19	4	88
K23C09	Bromodichloroacetamide	0.998		14	3.1	67
K23C11	Dibromochloroacetamide	0.998		14	3.1	67
K23C15	2-Bromopropamide	0.990		641	70	5900
K23C17	2-Chloropropamide	0.990		127	14.2	1100
K23C05	3,3-Dichloropropamide	0.981		1.2	0.18	7.3
K23C03	3,3-Dibromopropamide	0.981		3.6	0.64	20
K23C19	2,3-Dibromopropamide	0.956		1.8	0.32	9.7
K23C21	2-Bromo-3-chloropropamide	0.956		2.2	0.38	12.7
K23C20	2,3-Dichloropropamide	0.956		0.50	0.08	3.2
K23C01	3-Bromo-3-Chloropropamide	0.949		0.62	0.10	3.7
K28	**Halothiophenes**					
K28C07	Tetrachlorothiophene	1		7.7	1.3	45
K28C05	Bromotrichlorothiophene	1		18	3.5	88
K28C03	2,3-Dibromo-4,5-dichlorothiophene	1		39	8.6	175
K28C04	2,4-Dibromo-3,5-dichlorothiophene	1		39	8.6	175
K28C01	2,3,4-Tribromothiophene	1		150	36	627
K28C02	2,3,5-Tribromothiophene	1		150	36	627
K28C06	Tetrabromothiophene	1	yes	178	43	742

* OPS = optimal predictive space of the model

Table E.28
Halodicarboxylic and halotricarboxylic acids predicted to be carcinogenic by at least one TOPKAT® model and have a chronic LOAEL < 10 mg/kg day^{-1}

Index Name	Chemical	Pred. Ames' Mut.	No. Pos. Cancer Pred.	No. outside OPS*	LOAEL mg/kg	LOAEL Lower Limit mg/kg	LOAEL Upper Limit mg/kg
K11	**Saturated halodicarboxylic acids**						
K11C23	Chlorobutanedioic	+	2	1	0.9	0.1	5.1
K11C13	2,3-Dichloropentanedioic	+	3	1	0.9	0.1	5.4
K11C22	3-Chlorohexanedioic	+	3	0	2.1	0.4	12
K11C17	2-Chloroheptanedioic	+	2	1	2.9	0.5	16
K11C18	2-Chlorohexanedioic	+	2	1	2.9	0.5	16
K11C21	2-Chlorooctanedioic	+	2	1	3.5	0.7	19
K11C14	2,4-Dichloropentanedioic	+	4	1	3.7	0.6	22
K11C20	2-Chlorononanedioic	+	2	1	4.2	0.8	23
K11C03	2,2,3-Trichloro-4-hydroxypentanedioic	+	4	3	4.3	0.8	25
K11C24	Trichlorobutanedioic	+	4	3	5.0	0.9	27
K11C06	2,2-Dichloro-4-hydroxypentanedioic	-	1	0	5.5	0.9	34
K11C15	2-(Dichloromethyl)-2,3-dichloro-3-hydroxybutanedioic	-	3	3	6.2	1.1	34
K16	**Unsaturated halodicarboxylic acids**						
K16C04	2-Chloro-3-pentenedioic	+	2	1	0.2	0.0	1.6
K16C09	trans-2-bromobutendioic	+	4	1	1.0	0.2	6.3
K16C10	trans-2-chlorobutenedioic	+	4	2	1.0	0.2	6.2
K16C15	cis-2-Bromobutenedioic	+	4	1	1.0	0.2	6.3
K16C16	cis-2-Chlorobutenedioic	+	4	2	1.0	0.2	6.2
K16C06	3-Chloro-2-pentenedioic	+	3	0	1.7	0.3	9.8
K16C18	Trichloropentenedioic	+	5	3	2.8	0.4	17

* OPS = optimal predictive space of the model

Table E.29
TOPKAT® predictions of carcinogenicity of halodicarboxylic acids in mice and rats

Index Number	Chemical	Prob. Ames' Mut.	Male rat	Female rat	Male mice	Female mice	LOAEL mg/kg
K11	**Saturated halodicarboxylic acids**						
K11C22	3-Chlorohexanedioic	+	pos	pos	pos		2.1
K11C14	2,4-Dichloropentanedioic	+	pos	pos	pos		3.7
K16	**Unsaturated halodicarboxylic acids**						
K16C09	trans-2-bromobutendioic	+	pos*	pos		pos	1
K16C10	trans-2-chlorobutenedioic	+	pos*	pos		pos	1
K16C15	cis-2-Bromobutenedioic	+	pos*	pos		pos	1
K16C16	cis-2-Chlorobutenedioic	+	pos*	pos		pos	1
K16C06	3-Chloro-2-pentenedioic	+	pos	pos		pos	1.7

* Outside the optimal predictive space of the model, but within permissible limits
† Outside the optimal predictive space of the model and exceeds permissible limits

Table E.30
Probability scores and similarity indices for halodicarboxylic acids predicted to be carcinogenic by TOPKAT®

Rats

Index Number	Chemical	NTP Male rat	Simil. index	FDA Male rat	Simil. index	NTP Female rat	Simil. index	FDA Female rat	Simil. index
K11	**Sat. halodicarboxylic acids**								
K11C22	3-Chlorohexanedioic	0.009	0.140	0.922	0.384	0.973	0.513	0.0*	0.490
K11C14	2,4-Dichloropentanedioic acid	0.341	0.092	0.915	0.372	1.0	0.408	0.0*	0.327
K16	**Unsat. halodicarboxylic acids**								
K16C09	trans-2-Bromobutendioic acid	0.033	0.07	1.0†	0.317	1.000	0.323	0.986	0.347
K16C10	trans-2-Chlorobutenedioic acid	0.004	0.149	1.0†	0.337	1.0*	0.437	0.899	0.373
K16C15	cis-2-Bromobutenedioic acid	0.033	0.070	1.0†	0.317	1.000	0.323	0.986	0.347
K16C16	cis-2-Chlorobutenedioic acid	0.004	0.149	1.0†	0.337	1.0*	0.437	0.899	0.373
K16C06	3-Chloro-2-pentenedioic acid	0.003	0.214	0.996	0.384	1.000	0.587	0.004	0.370

Mice

Index Number	Chemical	NTP Male mice	Simil. index	FDA Male mice	Simil. index	NTP Female mice	Simil. index	FDA Female mice	Simil. index
K11	**Sat. halodicarboxylic acids**								
K11C22	3-Chlorohexanedioic	0.973	0.378	0.001*	0.541	0.133†	0.123	0.0†	0.461
K11C14	2,4-Dichloropentanedioic acid	0.967	0.431	1.0†	0.494	0.0	0.154	0.363*	0.437
K16	**Unsat. halodicarboxylic acids**								
K16C09	trans-2-Bromobutendioic acid	0.001	0.193	0.000	0.214	0.999	0.260	0.0	0.173
K16C10	trans-2-Chlorobutenedioic acid	0.000	0.206	0.0†	0.422	1.000	0.261	0.0*	0.386
K16C15	cis-2-Bromobutenedioic acid	0.001	0.193	0.000	0.214	0.999	0.260	0.000	0.173
K16C16	cis-2-Chlorobutenedioic acid	0.000	0.206	0.0†	0.422	1.000	0.261	0.0*	0.386
K16C06	3-Chloro-2-pentenedioic acid	0.002	0.207	0.0*	0.469	0.999	0.219	0.0*	0.376

* Outside the optimal prediction space of the model, but within permissible limite

† Outside the optimal prediction space of the model and exceeds permissible limits

Table E.31
Halodicarboxylic and halotricarboxylic acids predicted by TOPKAT® to be developmental toxicants

Index number	Chemical	Prob. Develop. Toxicity	Pred. outside OPS?*	LOAEL mg/kg	LOAEL Lower Limit mg/kg	LOAEL Upper Limit mg/kg
K11	**Saturated halodicarboxylic acids**					
K11C19	2-Chloromethylpropanedioic	1	yes	13	2.3	70
K11C10	2,2-Dichloropentanedioic	1		35	7.5	158
K11C16	2-Chloro-2-methylpropanedioic	1		62	14	268
K11C11	2,2-Dichloropropanedioic	1		68	16	300
K11C07	2,2-Dichlorobutanedioic	0.958		26	5.5	123
K11C06	2,2-Dichloro-4-hydroxypentanedioic	0.834	yes	5.5	0.88	34
K16	**Unsaturated halodicarboxylic acids**					
K16C05	2-Chloro-3,4-dicarboxy-2-pentendioic	1	yes	25	5	122
K16C03	2-Chloro-3-(trichloroacetyl)butenedioc	1		34	7.1	165
K16C13	Tetrachloropentenedioic	0.979		21	4.1	107
K16C14	trans-2,3-Dibromobutenedioic	0.968		475	10.1	215
K16C11	cis-Dichlorobutenedioic	0.968		52	11.1	240
K16C17	trans-Dichlorobutenedioic	0.968		52	11.1	240
K16C09	trans-2-bromobutendioic	0.947		1	0.17	6.3
K16C10	trans-2-chlorobutenedioic	0.947		1	0.17	6.2
K16C15	cis-2-Bromobutenedioic	0.947		1	0.17	6.3
K16C16	cis-2-Chlorobutenedioic	0.947		1	0.17	6.2
K16C12	Dichloropentenedioic	0.749		16	3	80

* OPS = optimal predictive space of the model

Table E.32
Haloaromatic chemicals predicted to be developmental toxicants

Index number	Chemical	Prob. Develop. Toxicity	Pred. outside OPS?*	LOAEL mg/kg	LOAEL Lower Limit mg/kg	LOAEL Upper Limit mg/kg
K08	**Halophenols**					
K08C17	Trichloro-trihydroxybenzene	0.966		12	2	68
K08C16	Pentachlorophenol	0.871		12	1.9	69
K24	Haloaromatic acids					
K24C24	4-Chlorophenylacetic acid	0.945		21	3.6	118
K29	**Other haloaromatics**					
K29C09	3,5-Dichloroethylbenzene	0.836		76	15	382
K29C10	3,5-Dichlorotoluene	0.836		76	15	382

* OPS = optimal predictive space of the model

Table E.33
Analogs of MX that are predicted to be carcinogenic in at least one TOPKAT® carcinogenesis model and with a predicted chronic LOAEL below 10 mg/kg day^{-1}

Index number	Chemical	Pred. Ames' Mut.	# Pos. Cancer Pred. (8 poss.)	No. outside OPS*	LOAEL mg/kg	LOAEL Lower Limit mg/kg	LOAEL Upper Limit mg/kg
K21C05	CMCF	-	3	1	0.03	0.004	0.2
K21C11	Red-MX	+	3	3	0.07	0.013	0.4
K21C08	MX	-	2	2	0.10	0.023	0.4
K21C01	4-Bromochloromethyl-MX	-	5	3	0.16	0.04	0.7
K21C03	5-Hydroxy-4-trichloromethyl-2-furanone	-	4	3	0.21	0.04	1.2
K21C02	4-Dibromomethyl-MX	-	3	2	0.25	0.06	1.1
K21C12	Tribromomethyl-MX	-	3	1	0.41	0.09	1.8
K21C06	E-MX	+	4	3	0.67	0.11	4.1
K21C07	MCA	-	2	1	0.79	0.22	2.8
K21C10	Ox-EMX	-	4	3	5.2	0.94	29

* OPS = optimal predictive space of the model

Table E.34
TOPKAT® predictions of carcinogenicity of MX and its analogs in rats and mice.

Index Name	Chemical	Pred. Ames' Mut.	Male rat	Female rat	Male mice	Female mice	LOAEL mg/kg
K21C05	CMCF	-		pos	pos	pos*	0.03
K21C11	Red-MX	+	pos*		pos*	pos*	0.07
K21C08	MX	-			pos*	pos*	0.10
K21C01	4-Bromochloromethyl-MX	-		pos*	pos	pos	0.16
K21C03	5-Hydroxy-4-trichloromethyl-2-furanone	-		pos*	pos*	pos	0.21
K21C02	4,4-Dibromomethyl-MX	-		pos*	pos	pos*	0.25
K21C12	Tribromomethyl-MX	-			pos	pos*	0.41
K21C06	E-MX	+	pos*	pos*	pos*	pos	0.67
K21C07	MCA	-			pos	pos*	0.79
K21C10	Ox-EMX	-	pos*	pos*	pos		5.2

* Outside the optimal predictive space of the model, but within permissible limits
† Outside the optimal predictive space of the model and exceeds permissible limits.

Table E.35

Probability scores and similarity indices for TOPKAT® predictions of carcinogenicity of MX and CMCF

Index Number	Chemical	NTP Male rat	Simil. index	FDA Male rat	Simil. index	NTP Female rat	Simil. index	FDA Female rat	Simil. index	NTP Male mice	Simil. index	FDA Male mice	Simil. index	NTP Female mice	Simil. index	FDA Female mice	Simil. index
K21C05	CMCF	0.0	0.205	0.0*	0.459	0.997	0.433	0.0	0.411	0.084	0.219	1.0	0.193	0.14	0.221	1.0†	0.457
K21C08	MX	0.002	0.169	0.0†	0.472	0.090†	0.453	0.0†	0.460	0.431	0.399	1.0†	0.242	0.032	0.111	1.0†	0.500

* Outside the optimal prediction space of the model, but within permissible limits

† Outside the optimal prediction space of the model and exceeds permissible limits

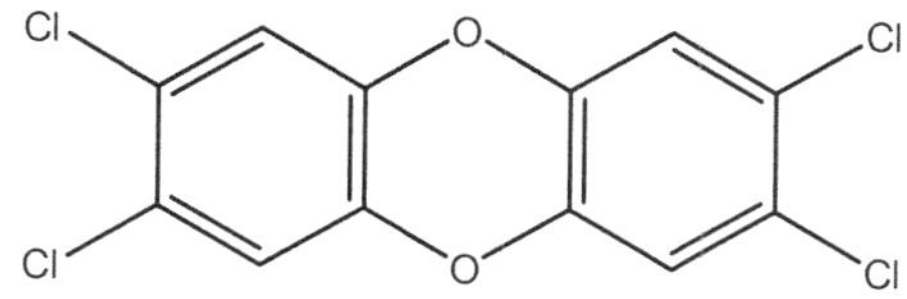

2,3,7,8-Tetrachlorodibenzo-*p*-dioxin
TD_{50} = 0.0000457 mg/kg

Arochlor 1260
Mixture of Polychlorinated Biphenyls
TD_{50} = 1.74

Mixture of Polybrominated biphenyls
TD_{50} = 0.322

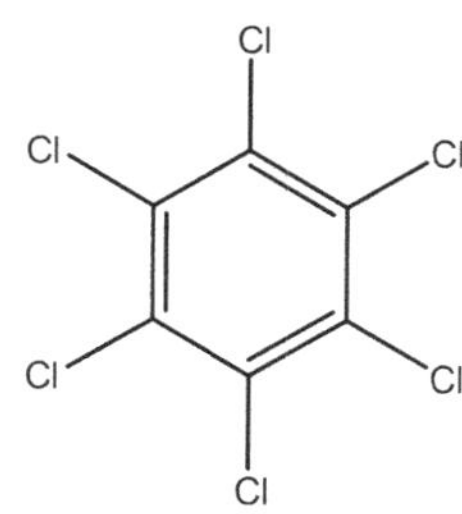

Hexachlorobenzene
TD_{50} = 3.86

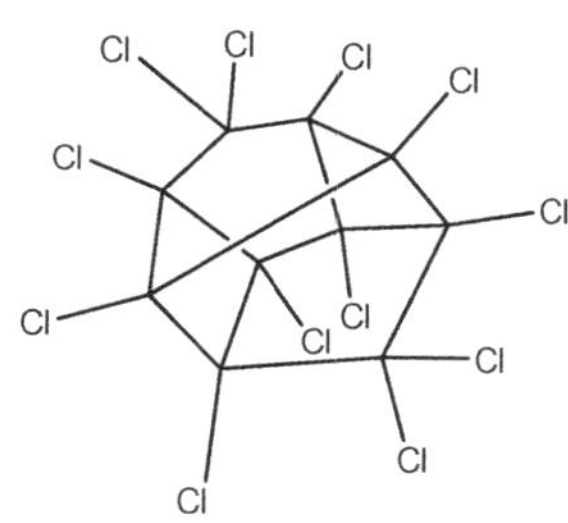

Mirex
TD_{50} = 1.77
Photo-Mirex
TD_{50} = 1.46

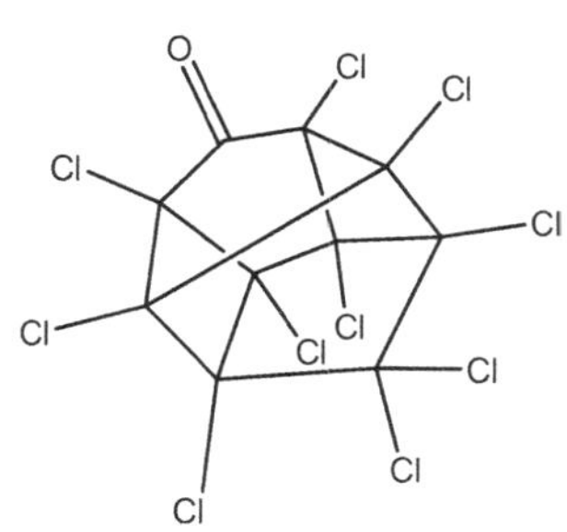

Kepone
TD_{50} = 2.96

Figure E.1 Structures of halogenated compounds that are potent carcinogens

(continued)

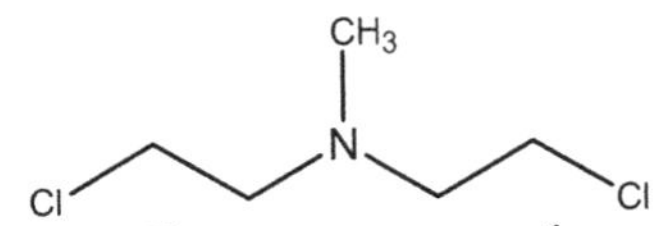

Nitrogen mustard
$TD_{50} = 0.0114$

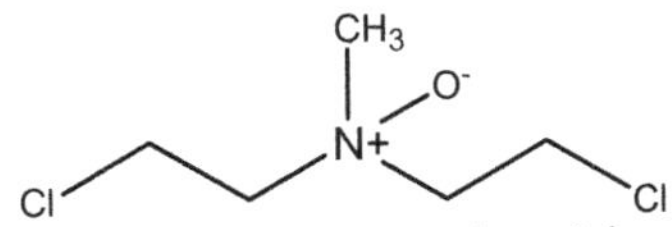

Nitrogen mustard oxide
$TD_{50} = 0.764$

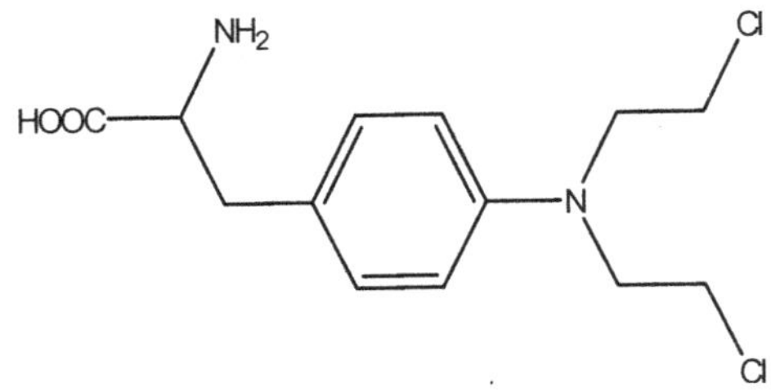

Mephalan
$TD_{50} = 0.0938$

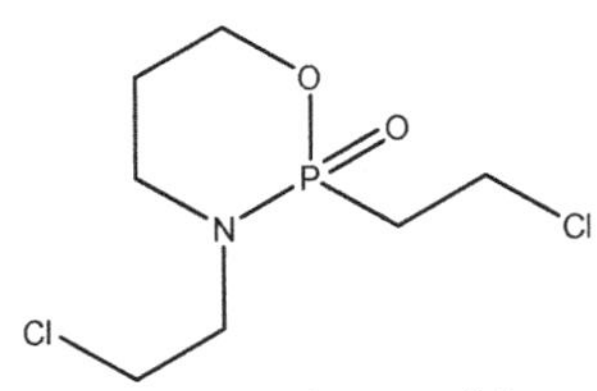

Isophosphoramide
$TD_{50} = 0.739$

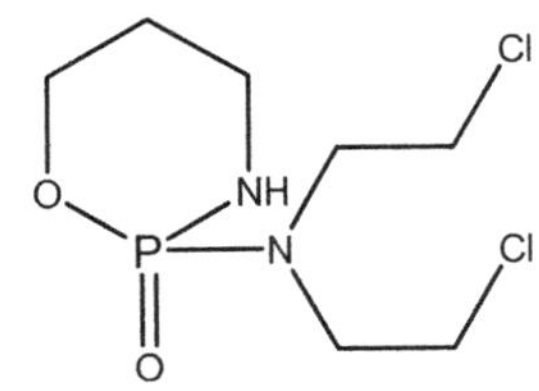

Cyclophosphamide
$TD_{50} = 2.23$

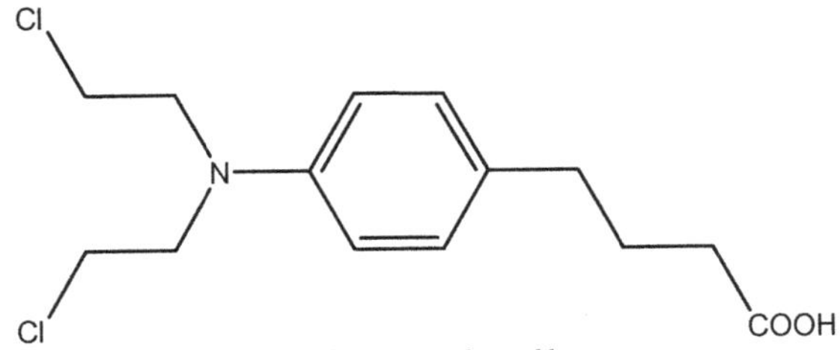

Chlorambucil
0.896

Figure E.1. (Continued)

(continued)

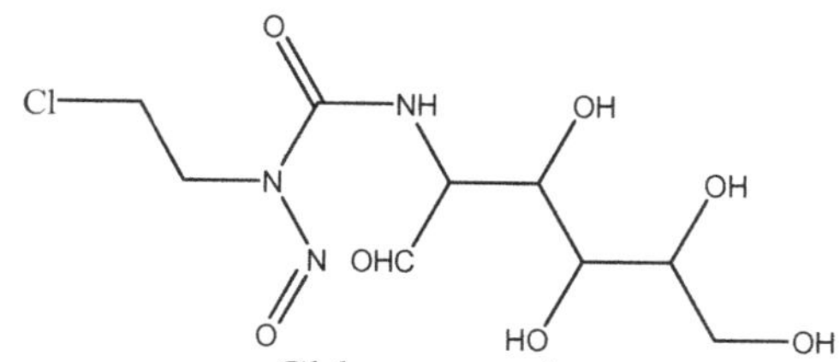

Chlorozotocin
$TD_{50} = 0.0375$

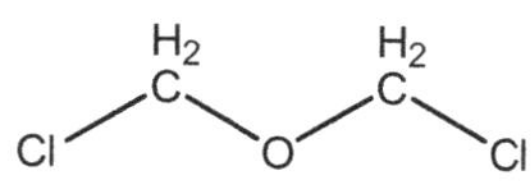

Bis(chloromethyl)ether
$TD_{50} = 0.00357$

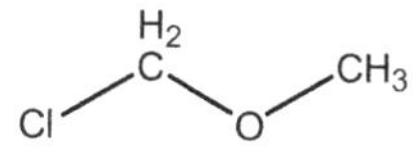

Chloromethyl methyl ether
$TD_{50} = 5.5$

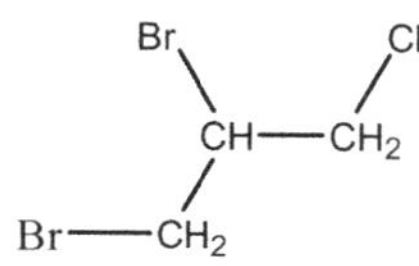

1,2-Dibromo-3-chloropropane
$TD_{50} = 0.259$

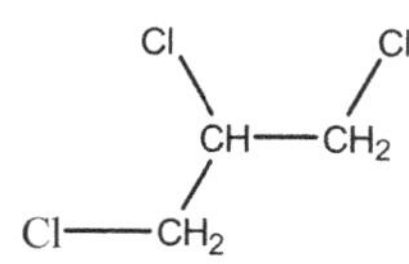

1,2,3-Trichloropropane
$TD_{50} = 1.35$

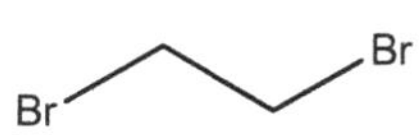

1,2-Dibromoethane
$TD_{50} = 1.52$

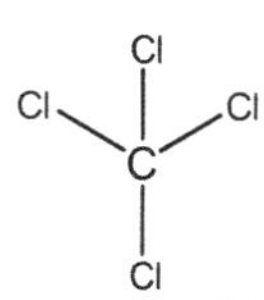

Carbon tetrachloride
$TD_{50} = 2.29$

Figure E.1 (Continued)

(continued)

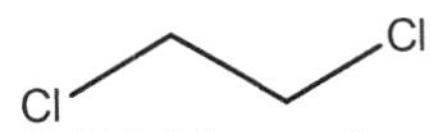

1,2-Dichloroethane
TD_{50}= 8.04

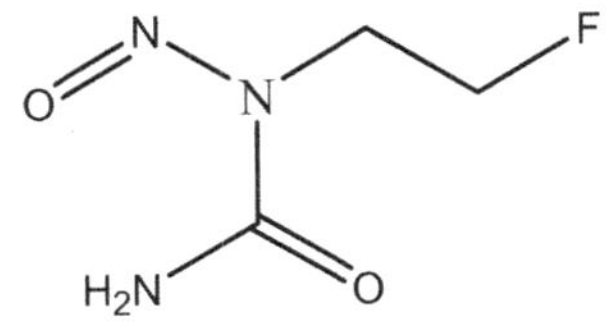

2-Fluoroethylnitrosourea
TD_{50} = 0.125

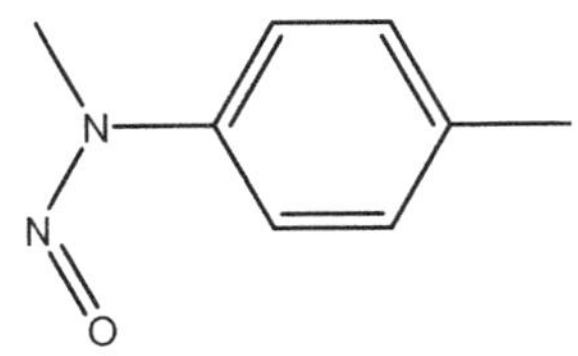

N-Nitroso-N-methyl-4-fluoroaniline
TD_{50} = 0.255

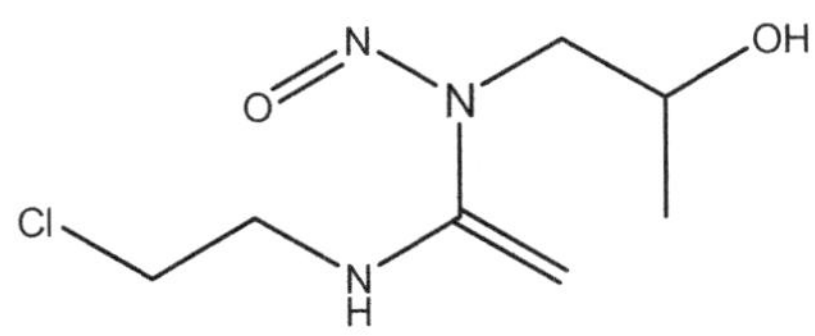

1-Nitroso-1-(2-hydroxypropyl)-3-chloroethylurea
TD_{50} = 0.873

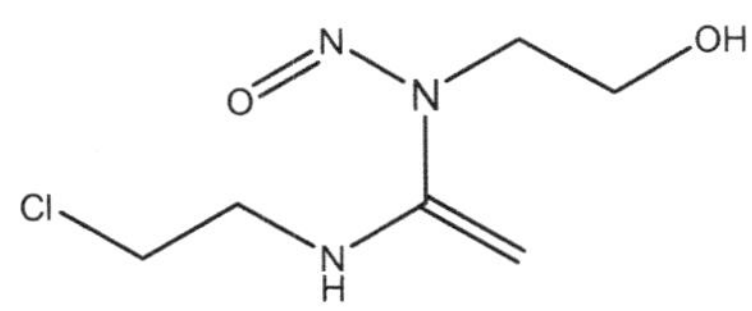

1-Nitroso-1-hydroxyethyl-3-chloroethylurea
TD_{50} = 0.356

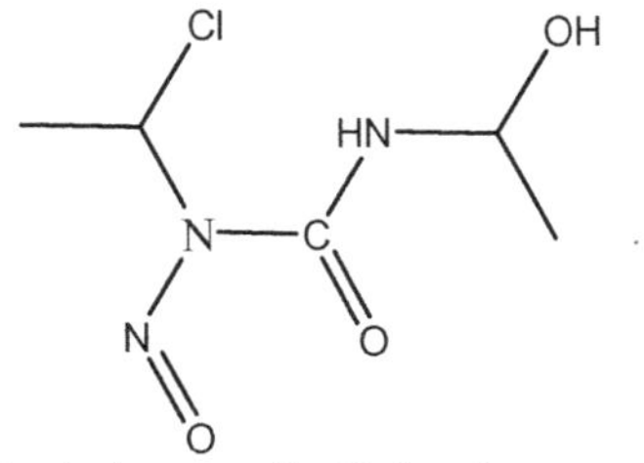

2-Chloroethylnitroso-3-(2-hydroxypropyl)urea
TD_{50} = 0.124

Figure E.1 (Continued)

(continued)

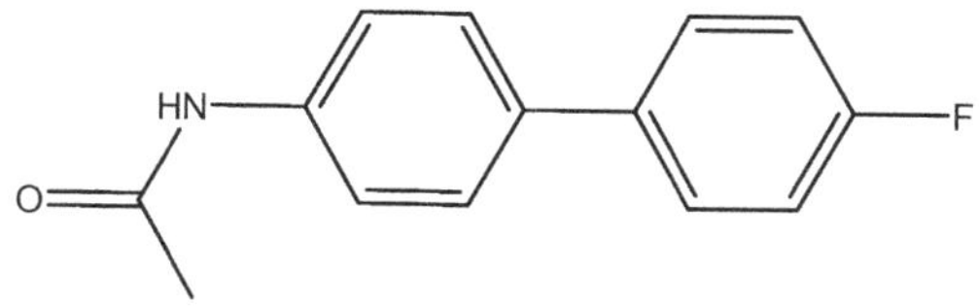

N-4-(4'-Fluorobiphenyl)acetamide
$TD_{50} = 1.01$

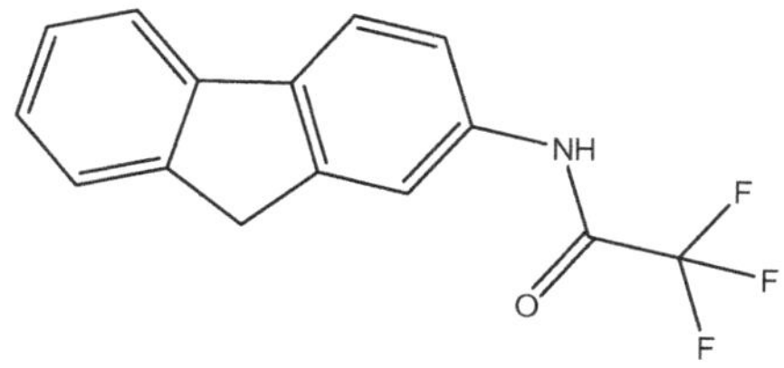

N-(2-Fluorenyl)-2,2,2-trifluoroacetamide
$TD_{50} = 1.62$

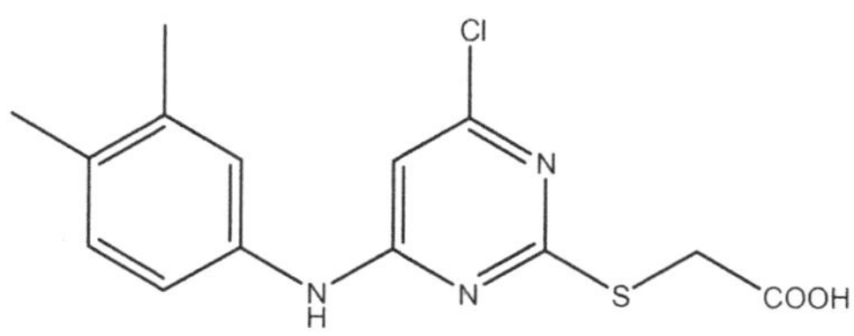

4-Chloro-6-(2,3-xylidino)-2-pyrimidinylthio(N-β-hydroxyethyl)acetamide
$TD_{50} = 6.49$

2,2,2-Trifluoro-N-[4-(5-nitro-2-furyl)-2-thiazolyl]acetamide
$TD_{50} = 6.79$

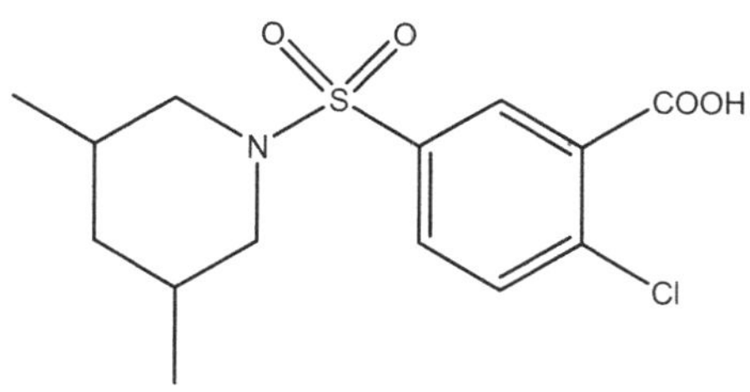

2-Chloro-5-(3,5-dimethylpiperidineosulfonyl)benzoic acid
$TD_{50} = 4.85$

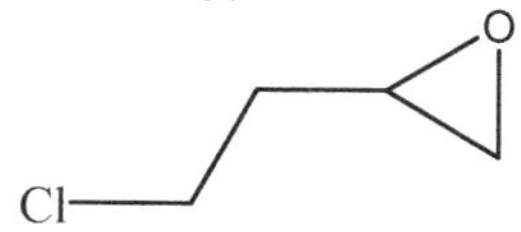

Epichlorohydrin
$TD_{50} = 2.86$

Figure E.1 (Continued)

(continued)

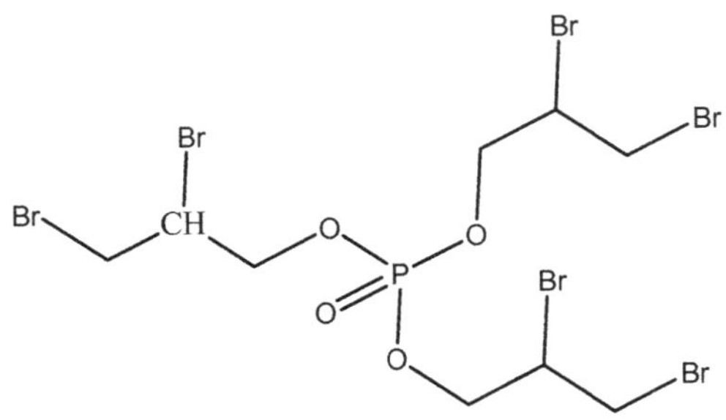

Tris(2,3-dibromopropyl)phosphate
$TD_{50} = 3.83$

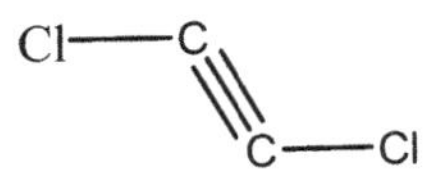

Dichloroacetylene
$TD_{50} = 3.58$

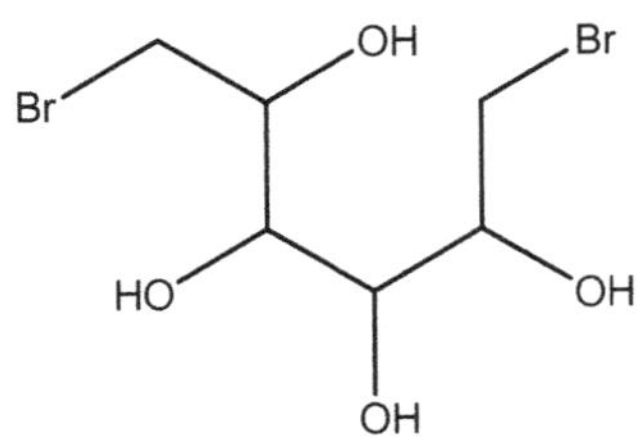

Dibromodulcitol
$TD_{50} = 8.37$

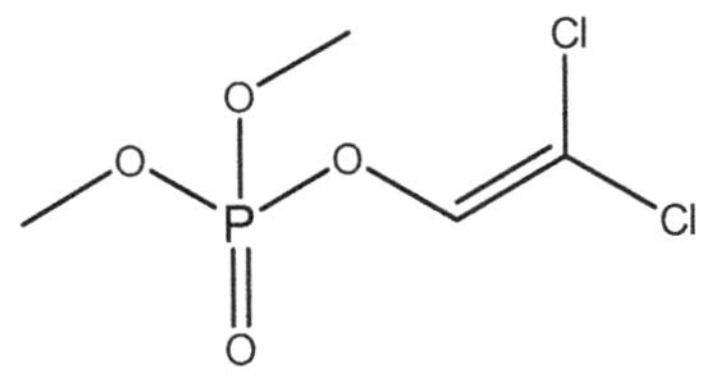

Dichlorvos
$TD_{50} = 4.16$

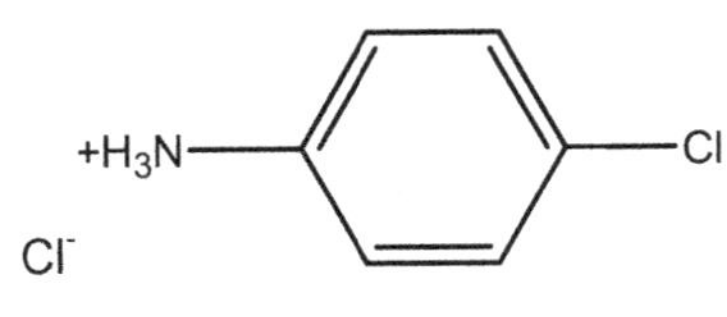

p-Chloroaniline.HCl
$TD_{50} = 7.62$

Figure E.1 (Continued)

(continued)

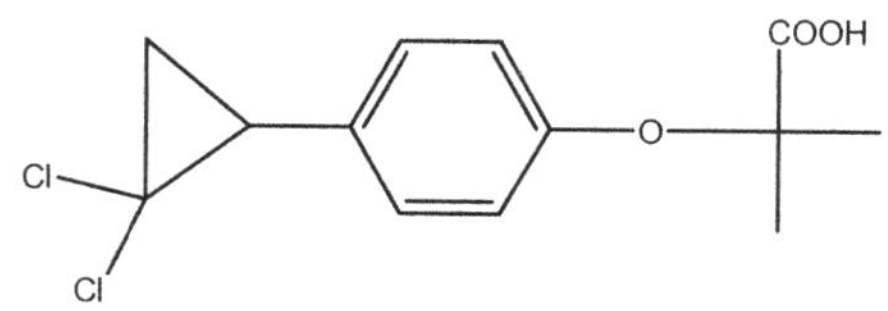

Ciprofibrate
$TD_{50} = 2.48$

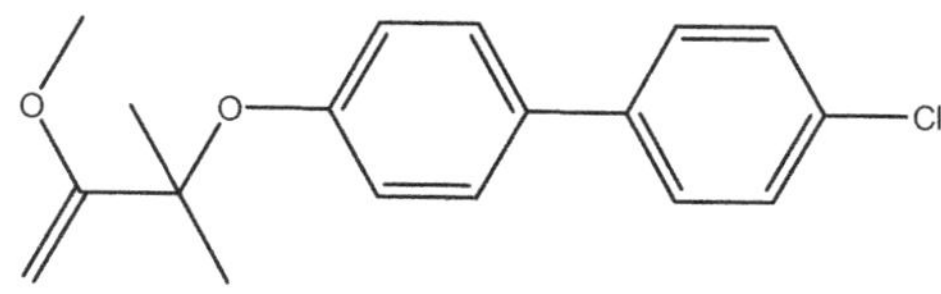

Methyl clofenapate
$TD_{50} = 4.78$

Figure E.1 (Continued)

REFERENCES

Abia, L., X.L. Armesto, M. Canle, M.V. Garcia, and J.A. Santaballa. 1998. Oxidation of aliphatic amines by aqueous chlorine. *Tetrahedron,* 54 (3-4):521-530.

Accelrys. 2001. TOPKAT® User Guide. Version 6.1. Accelrys, Inc., San Diego, Calif.

Ambrosone, C.B. and F.F. Kadlubar. 1998. Acetyltransferase polymorphisms and disease susceptibility. In *Biomarkers: Medical and Workplace Applications*, edited by M. L. Mendelsohn, L. C. Mohr and J. P. Peeters. Washington DC: Joseph Henry Press.

American Cancer Society. 2005. Cancer Facts [Online]. Available: <http://www.cancer.org.> [cited December 15, 2005].

Antelo, J.M., F. Arce, and M. Parajo. 1995. Kinetic study of the formation of N-chloramines. *Journal of Chemical Kinetics,* 27 (7):637-647.

Archer, C. L., P. Morse, R. F. Jones, T. Shirai, G. P. Haas, and C. Y. Wang. 2000. Carcinogenicity of the N-hydroxy derivative of 2-amino-1-methyl-6-phenylimidazo[4,5-b]pyridine, 2-amino-3, 8-dimethyl-imidazo[4,5-f]quinoxaline and 3, 2'-dimethyl-4-aminobiphenyl in the rat. *Cancer Lett.,* 155 (1):55-60.

Armesto, X. L., L. M. Canle, M. I. Fernandez, M. V. Garcia, S. Rodriguez, and J. A. Santaballa. 2001. Intracellular oxidation of dipeptides. Base-promoted elimination from N-halodipeptides to 2-[N-alkyl-N-(2-N-alkylimino-2-alkyl-ethanoyl)amino]-2,2-dialkylethanoic acids. *J. Org. Chem.,* 66 (17):5692-5700.

Armesto, X.L., L.M. Canle, M. Losada, and J.A. Santaballa. 1993. Nitrenium ions in N-chloro-α-amino acids decomposition. *International Journal of Chemical Kinetics,* 25:1-8.

Armesto, X.L., M. Canle, A.M. Gamper, M. Losada, and J.A. Santaballa. 1994. Acid-base equilibria and decomposition of secondary (N-Cl)-alpha-amino acids. *Tetrahedron,* 50 (35):10509-10520.

Armesto, X.L., M. Canle, M.V. Garcia, and J.A. Santaballa. 1998. Aqueous chemistry of N-halo-compounds. *Chemical Society Reviews,* 27 (6):453-460.

Awad, R., A., A. Hussain, and P. A. Crooks. 1990. Kinetics and mechanism of decomposition of N-chloro-alpha-amino acids. *Journal of the Chemical Society, Perkin Transactions 2*:1233-1238.

Barua, M., Y. Liu, and M. R. Quinn. 2001. Taurine chloramine inhibits inducible nitric oxide synthase and TNF-alpha gene expression in activated alveolar macrophages: Decreased NF-kappaB activation and I-kappaB kinase activity. *J. Immunol.,* 167 (4):2275-2281.

Bedner, M., W. A. MacCrehan, and G. R. Helz. 2004. Production of macromolecular chloramines by chlorine-transfer reactions. *Environ. Sci. Technol.,* 38 (6):1753-1758.

Belman, S., W. Troll, G. Teebor, and F. Mukai. 1968. The carcinogenic and mutagenic properties of N-hydroxy-aminonaphthalenes. *Cancer Res.,* 28 (3):535-542.

Benigni, R. 2003. *Quantitative Structure-Activity Relationship (QSAR): Models of Mutagens and Carcinogens*. Boca Raton, Fla.: CRC Press.

Benigni, R., A. Giuliani, A. Gruska, and R. Franke. 2003. QSARs for the mutagenicity and carcinogenicity of the aromatic amines. In *Quantitative Structure-Activity Relationship (QSAR). Models of Mutagens and Carcinogens*, edited by R. Benigni. Boca Raton, Fla.: CRC Press.

Benner, R., J.D. Pakulshi, M. McCarthy, J.I. Hedges, and P.G. Hatcher. 1992. Bulk chemical characteristics of dissolved organic matter in the ocean. Science 255:1561-1564.

Berne, F. D., B. Panais, N. Merlet, B. Cauchi, and B. Legube. 1994. Total dissolved amino-acid-analysis in natural and drinking waters. *Environmental Technology,* 15:901-916.

Berne, F.D., N. Merlet, B. Cauchi and B. Legube. 1996. Evolution des acides amines et de la matiere organique dissoute dans une filiere de production d'eau potable: Correlations avec le carbon organique dissous biodegradable et le potential de demande en chlore a long terme. *Revue des Sciences de l'Eau*, 9 (1):115-133.

Bernofsky, C. 1991. Nucleotide chloramines and neutrophil-mediated cytotoxicity. *FASEB J.,* 5 (3):295-300.

Biaudet, H., T. Mavelle, and G. Debry. 1994. Mean daily intake of N-nitrosodimethylamine from foods and beverages in France in 1987-1992. *Food and Chemical Toxicology,* 32:417-421.

Bignami, M., G. Conti, L. Conti, R. Crebelli, F. Misuraca, A.M. Puglia, R. Randazzo, G. Sciandrello, and A. Carere. 1980. Mutagenicity of halogenated aliphatic hydrocarbons in Salmonella typhimurium, Streptomyces coelicolor, and Aspergilllus nidulans. *Chemical-Biological Interactions,* 30:9-23.

Birch, A.J. 1967. Biosynthesis of polyketides and related compounds. *Science,* 156 (3772):202-206.

Birch, A.J. 1968. Polyketide metabolism. *Annual Review of Plant Physiology,* 19:321-332.

Bolton, J. L., M. A. Trush, T. M. Penning, G. Dryhurst, and T. J. Monks. 2000. Role of quinones in toxicology. *Chem. Res. Toxicol.,* 13 (3):135-160.

Boyce, S. D. and J. F. Hornig. 1983. Reaction pathways of trihalomethane formation from the halogenation of dihydroxyaromatic model compounds for humic acid. *Environmental Science & Technology,* 17:202-211.

Bronk, D.A. 2002. Dynamics of DON. In *Biogeochemistry of Marine Dissolved Organic Matter*, edited by D. A. Hansell and C. A. Carlson. London: Academic Press.

Bruchet, A., E. Costentin, M. F. Legrand, and J. Mallevialle. 1992. Influence of the chlorination of natural nitrogenous organic-compounds on tastes and odors in finished drinking waters. *Water Science and Technology,* 25:323-333.

Bubnis, B. 1993. Personal communication. Oxford, OH, Novatek.

Buffle, J. 1988. *Complexation Reactions in Aquatic Systems: An Analytical Approach*. London: Ellis Horwood, Ltd.

Bull, R. J., J. R. Meier, M. Robinson, H. P. Ringhand, R. D. Laurie, and J. A. Stober. 1985. Evaluation of mutagenic and carcinogenic properties of brominated and chlorinated acetonitriles: By-products of chlorination. *Fundam. Appl. Toxicol.,* 5 (6 Pt 1):1065-1074.

Bull, R.J. 2000. Mode of action of liver tumor induction by trichloroethylene and its metabolites, trichloroacetate and dichloroacetate. *Environmental Health Perspectives,* 108 (Supp. 2):241-259.

Bull, R.J., S.W. Krasner, P.A. Daniel, and R.D. Bull. 2001. *Health Effects and Occurrence of Disinfection By-Products*. Denver, Colo.: AWWA Research Foundation.

Candeias, L.P., M.R.L. Stratford, and P. Wardman. 1994. Formation of hydroxyl radicals on reaction of hypochlorous acid with ferrocyanide, a model iron(Ii) complex. *Free Radical Research,* 20:241-249.

Cantor, K. P., C. F. Lynch, M. E. Hildesheim, M. Dosemeci, J. Lubin, M. Alavanja, and G. Craun. 1999. Drinking water source and chlorination byproducts in Iowa. III. Risk of brain cancer. *Am. J. Epidemiol.,* 150 (6):552-560.

Caraco, N.F. and J.J. Cole. 2002. The importance of organic nitrogen production in aquatic systems: A landscape perspective. In *Aquatic Ecosystems: Interactivity of Dissolved Organic Matter*, edited by S. Findlay and R. L. Sinsabaugh. New York, N.Y.: Academic Press.

Carney, E.W., C.L. Zablotny, M.S. Marty, J.W. Crissman, P. Anderson, M. Woolhiser, and M. Holsapple. 2004. The effects of feed restriction during in utero and postnatal development in rats. *Toxicol. Sci.,* 82 (1):237-249.

Carr, A. C., C. L. Hawkins, S. R. Thomas, R. Stocker, and B. Frei. 2001. Relative reactivities of N-chloramines and hypochlorous acid with human plasma constituents. *Free Radic. Biol. Med.,* 30 (5):526-536.

Charrois, J. W., M. W. Arend, K. L. Froese, and S. E. Hrudey. 2004. Detecting N-nitrosamines in drinking water at nanogram per liter levels using ammonia positive chemical ionization. *Environ. Sci. Technol.,* 38 (18):4835-4841.

Chevrier, C., B. Junod, and S. Cordier. 2004. Does ozonation of drinking water reduce the risk of bladder cancer? *Epidemiology,* 15 (5):605-614.

Choi, J. and R. L. Valentine. 2002. Formation of N-nitrosodimethylamine (NDMA) from reaction of monochloramine: A new disinfection by-product. *Water Res.,* 36 (4):817-824.

Choi, J. and R. L. Valentine. 2003. N-nitrosodimethylamine formation by free-chlorine-enhanced nitrosation of dimethylamine. *Environ. Sci. Technol.,* 37 (21):4871-4876.

Christman, R.F., D.L. Norwood, and J.D. Johnson. 1985. New directions in oxidant by-product research - identification and significance. *Science of the Total Environment,* 47:195-210.

Christman, R.F., D.L. Norwood, D.S. Millington, J.D. Johnson, and A.A. Stevens. 1983. Identity and yields of major halogenated products of aquatic fulvic-acid chlorination. *Environmental Science & Technology,* 17:625-628.

Cleverly, D., M. Monetti, L. Phillips, P. Cramer, M. Heit, S. McCarthy, K. O'Rourke, J. Stanley, and D. Winters. 1996. A time-trends study of the occurrence of CDDs, CDFs, and dioxin-like PCBs in sediment cores from geographically distributed lakes in the United States. *Organohalogen Compounds* 28:77-82.

Cohen, S.M. 1999. Calcium phosphate-containing urinary precipate in rat urinary bladder carcinogenesis. In *Species Differences in Thryoid, Kidney and Urinary Bladder Carcinogenesis*, edited by C. C. Capen, E. Dybing, J. M. Rice and J. D. Wilbourn. Lyon, France: IARC Scientific Publications.

Collette, T.W., S.D. Richardson, and A.D. Thurston, Jr. 1994. Identification of bromohydrins in ozonated waters. *Applied Spectroscopy,* 48:1181-1192.

Daumer, K. M., A. U. Khan, and M. J. Steinbeck. 2000. Chlorination of pyridinium compounds. Possible role of hypochlorite, N-chloramines, and chlorine in the oxidation of pyridinoline cross-links of articular cartilage collagen type II during acute inflammation. *J. Biol. Chem.,* 275 (44):34681-34692.

Davies, J. M. S., D. A. Horwitz, and K. J. A. Davies. 1994. Inhibition of collagenase activity by N-chlorotaurine, a product of activated neutrophils. *Arthritis Rheum.,* 37 (3):424-427.

De Laat, J. 1981. *Contribution à L'Étude Du Mécanisme De Formation Des Trihalométhanes: Incidence De L'Azote Ammoniacal Et Des Traitements De Préoxydation*. Poitiers: Universite de Poitiers.

Dennis, W.H., V.P. Olivieri, and C.W. Kruse. 1979. Reaction of nucleotides with aqueous hypochlorous acid. *Water Research,* 13:357-362.

Dooley, K. L., F. A. Beland, T. J. Bucci, and F. F. Kadlubar. 1984. Local carcinogenicity, rates of absorption, extent and persistence of macromolecular binding, and acute histopathological effects of N-hydroxy-1-naphthylamine and N-hydroxy-2-naphthylamine. *Cancer Res.,* 44 (3):1172-1177.

Dossier-Berne, F., N. Merlet, B. Cauchi, and B. Legube. 1996. Evolution des acides aminés et de la matiere organique dissoute dans une filiere de production d'eau potable: Corrélations avec le carbone organique dissous biodégradable et le potentiel de demande en chlore a long terme. *Revue Des Sciences De L'Eau,* 1:115-133.

Dossier-Berne, F., B. Panais, N. Merlet, B. Cauchi, and B. Legube. 1994. Total dissolved amino-acid-analysis in natural and drinking waters. *Environmental Technology,* 15 (10):901-916.

Ellis, A.J. and F.G. Soper. 1954. Studies of N-halogeno compounds. VI. The kinetics of chlorination of tertiary amines. *Journal of the Chemical Society*:1750-1755.

Englert, R. P. and E. Shacter. 2002. Distinct modes of cell death induced by different reactive oxygen species: Amino acyl chloramines mediate hypochlorous acid-induced apoptosis. *J. Biol. Chem.,* 277 (23):20518-20526.

Folkes, L.K., L.P. Candeias, and P. Wardman. 1995. Kinetics and mechanisms of hypochlorous acid reactions. *Archives of Biochemistry and Biophysics,* 323:120-126.

Fox, T.C., D.J. Keefe, F.E. Scully, and A. Laikhter. 1997. Chloramines.7. Chlorination of alanylphenylalanine in model solutions and in a wastewater. *Environmental Science & Technology,* 31:1979-1984.

Friend, A.G. 1956. Rates of N-Chlorination of Amino Acids, Ph. D. dissertation, Harvard University, Cambridge, MA.

Furst, S. M. and J. P. Uetrecht. 1993. Carbamazepine metabolism to a reactive intermediate by the myeloperoxidase system of activated neutrophils. *Biochem. Pharmacol.,* 45 (6):1267-1275.

Gallard, H. and U. von Gunten. 2002. Chlorination of phenols: Kinetics and formation of chloroform. *Environmental Science & Technology,* 36:884-890.

Gaylor, D. W. and L. S. Gold. 1995. Quick estimate of the regulatory virtually safe dose based on the maximum tolerated dose for rodent bioassays. *Regul. Toxicol. Pharmacol.,* 22 (1):57-63.

Glaze, W.H. and H.S. Weinberg. 1993. *Identification and Occurrence of Ozonation By-Products in Drinking Water.* Denver, CO: American Water Works Association Research Foundation.

Gold, L. S. 2005. *Carcinogen Potency Database (CPDB)* [Online]. Available: <http://potency.berkeley.edu/pub/tables/chemicalsummary> April 7, 2005 [cited 2005].

Goldschmidt, S., E. Wiberg, F. Nagel, and K. Martin. 1927. About proteins. *Justus Liebigs Ann. Chem.,* 456:1-38.

Gordon, G., L. Adam, and B. Bubnis. 1995. *Minimizing Chlorate Ion Formation in Drinking Water When Hypochlorite Ion is the Chlorination Agent.* Denver, CO: Awwa Research Foundation and American Water Works Association.

Gould, J.P. and T.R. Hay. 1982. The nature of the reactions between chlorine and purine and pyrimidine bases: Products and kinetics. *Wat. Sci. Tech.,* 14:629-640.

Gould, J.P., J.T. Richards, and M.G. Miles. 1984. The kinetics and primary products of uracil chlorination. *Water Research,* 18:205-212.

Guengerich, F.P. and M. Persmark. 1994. Mechanism of formation of ethenoguanine adducts from 2-haloacetaldehydes: 13C-labeling patterns with 2-bromoacetaldehyde. *Chemical Research in Toxicology,* 7 (2):205-208.

Halliwell, B., M. Wasil, and M. Grootveld. 1987. Biologically significant scavenging of the myeloperoxidase-derived oxidant hypochlorous acid by ascorbic acid. Implications for antioxidant protection in the inflamed rheumatoid joint. *FEBS Lett.,* 213 (1):15-18.

Hartman, C. and D. A. Reckhow. 2005. Chlorination and Chloramination of Ozonation By-products. Model Compound Studies. Unpublished manuscript.

Hawkins, C. L. and M. J. Davies. 1998. Degradation of hyaluronic acid, poly- and monosaccharides, and model compounds by hypochlorite: Evidence for radical intermediates and fragmentation. *Free Radic. Biol. Med.,* 24 (9):1396-1410.

Hawkins, C.L. and M.J. Davies. 1999. Hypochlorite-induced oxidation of proteins in plasma: Formation of chloramines and nitrogen-centered radicals and their role in protein fragmentation. *Biochemical Journal,* 340:539-548.

Hawkins, C. L. and M. J. Davies. 2001. Hypochlorite-induced damage to nucleosides: Formation of chloramines and nitrogen-centered radicals. *Chem. Res. Toxicol.,* 14 (8):1071-1081.

Hawkins, C.L. and M.J. Davies. 2002. Hypochlorite-induced damage to DNA, RNA, and polynucleotides: Formation of chloramines and nitrogen-centered free radicals. *Chem. Res. Toxicol.,* 15 (1):83-92.

Hawkins, C.L., D.I. Pattison, and M.J. Davies. 2002. Reaction of protein chloramines with DNA and nucleosides: Evidence for the formation of radicals, protein-DNA cross-links and DNA fragmentation. *Biochemical Journal,* 365 (Pt 3):605-615.

Hayatsu, H., S.K. Pan, and T. Ukita. 1971. Reaction of sodium hypochlorite with nucleic acids and their constituents. *Chemical & Pharmaceutical Bulletin,* 19:2189-2192.

Hayes, J. R., L. W. Condie, Jr., and J. F. Borzelleca. 1986. Toxicology of haloacetonitriles. *Environ. Health Perspect.,* 69:183-202.

Health Canada. 2003. Findings from the Health Canada Workshop held on 18 and 19 September 2002 to identify critical end points for assessment of health risks related to trihalomethanes (THMs) in drinking water, at Ottawa, Canada.

Heasley, V.L., A.M. Fisher, E.E. Herman, F.E. Jacobsen, E.W. Miller, A.M. Ramirez, N.R. Royer, J.M. Whisenand, D.L. Zoetewey, and D.F. Shellhamer. 2004. Investigations of the reactions of monochloramine and dichloramine with selected phenols: Examination of humic acid models and water contaminants. *Environ. Sci. Technol.,* 38:5022-5029.

Hecht, S. S. 1998. Biochemistry, biology, and carcinogenicity of tobacco-specific N-nitrosamines. *Chem. Res. Toxicol.,* 11 (6):559-603.

Hedges, J.I. and J.R. Ertel. 1982. Characterization of lignin by gas capillary chromatography of cupric oxide oxidation-products. *Analytical Chemistry,* 54:174-178.

Henderson, J.P., J. Byun, and J.W. Heinecke. 1999. Molecular chlorine generated by the myeloperoxidase-hydrogen peroxide-chloride system of phagocytes produces 5-chlorocytosine in bacterial RNA -olecular chlorine generated by the myeloperoxidase-hydrogen peroxide-chloride system of phagocytes produces 5-chlorocytosine in bacterial RNA. *Journal of Biological Chemistry,* 274:33440-33448.

Hernandez, O., L. Rhomberg, K. Hogan, C. Siegel-Scott, D. Lai, G. Grindstaff, M. Henry, and J.A. Cotruvo. 1994. Risk assessment of formaldehyde. *J. Hazardous Materials,* 39:161-172.

Hoyano, Y., V. Bacon, R.E. Summers, W.E. Pereira, B. Halpern, and A.M. Duffield. 1973. Chlorination studies IV: The reaction of aqueous hypochlorous acid with pyrimidine and purine bases. *Biochemical and Biophysics Research Communication,* 53:1195-1199.

Hrudey, S.E., A. Gac, and S.A. Daignault. 1988. Potent odor-causing chemicals arising from drinking-water disinfection. *Water Science and Technology,* 20:55-61.

Hureiki, L., J-P. Croué, and B. Legube. 1994. Chlorination studies of free and combined amino-acids. *Water Research,* 28:2521-2531.

Hurni, H. and H. Ohder. 1973. Reproduction study with formaldehyde and hexamethylenetetramine in Beagle dogs. *Food Cosmet. Toxicol.,* 11 (3):459-462.

Hussain, A., A. Hurwitz, T. Higuchi, and I.H. Pitman. 1972. Rates of hydrolysis of N-chlorinated molecules. *Journal of Pharmaceutical Sciences,* 61 (3):371-374.

Hwang, S.C., R.A. Larson, and V.L. Snoeyink. 1990. Reactions of free chlorine with substituted anilines in aqueous-solution and on granular activated carbon. *Water Research,* 24:427-432.

Hwang, C.J., S.W. Krasner, M.J. Sclimenti, G.L. Amy, E. Dickensen, A. Bruchet, C. Prompsy, G. Filippi, J.P. Croué, D. Violleau, and J.A. Leenheer. 2001. *Polar NOM: Characterization, DBPs, Treatment.* Denver, CO: Awwa Research Foundation.

IARC. 1999. *Species Differences in Thyroid, Kidney and Urinary Bladder Carcinogenesis.* Edited by C. C. Capen, E. Dybing, J. M. Rice and J. D. Wilbourn. Vol. Appendix 1 and 2, 191-225. Lyon, France: IARC Press.

IARC. 2004. *Some Drinking Water Disinfectants and Contaminants, including Arsenic.* Vol. 84. Lyon, France: IARC Press.

Imafidon, G.I. and F.W. Sosulski. 1990. Nucleic-acid nitrogen of animal and plant foods. *Journal of Agricultural and Food Chemistry,* 38:118-120.

Isaac, R.A. and J.C. Morris. 1983. Transfer of active chlorine from chloramine to nitrogenous organic compounds. 1. Kinetics. *Environ. Sci. Technol.,* 17:738-742.

Jacob, J.S., D.P. Cistola, F.F. Hsu, S. Muzaffar, D.M. Mueller, S.L. Hazen, and J.W. Heinecke. 1996. Human phagocytes employ the myeloperoxidase-hydrogen peroxide system to synthesize dityrosine, trityrosine, pucherosine, and isodityosine by a tyrosyl radical-dependent pathway. *Journal of Biological Chemistry,* 271 (33):19950-19956.

Jersey, J.A. and J.D. Johnson. 1992. *Development of Analytical Methods for the Identification of Organic Nitrogen Chlorination By-Products in North Carolina Surface Waters.* Raleigh, N.C.: North Carolina Water Resources Research Institute.

Jesaitis, A.J. and E.A. Dratz. 1992. *The Molecular Basis of Oxidative Damage by Leukocytes.* Boca Raton, Fla.: CRC Press.

Jorgenson, T. A., E. F. Meierhenry, C. J. Rushbrook, R. J. Bull, and M. Robinson. 1985. Carcinogenicity of chloroform in drinking water to male Osborne-Mendel rats and female B6C3F1 mice. *Fundam. Appl. Toxicol.,* 5 (4):760-769.

Kaminski, J.J., N. Bodor, and T. Higuchi. 1976. N-Halo derivatives III: Stabilization of nitrogen-chlorine bond in N-chloroamino acid derivatives. *Journal of Pharmaceutical Sciences,* 65 (4):553-558.

Kang, J.I., A. Burdzy, P.F. Liu, and L.C. Sowers. 2004. Synthesis and characterization of oligonucleotides containing 5-Chlorocytosine. *Chemical Research in Toxicology,* 17:1236-1244.

Karimi, A.A. and P.C. Singer. 1991. Trihalomethane formation in open reservoirs. *Journal of the American Water Works Association,* 83 (3):84-88.

Kaufman, D.D., J.R. Plimmer, and U.I. Klingebi. 1973. Microbial oxidation of 4-chloroaniline. *Journal of Agricultural and Food Chemistry,* 21:127-132.

Kawai, Y., H. Morinaga, H. Kondo, N. Miyoshi, Y. Nakamura, K. Uchida, and T. Osawa. 2004. Endogenous formation of novel halogenated 2 '-Deoxycytidine - hypohalous acid-mediated DNA modification at the site of inflammation. *Journal of Biological Chemistry,* 279:51241-51249.

King, W.D. 1999. *Chlorination By-products and Cancer: Relative Risk, Population Attributable Risk, and Individual Risk.* Contract # H1021-8-0282: Health Canada.

Kirmeyer, G.J., G.W. Foust, G.L. Pierson, J.J. Simmler, and M.W. LeChevallier. 1993. *Optimizing Chloramine Treatment.* Denver, CO: Awwa Research Foundation.

Knize, M. G., K. S. Kulp, C. P. Salmon, G. A. Keating, and J. S. Felton. 2002. Factors affecting human heterocyclic amine intake and the metabolism of PhIP. *Mutat. Res.,* 506-507:153-162.

Komulainen, H., V.-M. Kosma, S.-L. Vaittinen, T. Vartiainen, E. Kaliste-Korhonen, S. Lotjonen, R. K. Tuominen, and J. Tuomisto. 1997. Carcinogenicity of the drinking water mutagen 3-chloro-4-(dichloromethyl)-5-hydroxy-2(5H)-furanone in the rat. *J. Natl. Cancer Inst.,* 89 (12):848-856.

Kozubek, A. and J. H. P. Tyman. 1999. Resorcinolic lipids, the natural non-isoprenoid phenolic amphiphiles and their biological activity. *Chemical Reviews,* 99 (1):1-25.

Krasner, S.W., M.J. McGuire, J.G. Jacangelo, N.J. Patania, K.M. Reagan, and E.M. Aieta. 1989. The occurrence of disinfection by-products in U.S. drinking water. *Journal of American Water Works Association,* 81:41-53.

Kumar, R., P. Mende, C.-D. Wacker, B. Spiegelhalder, R. Preussmann, and M. Siddiqi. 1992. Caffeine-derived N-nitroso compounds. I. Nitrosatable precursors from caffeine and their potential relevance in the etiology of oesophageal and gastric cancers in Kashmir, India. *Carcinogenesis,* 13 (11):2179-2182.

Kumar, R., C.-D. Wacker, P. Mende, B. Spiegelhalder, R. Preussmann, and M. Siddiqi. 1993. Caffeine-derived N-nitroso compounds. II. Synthesis and characterization of nitrosation products from caffeidine and caffeidine acid. *Chem. Res. Toxicol.,* 6 (1):50-58.

Kusek, J.W. and L.M. Nyberg. 2001. The epidemiology of interstitial cystitis: Is it time to expand our definition? *Urology,* 57 (Suppl 6A):95-99.

La Vecchia, C. and L. Airoldi. 1999. Human bladder cancer: Epidemiological, pathological and mechanistic aspects. In *Species Differences in Thyroid, Kidney and Urinary Bladder Carcinogenesis*, edited by C. C. Capen, E. Dybing, J. M. Rice and J. D. Wilbourn. Lyon, France: IARC Scientific Publications.

Lakshmi, V. M., F. F. Hsu, B. B. Davis, and T. V. Zenser. 2000a. N-Acetylbenzidine-DNA adduct formation by phorbol 12-myristate-stimulated human polymorphonuclear neutrophils. *Chem. Res. Toxicol.,* 13 (8):785-972.

Lakshmi, V. M., F. F. Hsu, A. E. McGarry, B. B. Davis, and T. V. Zenser. 2000b. Hypochlorous acid-mediated activation of N-acetylbenzidine to form N'-(3'-monophospho-deoxyguanosin-8-yl)-N-acetylbenzidine. *Toxicol. Sci.,* 53 (2):202-212.

Lakshmi, V. M., F. F. Hsu, and T. V. Zenser. 2002. Nitrosation and nitration of 2-amino-3-methylimidazo[4,5-f]quinoline by reactive nitrogen oxygen species. *Chem. Res. Toxicol.,* 15 (8):1059-1068.

LaLonde, R. T. and H. R. Leo. 1994. Interactive chlorine-by-bromine and hydrogen-by-hydroxyl group replacement effects in 2(5H)-furanone mutagenicity. *Chem. Res. Toxicol.,* 7 (6):779-783.

Lampert, M. B. and S. J. Weiss. 1983. The chlorinating potential of the human monocyte. *Blood,* 62 (3):645-651.

Larson, R.A. and A.L. Rockwell. 1979. Chloroform and chlorophenol production by decarboxylation of natural acids during aqueous chlorination. *Environmental Science and Technology,* 13:325-329.

Lau, S. S., T. J. Monks, J. I. Everitt, E. Kleymenova, and C. L. Walker. 2001. Carcinogenicity of a nephrotoxic metabolite of the "nongenotoxic" carcinogen hydroquinone. *Chem. Res. Toxicol.,* 14 (1):25-33.

Le Cloirec, C. and G. Martin. 1985. Evolution of Amino Acids in Water Treatment Plants and the Effect of Chlorination on Amino Acids. In *Water Chlorination: Environmental Impact and Health Effects*, edited by R. L. Jolley, R. J. Bull, W. P. David, S. Katz, M. H. Roberts, Jr. and V. A. Jacobs. Chelsea, MI: Lewis Publishers.

Lee, G.F. and J.C. Morris. 1962. Kinetics of chlorination of phenol - Chlorophenolic tastes and odors. *International Journal of Air and Water Pollution,* 6:419-431.

Liu, Z. C. and J. P. Uetrecht. 1995. Clozapine is oxidized by activated human neutrophils to a reactive nitrenium ion that irreversibly binds to the cells. *J. Pharmacol. Exp. Ther.,* 275 (3):1476-1483.

Loew, G. H., M. Poulsen, E. Kirkjian, J. Ferrell, B. S. Sudhindra, and M. Rebagliati. 1985. Computer-assisted mechanistic structure-activity studies: Application to diverse classes of chemical carcinogens. *Environ. Health Perspect.,* 61:69-96.

MacCarthy, P. 2001. The Principles of Humic Substances: An Introduction to the First Principle. In *Humic Substances: Structures, Models and Functions*, edited by E. A. Ghabbour and G. Davies. Cambridge, U.K.: Royal Society of Chemistry.

Mann, J. 1994. Alkaloids. In *Natural Products: Their Chemistry and Biological Significance*, edited by J. Mann, R. S. Davidson, J. B. Hobbs, D. V. Banthorpe and J. B. Harborne. New York: Prentice Hall.

Marks, T. A., W. C. Worthy, and R. E. Staples. 1980. Influence of formaldehyde and Sonacide® (potentiated acid glutaraldehyde) on embryo and fetal development in mice. *Teratology,* 22 (1):51-58.

Martin, W. J. 1990. A teratology study of inhaled formaldehyde in the rat. *Reprod. Toxicol.,* 4 (3):237-239.

Masuda, M., T. Suzuki, M.D. Friesen, J.L. Ravanat, J. Cadet, B. Pignateli, H. Nishino, and H. Ohshima. 2001. Chlorination of guanosine and other nucleosides by hypochlorous acid and myeloperoxidase of activated human neutrophils - Catalysis by nicotine and trimethylamine. *Journal of Biological Chemistry,* 276:40486-40496.

McDorman, K.S., S. Chandra, M.J. Hooth, S.D. Hester, R. Schoonhoven, and D.C. Wolf. 2003. Induction of transitional cell hyperplasia in the urinary bladder and aberrant crypt foci in the colon of rats treated with individual and a mixture of drinking water disinfection by-products. *Toxicological Pathology,* 31 (2):235-242.

McGeehin, M.A., J.S. Reif, J.C. Becher, and E.J. Mangione. 1993. Case-control study of bladder cancer and water disinfection methods in Colorado. *American Journal of Epidemiology,* 138 (7):492-501.

McGuire, M.J., J.L. McLain, and A. Obolensky. 2002. *Infomation Collection Rule Data Analysis*. Denver: American Water Works Association Research Foundation.

Mills, C. J., R. J. Bull, K. P. Cantor, J. Reif, S. E. Hrudey, and P. Huston. 1998. Health risks of drinking water chlorination by-products: Report of an expert working group. *Epidemiology,* 19 (3):91-102.

Mitch, W. A. and D. L. Sedlak. 2002. Formation of N-nitrosodimethylamine (NDMA) from dimethylamine during chlorination. *Environ. Sci. Technol.,* 36 (4):588-595.

Miyamoto, G., N. Zahid, and J.P. Uetrecht. 1997. Oxidation of diclofenac to reactive intermediates by neutrophils, myeloperoxidase, and hypochlorous acid. *Chem. Res. Toxicol.,* 10:414-419.

Monks, T. J. and S. S. Lau. 1997. Biological reactivity of polyphenolic-glutathione conjugates. *Chem. Res. Toxicol.,* 10 (12):1296-1313.

Monks, T. J. and S. S. Lau. 1998. The pharmacology and toxicology of polyphenolic-glutathione conjugates. *Annu. Rev. Pharmacol. Toxicol.,* 38:229-255.

Moore, M.M. and K. Harrington-Brock. 2000. Mutagenicity of trichloroethylene and its metabolites: Implications for the risk assessment of trichloroethylene. *Environmental Health Perspectives,* 108 (Suppl. 2):215-223.

Moriarty, S. E., J. H. Shah, M. Lynn, S. Jiang, K. Openo, D. P. Jones, and P. Sternberg, Jr. 2003. Oxidation of glutathione and cysteine in human plasma associated with smoking. *Free Radic. Biol. Med.,* 35 (12):1582-1588.

Morris, J.C. 1967. Kinetics of reactions between aqueous chlorine and nitrogenous compounds. In *Principles and Applications of Water Chemistry*, edited by S. D. Faust and J. V. Hunter. New York: John Wiley and Sons.

Morris, R. D., A.-M. Audet, I. F. Angelillo, T. C. Chalmers, and F. Mosteller. 1992. Chlorination, chlorination by-products, and cancer: A meta-analysis. *Am. J. Public Health,* 82 (7):955-963.

Moudgal, C. J., J. C. Lipscomb, and R. M. Bruce. 2000. Potential health effects of drinking water disinfection by-products using quantitative structure toxicity relationship. *Toxicology,* 147 (2):109-131.

Napolitano, M.J., B.J. Green, J.S. Nicoson, and D.W. Margerum. 2005. Chlorine dioxide oxidations of tyrosine, N-acetyltyrosine, and Dopa. *Chem. Res. Toxicol.,* 18:501-508.

NRC. 1999. *Hormonally Active Agents in the Environment*. Washington, D.C.: National Academy Press.

NRC. 2001. *Classifying Drinking Water Contaminants for Regulatory Consideration*. Washington, D.C.: National Academy Press.

NRC. 2005. *Health Effects of Perchlorate Ingestion*. Washington D.C.: National Academy Press.

NTP. 1990a. *Toxicology and Carcinogenesis Studies of Monochloroacetic Acid in F344/N Rats and B6C3F1 Mice*. Research Triangle Park, N.C.: NIEHS; National Toxicology Program.

NTP. 1990b. *Toxicology and Carcinogenesis Studies of Furfural in F344/N Rats and B6C3F1 Mice (gavage studies)*. NTP Technical Report TR 382, NIH Publication No. 90-2837. Research Triangle Park, N.C.: NIEHS; National Toxicology Program.

NTP. 1993. *Toxicology and Carcinogenesis Studies of Furan in F344/N Rats and B6C3F1 Mice (gavage studies)*. NTP Technical Report TR 402, NIH Publication No. 93-2857. Research Triangle Park, N.C.: NIEHS; National Toxicology Program.

NTP. 1994. *Toxicology and Carcinogenesis Studies of Hexachlorocyclopentadiene in F344/N Rats and B6C3F1 Mice (Inhalation studies).* NTP Technical Report TR 437, NIH Publication No. 94-3168. Research Triangle Park, N.C.: NIEHS; National Toxicology Program.

NTP. 1996. *Short-Term Reproductive & Developmental Toxicity of Chlorodibromomethane (CAS No. 124-48-1) Administered in Drinking Water to Sprague-Dawley Rats* ((NTP Study Number: RDGT94003)) [Online]. Available: <http://ntp.niehs.nih.gov/ntpweb/index.cfm?objectid=4B4F9501-F1F6-975E-70DDA4D64E8A65AF> 1996a [cited January 17, 2005 1996a].

NTP. 1997a. *Short Term Reproductive and Developmental Toxicity of Bromoacetonitrile (CAS No. 590-17-0) Administered in Drinking Water to Sprague-Dawley Rats* (NTP Study Number: RDGT94013) [Online]. Available: <http://ntp.niehs.nih.gov/ntpweb/index.cfm?objectid=4B4F9501-F1F6-975E-70DDA4D64E8A65AF> 1997a [cited January 17, 2005 1997a].

NTP. 1997b. *Reproductive Toxicity of Dibromoacetonitrile (CAS No. 3252-43-5) Administered to Sprague-Dawley Rats in the Drinking Water* (NTP Study Number: RDGT94014) [Online]. Available: <http://ntp.niehs.nih.gov/ntpweb/index.cfm?objectid=4B4F9501-F1F6-975E-70DDA4D64E8A65AF> 1997b [cited January 17, 2005 1997b].

NTP. 1998a. *Toxicology and Carcinogenesis Studies of Tetrahydrofuran in F344/N Rats and B6C3F1 Mice (Inhalation Studies).* NTP Technical Report TR 437, NIH Publication No. 94-3168. Research Triangle Park, N.C.: NIEHS; National Toxicology Program.

NTP. 1998b. *Short Term Reproductive and Developmental Toxicity of Bromochloroacetic Acid (CAS No. 5589-96-8) Administered in the Drinking Water to Sprague-Dawley Rats* (NTP Study No.RDGT96001) [Online]. Available: <http://ntp.niehs.nih.gov/ntpweb/index.cfm?objectid=4B4F9501-F1F6-975E-7099A4D64E8A65AF> 1998b [cited January 15, 2005 1998b].

NTP. 1998c. *Short Term Reproductive and Developmental Toxicity of Bromodichloromethane (CAS No. 75-27-4) Administered in the Drinking Water to Sprague-Dawley Rats* (NTP Study Number: RDGT94017) [Online]. Available: <http://ntp.niehs.nih.gov/ntpweb/index.cfm?objectid=4B4F9501-F1F6-975E-70DDA4D64E8A65AF> 1998c [cited January 17, 2005 1998c].

NTP. 1998d. *Short Term Reproductive and Developmental Toxicity of Tribromoacetic Acid (CAS No. 75-96-7) Administered in Drinking Water to Sprague-Dawley Rats* (NTP Study Number: RDGT94009) [Online]. Available: <http://ntp.niehs.nih.gov/ntpweb/index.cfm?objectid=4B4F9501-F1F6-975E-70DDA4D64E8A65AF> 1998d [cited January 17, 2005 1998d].

NTP. 1999. *Toxicology and Carcinogenesis Studies of Furfuryl Alcohol in F344/N Rats and B6C3F1 Mice (Inhalation studies).* NTP Technical Report TR 482, NIH Publication 99-3972. Research Triangle Park, N.C.: NIEHS; National Toxicology Program.

NTP. 2000a. *Toxicology and Carcinogenesis Studies of Methyleugenol in F344/N Rats and B6C3F1 Mice (Gavage studies).* NTP Technical Report TGR 491, NIH Publication No. 00-3950. Research Triangle Park, N.C.: NIEHS; National Toxicology Program.

NTP. 2000b. *Short Term Reproductive and Developmental Toxicity Study of Dibromochloroacetic Acid (CAS No. 5278-95-5) Administered to Sprague-Dawley Rats in the Drinking Water* (NTP Study Number: RDGT98001) [Online]. Available: http://ntp.niehs.nih.gov/ntpweb/index.cfm?objectid=4B4F9501-F1F6-975E-70DDA4D64E8A65AF 2000b [cited January 17, 2005 2000b].

NTP. 2005a. *Toxicology and Carcinogenesis Studies of Sodium Chlorate (CAS No. 7775-09-9) in F344/N rates and B6C3F1 Mice* [Online]. Available: <http://ntp.niehs.nih.gov/index.cfm?objectid=00132319-F1F6-975E-778A4E6504EB919> 2005a [cited January 17, 2005 2005a].

NTP. 2005b. *Toxicology and Carcinogenesis Studies of Bromodichloromethane (CAS No. 75-27-4) in Male F344/N Rats and Female B6C3F1 Mice* (Drinking Water Studies. TR-532) [Online]. Available: <http://ntp.niehs.nih.gov/index.cfm?objectid=00271EF5-F1F6-975E-73E6FE7AEE1A1A31> 2005b [cited January 17, 2005 2005b].

Nukaya, H., J. Yamashita, K. Tsuji, Y. Terao, T. Ohe, H. Sawanishi, T. Katsuhara, K. Kiyokawa, M. Tezuka, A. Oguri, T. Sugimura, and K. Wakabayashi. 1997. Isolation and chemical-structural determination of a novel aromatic amine mutagen in water from the Nishitakase River in Kyoto. *Chem. Res. Toxicol.,* 10 (10):1061-1066.

Nweke, A. and F.E. Scully, Jr. 1989. Stable N-chloraldimines and other products of the chlorination of isoleucine in model solutions and in a wastewater. *Environ. Sci. Technol.,* 23:989-994.

Onodera, S., K. Yamada, Y. Yamaji, and S. Ishikura. 1984. Chemical changes of organic compounds in chlorinated water. IX. Formation of polychlorinated phenoxyphenols during the reaction of phenol with hypochlorite in dilute aqueous solution. *Journal of Chromatography,* 288:91-100.

Passerini, L. 2003. QSARs for Iindividual Classes of Chemical Mutagens and Carcinogens. In *Quantitative Structure-Activity Relationship (QSAR): Models of Mutagens and Carcinogens*, edited by R. Benigni. Boca Raton, Fla.: CRC Press.

Pattison, D. I., C. L. Hawkins, and M. J. Davies. 2003. Hypochlorous acid-mediated oxidation of lipid components and antioxidants present in low-density lipoproteins: absolute rate constants, product analysis, and computational modeling. *Chem. Res. Toxicol.,* 16 (4):439-49.

Patton, W., W. Pereira, A.M. Duffield, V. Bacon, B. Halpern, Y. Hoyano, and J. Lederber. 1972. Chlorination studies.1. Reaction of aqueous hypochlorous acid with cytosine. *Biochemical and Biophysical Research Communications,* 48:880-884.

Peake, E., B.L. Baker, and G.W. Hodgson. 1972. Hydrogeochemistry of the surface waters of the Mackenzie River Drainage Basin, Canada - II. The contribution of amino acids, hydrocarbons and chlorins to the Beaufort Sea by the Mackenzie River system. *Geochim. Cosmochim. Acta.,* (36):867-883.

Perdue, E.M. and J.D. Ritchie. 2004. Dissolved Organic Matter in Fresh Waters. In *Treatise on Geochemistry; Volume 5: Surface and Ground Water, Weathering, Erosion and Soils*, edited by J. I. Drever.

Pereira, M. A. 1994. Route of administration determines whether chloroform enhances or inhibits cell proliferation in the liver of B6C3F1 mice. *Fundam. Appl. Toxicol.,* 23 (1):87-92.

Pereira, M. A. and M. Grothaus. 1997. Chloroform in drinking water prevents hepatic cell proliferation induced by chloroform administered by gavage in corn oil to mice. *Fundam. Appl. Toxicol.,* 37 (1):82-87.

Pereira, W.E., Y. Hoyano, R.E. Summons, V.A. Bacon, and A.M. Duffield. 1973. Chlorination studies. II. The Reaction of aqueous hypochlorous acid with alpha-amino acids and dipeptides. *Biochimica Et Biophysica Acta,* 313:170-180.

Peskin, A. V. and C. C. Winterbourn. 2001. Kinetics of the reactions of hypochlorous acid and amino acid chloramines with thiols, methionine, and ascorbate. *Free Radic. Biol. Med.,* 30 (5):572-579.

Peters, M. M., T. W. Jones, T. J. Monks, and S. S. Lau. 1997. Cytotoxicity and cell-proliferation induced by the nephrocarcinogen hydroquinone and its nephrotoxic metabolite 2,3,5-(tris-glutathion-S-yl)hydroquinone. *Carcinogenesis,* 18 (12):2393-2401.

Peters, M.M., T.W. Jones, T.J. Monks, and S.S Lau. 1996. Glutathione conjugates of 2-*tert*-butylhydroxyquinone, metabolites of the renal tumor promoter *tert*-butylhydroxyanisole, are toxic to kidney and bladder. *Cancer Research,* 56:1006-1011.

Pitman, I.H., H. Dawn, T. Higuchi, and A. Hussain. 1969. Prediction of chlorine potentials of N-chlorinated organic molecules. *Journal of the Chemical Society. B Physical Organic Chemistry*:1230-1232.

Plewa, M. J., E. D. Wagner, S. D. Richardson, A. D. Thruston, Jr., Y. T. Woo, and A. B. McKague. 2004a. Chemical and biological characterization of newly discovered iodoacid drinking water disinfection byproducts. *Environ. Sci. Technol.,* 38 (18):4713-4722.

Plewa, M. J., E. D. Wagner, P. Jazwierska, S. D. Richardson, P. H. Chen, and A. B. McKague. 2004b. Halonitromethane drinking water disinfection byproducts: Chemical characterization and mammalian cell cytotoxicity and genotoxicity. *Environ. Sci. Technol.,* 38 (1):62-68.

Plummer, J.D. and J.K. Edzwald. 2001. Effect of ozone on algae as precursors for trihalomethane and haloacetic acid production. *Environmental Science & Technology,* 35 (18):3661-3668.

Poon, R., I. Chu, G. LeBel, A. Yagminas, and V. E. Valli. 2003. Effects of dibromoacetonitrile on rats following 13-week drinking water exposure. *Food Chem. Toxicol.,* 41 (8):1051-1061.

Prutz, W. A. 1998. Interactions of hypochlorous acid with pyrimidine nucleotides, and secondary reactions of chlorinated pyrimidines with GSH, NADH, and other substrates. *Arch. Biochem. Biophys.,* 349 (1):183-91.

Radomski, J. L. 1979. The primary aromatic amines: Their biological properties and structure-activity relationships. *Annu. Rev. Pharmacol. Toxicol.,* 19:129-157.

Rapp, T. and D.A. Reckhow. 2005. Formation of dichloroacetamide in drinking waters. *(Manuscript in Preparation).*

Razdan, R., E. Frei, B. Spiegelhalder, and M. Siddiqi. 1994. Caffeine-derived N-nitroso compounds IV: Kinetics of mononitrosocaffeidine demethylation by rat liver microsomes. *Cancer Letters,* 79:117-122.

Rebenne, L.M., A.C. Gonzalez, and T.M. Olson. 1996. Aqueous chlorination kinetics and mechanism of substituted dihydroxybenzenes. *Environmental Science & Technology,* 30:2235-2242.

Reckhow, D. A., J.N. McClellan, B. Mousset, and B.K. Bezbarua. 1999. *A Comprehensive Look at DBP Precursors and Their Chemical Properties*. Washington, D.C.: American Chemical Society, Preprint of extended abstracts. 39(1), 217-218.

Reckhow, D.A., T.L. Platt, A.L. MacNeill, and J.N. McClellan. 2001. Formation and degradation of dichloroacetonitrile in drinking waters. *Journal of Water Supply: Research and Technology - AQUA,* 50 (1):1-13.

Reckhow, D.A. and P.C. Singer. 1985. Mechanisms of Organic Halide Formation During Fulvic Acid Chlorination and Implications With Respect to Preozonation. In *Water Chlorination: Environmental Impact and Health Effects*, edited by R. L. Jolley, R. J. Bull, W. P. David, S. Katz, M. H. Roberts Jr. and V. A. Jacobs. Chelsea, Mich.: Lewis Publishers.

Reckhow, D.A., P.C. Singer, and R.L. Malcolm. 1990. Chlorination of humic materials. Byproduct formation and chemical interpretations. *Environmental Science and Technology,* 24:1655-1664.

Reif, J. S., M. C. Hatch, M. Bracken, L. B. Holmes, B. A. Schwetz, and P. C. Singer. 1996. Reproductive and developmental effects of disinfection by-products in drinking water. *Environ. Health Perspect.,* 104 (10):1056-1061.

Repeta, D.J., T.M. Quan, L.I. Aluwihare, and A. Accardi. 2002. Chemical characterization of high molecular weight dissolved organic matter in fresh and marine waters. *Geochimica Et Cosmochimica Acta.,* 66:955-962.

Reynolds, G.L., H.A. Filaderli, A.E. McIntyre, N.J.D. Graham, and R. Perry. 1988. Isolation and identification of reaction products arising from the chloriation of cytosine in aqueous solution. *Environmental Science and Technology,* 22:1425-1429.

Richardson, S.D. 1998. Drinking Water Disinfection By-products. In *Encyclopedia of Environmental Analysis and Remediation*, edited by R. A. Meyers. New York: John Wiley & Sons, Inc.

Richardson, S.D., A.D. Thruston, Jr., T. V. Caughran, P.H. Chen, T.W. Collette, T.L. Floyd, K.M. Schenck, B.W. Lykins, Jr., G.-R. Sun, and G. Majetich. 1999a. Identification of new ozone disinfection byproducts in drinking water. *Environ. Sci. Technol.,* 33:3368-3377.

Richardson, S.D., A.D. Thruston, Jr., T.W. Caughran, P.H. Chen, T.W. Collette, T.L. Floyd, K.M. Schenck, B.W. Lykins, Jr., G.-R. Sun, and G. Majetich. 1999b. Identification of new drinking water disinfection byproducts formed in the presence of bromide. *Environ. Sci. Technol.,* 33:3378-3383.

Richardson, S. D., A. D. Thruston, Jr., C. Rav-Acha, L. Groisman, I. Popilevsky, O. Juraev, V. Glezer, A. B. McKague, M. J. Plewa, and E. D. Wagner. 2003. Tribromopyrrole, brominated acids, and other disinfection byproducts produced by disinfection of drinking water rich in bromide. *Environ. Sci. Technol.,* 37 (17):3782-3793.

Ritter, C. L., R. W. Decker, and D. Malejka-Giganti. 2000. Reductions of nitro and 9-Oxo groups of environmental nitrofluorenes by the rat mammary gland in vitro. *Chem. Res. Toxicol.,* 13 (8):793-800.

Ritter, C.L. and D. Malejka-Giganti. 1989. Oxidations of the carcinogen n-hydroxy-N-(2-fluorenyl)acetamide by enzymatically or chemically generated oxidants of chloride and bromide. *Chem. Res. Toxicol.,* 2 (5):325-333.

Robinson, M., R. J. Bull, G. R. Olson, and J. Stober. 1989. Carcinogenic activity associated with halogenated acetones and acroleins in the mouse skin assay. *Cancer Lett.,* 48 (3):197-203.

Robinson, T. 1991. *The Organic Constituents of Higher Plants*. 6th ed. North Amherst, Mass.: Cordus Press.

Rook, J.J. 1977. Chlorination reactions of fulvic acids in natural-waters. *Environmental Science & Technology,* 11:478-482.

Rosenkranz, H.S. 2004. SAR in the Assessment of Carcinogenesis: The MultiCASE Approach. In *Quantitative Structure-Activity Relationship (QSAR) Models of Mutagens and Carcinogens*, edited by R. Benigni. Boca Raton, Fla.: CRC Press.

Rostad, C.E., J.A. Leenheer, and S.R. Daniel. 1997. Organic carbon and nitrogen content associated with colloids and suspended particulates from the Mississippi River and some of its tributaries. *Environmental Science & Technology,* 31:3218-3225.

Sarkanen, K.V. and C.W. Dence. 1960. Reactions of para-hydroxybenzyl alcohol derivatives and their methyl ethers with molecular chlorine. *Journal of Organic Chemistry,* 25:715-720.

Schilling, J.D., M.A. Mulvey, and S.J. Hultgren. 2001. Dynamic interactions between host and pathogen during acute urinary tract infections. *Urology,* 57 (Suppl 6A):56-61.

Scully, F. E., Jr. and M. A. Bempong. 1982. Organic N-chloramines: Chemistry and toxicology. *Environ. Health Perspect.,* 46:111-116.

Scully, F. E., Jr., K. E. Mazina, H. P. Ringhand, E. K. Chess, J. A. Campbell, and J. D. Johnson. 1990. Identification of organic N-chloramines in vitro in stomach fluid from the rat after chlorination. *Chem. Res.Toxicol.,* 3 (4):301-306.

Shang, C., W.L. Gong, and E.R. Blatchley. 2000. Breakpoint chemistry and volatile byproduct formation resulting from chlorination of model organic-N compounds. *Environmental Science & Technology,* 34:1721-1728.

Shiozawa, T., K. Muraoka, H. Nukaya, T. Ohe, H. Sawanishi, A. Oguri, K. Wakabayashi, T. Sugimura, and Y. Terao. 1998. Chemical synthesis of a novel aromatic amine mutagen isolated from water of the Nishitakase River in Kyoto and a possible route of its formation. *Chem. Res. Toxicol.,* 11 (4):375-380.

Smith, J.G., S.F. Lee, and A. Netzer. 1975. Chlorination in dilute aqueous systems -2,4,6-trichlorophenol. *Environmental Letters,* 10 (1):47-52.

Smith, M. K., J. L. Randall, D. R. Tocco, R. G. York, J. A. Stober, and E. J. Read. 1988. Teratogenic effects of trichloroacetonitrile in the Long-Evans rat. *Teratology,* 38 (2):113-120.

Smith, M. K., J. L. Randall, J. A. Stober, and E. J. Read. 1989a. Developmental toxicity of dichloroacetonitrile: A by-product of drinking water disinfection. *Fundam. Appl. Toxicol.,* 12 (4):765-772.

Smith, M. K., J. L. Randall, E. J. Read, and J. A. Stober. 1989b. Teratogenic activity of trichloroacetic acid in the rat. *Teratology,* 40 (5):445-451.

Soffritti, M., C. Maltoni, F. Maffei, and R. Biagi. 1989. Formaldehyde: An experimental multipotential carcinogen. *Toxicol. Ind. Health,* 5 (5):699-730.

Soper, F.G. and G.F. Smith. 1926. The halogenation of phenols. *Journal of the Chemical Society*:1582-1591.

Stanbro, W.D. and W.D. Smith. 1979. Kinetics and mechanism of the decomposition of N-chloroalanine in aqueous solution. *Environmental Science and Technology,* 13:446-451.

Stepczuk, C., A.B. Martin, S.W. Effler, J.A. Bloomfield, and M.T. Auer. 1998. Spatial and temporal patterns of THM precursors in a eutrophic reservoir. *Journal of the Lake and Reservoir Management,* 14 (2-3):356-366.

Stiborova, M., M. Miksanonva, V. Havlicek, H.H. Schmeiser, and E. Frei. 2002. Mechanism of peroxidase-mediated oxidation of carcinogenic *o*-anisidine and its binding to DNA. *Mutation Research,* 500:49-66.

Struck, W.A. and P.J. Elving. 1965. Electrolytic oxidation of uric acid - Products and mechanism. *Biochemistry,* 4:1343-1353.

Suzuki, T., M. D. Friesen, and H. Ohshima. 2003. Identification of products formed by reaction of 3',5'-di-O-acetyl-2'-deoxyguanosine with hypochlorous acid or a myeloperoxidase-H2O2-Cl- system. *Chem. Res. Toxicol.,* 16 (3):382-389.

Suzuki, T., T. Nakano, M. Masuda, and H. Ohshima. 2004. Epigallocatechin gallate markedly enhances formation of 8-oxo-7,8-dihydro-2 '-deoxyguanosine in the reaction of 2-deoxyguanosine with hypochlorous acid. *Free Radical Biology and Medicine,* 36:1087-1093.

Thomas, E. L., M. M. Jefferson, J. J. Bennett, and D. B. Learn. 1987. Mutagenic activity of chloramines. *Mutat. Res.,* 188 (1):35-43.

Thomas, G. A. and E. D. Williams. 1999. Thyroid stimulating hormone (TSH)-associated follicular hypertrophy and hyperplasia as a mechanism of thyroid carcinogenesis in mice and rats. *IARC Sci. Publ.,* (147):45-59.

Thomas, J.D. 1997. The role of dissolved organic matter, particularly free amino acids and humic substances, in freshwater ecosystems. *Freshwater Biology,* 38:1-36.

Til, H. P., R. A. Woutersen, V. J. Feron, and J. J. Clary. 1988. Evaluation of the oral toxicity of acetaldehyde and formaldehyde in a 4-week drinking-water study in rats. *Food Chem. Toxicol.,* 26 (5):447-452.

Til, H. P., R. A. Woutersen, V. J. Feron, V. H. Hollanders, H. E. Falke, and J. J. Clary. 1989. Two-year drinking-water study of formaldehyde in rats. *Food Chem. Toxicol.,* 27 (2):77-87.

Tretyakova, N.Y., A.T. Lebedev, and V.S. Petrosyan. 1994. Degradative pathways for aqueous chlorination of orcinol. *Environmental Science & Technology,* 28:606-613.

Tricker, A. R. 1997. N-nitroso compounds and man: Sources of exposure, endogenous formation and occurrence in body fluids. *Eur. J. Cancer Prev.,* 6 (3):226-268.

Tyl, R.W. 2000. *Review of Animal Studies for Reproductive and Developmental Toxicity Assessment of Drinking Water Contaminants: Disinfection By-Products (DBPs).* Report to U.S. Environmental Protection Agency. RTI Project No. 07639. Research Triangle Park, N.C.: Research Triangle Institute.

Uetrecht, J. P., H. M. Ma, E. MacKnight, and R. McClelland. 1995. Oxidation of aminopyrine by hypochlorite to a reactive dication: Possible implications for aminopyrine-induced agranulocytosis. *Chem. Res. Toxicol.,* 8 (2):226-233.

USEPA. 1994. *Methods for Derivation of Inhalation Reference Concentrations and Application of Inhalation Dosimetry.* EPA/600/8-90/066F. Washington, D.C.: United States Environmental Protection Agency. October 1994.

USEPA. 1998. *Regulatory Impact Analysis of Final Disinfectant/Disinfection By-Products Regulations.* EPA Number 815-B-98-002. PB 99-111304. Washington, DC.

USEPA. 1999. *Chloroform Mode of Action Analysis.* [Online]. http://www.epa.gov/sab/chloro00.htm 1999 [cited January 3, 2002 1999].

USEPA. 2002. *A Review of the Reference Dose and Reference Concentration Processes.* EPA/630/P-02/002F. Washington, D.C.: United States Environmental Protection Agency. December 2002.

USEPA. 2003a. National Primary Drinking Water Regulations: Stage 2 Disinfectants and Disinfection Byproducts Rule; National Primary and Secondary Drinking Water Regulations: Approval of Analytical Methods for Chemical Contaminants. Proposed Rule: Parts 141, 142, 143:49548-49681.

USEPA. 2003b. *Health Risks to Fetuses, Infants, and Children*. EPA-822-R-03-010. Washington, D.C.: U. S. Environmental Protection Agency.

USEPA. 2005. *Guidelines for Carcinogenic Risk Assessment and Supplemental Guidance for Assessing Susceptibility from Early-Life Exposures to Carcinogens*. EPA/630/p-03/001B. Washington, D.C.: U.S. Environmental Protection Agency.

Vaittinen, S.-L., H. Komulainen, V.-M. Kosma, A. Julkunen, J. Mäki-Paakkanen, K. Jansson, T. Vartiainen, and J. Tuomisto. 1995. Subchronic toxicity of 3-chloro-4-(dichloromethyl)-5-hydroxy-2 (5H)-furanone (MX) in Wistar rats. *Fundamental Chemistry and Toxicity,* 33 (12):1027-1037.

Valentine R.L., J. Choi, Z. Chen, S.E. Barrett, C. Hwang, Y. Guo, M. Wehner, S. Fitzsimmons, S.A. Andrews, A. G. Werker, C. Brubacher, and K. Kohut. 2005. *Factors Affecting the Formation of NDMA and Occurrence.* Awwa Research Foundation and AWWA, Denver, CO.

Villaneueva, C.M., F. Fernandez, N. Malats, J.O. Grimalt, and M. Kogevinas. 2002. Meta-analysis of studies on individual consumption of chlorinated drinking water and bladder cancer. *J. Epidemiol. Community Health,* 57:166-173.

Walker, W.W., Jr. 1983. Significance of eutrophication in water supply reservoirs. *Journal of the American Water Works Association,* 75 (1):38-42.

Watanabe, T., H. Nukaya, Y. Terao, Y. Takahashi, A. Tada, T. Takamura, H. Sawanishi, T. Ohe, T. Hirayama, T. Sugimura, and K. Wakabayashi. 2001. Synthesis of 2-phenylbenzotriazole-type mutagens, PBTA-5 and PBTA-6, and their detection in river water from Japan. *Mutat. Res.,* 498 (1-2):107-115.

Watt, B.E., R.L. Malcolm, M.H.B. Hayes, N.W.E. Clark, and J.K. Chipman. 1996. Chemistry and potential mutagenicity of humic substances in waters from different watersheds in Britain and Ireland. *Water Res.,* 30:1502-1516.

Weinberg, H.S. 2000. *Bromate Occurrence in Chlorinated Waters*. Denver, Colo.: American Water Works Association - Government Affairs Office.

Weinberg, H.S., S.W. Krasner, S.D. Richardson, and A.D. Thruston, Jr. 2002. *The Occurrence of Disinfection By-Products (DBPs) of Health Concern in Drinking Water: Results of a Nationwide DBP Occurrence Study*. Athens, Ga.: U.S. Environmental Protection Agency; National Exposure Research Laboratory.

Westerhoff, P. and H. Mash. 2002. Dissolved organic nitrogen in drinking water supplies: A review. *Journal of Water Supply Research and Technology-Aqua,* 51:415-448.

Wetzel, R.G. 1975. *Limnology*. New York, N.Y.: Academic Press.

Wetzel, R.G. 2001. *Limnology: Lake and River Ecosystems*. 3rd ed. New York, N.Y.: Academic Press.

Whiteman, M., A. Jenner, and B. Halliwell. 1997. Hypochlorous acid-induced base modifications in isolated calf thymus DNA. *Chemical Research in Toxicology,* 10:1240-1246.

Whiteman, M., A. Jenner, and B. Halliwell. 1999. 8-Chloroadenine: A novel product formed from hypochlorous acid-induced damage to calf thymus DNA. *Biomarkers,* 4:303-310.

WHO. 2000. *Disinfectants and Disinfectant By-Products*. Geneva, Swizerland: International Programme on Chemical Safety.

Williams, G.M. and J.H. Weisburger. 1986. Chemical Carcinogens, Chapter 5. In *Casarett and Doull's Toxicology: The Basic Science of Poisons*, edited by C. D. Klaassen, M. O. Amdur and J. Doull. New York: MacMillan Publ. Co.

Woo, Y. T., D. Lai, J. L. McLain, M. K. Manibusan, and V. Dellarco. 2002. Use of mechanism-based structure-activity relationships analysis in carcinogenic potential ranking for drinking water disinfection by-products. *Environ. Health Perspect.,* 110 Suppl 1:75-87.

Wu, F.C., E. Tanoue, and C.Q. Liu. 2003. Fluorescence and amino acid characteristics of molecular size fractions of DOM in the waters of Lake Biwa. *Biogeochemistry,* 65 (2):245-257.

Yagi, N. and Y. Itokawa. 1979. Cleavage of thiamine by chlorine in tap water. *Journal of Nutritional Science and Vitaminology,* 25:281-287.

Yamashita, Y. and E. Tanoue. 2003. Distribution and alteration of amino acids in bulk DOM along a transect from bay to oceanic waters. *Marine Chemistry,* 82 (3-4):145-160.

Young, M.S. and P.C. Uden. 1994. Byproducts of aqueous chlorination of purines and pyrimidines. *Environmental Science and Technology,* 28:1755-1758.

Zenser, T. V., V. M. Lakshmi, F. F. Hsu, and B. B. Davis. 1999. Peroxygenase metabolism of N-acetylbenzidine by prostaglandin H synthase. Formation of an N-hydroxylamine. *J. Biol. Chem.,* 274 (21):14850-14856.

Zierler, S., L. Feingold, R. A. Danley, and G. Craun. 1988. Bladder cancer in Massachusetts related to chlorinated and chloraminated drinking water: A case-control study. *Arch. Environ. Health,* 43 (2):195-200.

Zou, H.X., J.H. Lu, Z. Chen, C.Y. Yang, J.Q. Zhang, and W. Zhou. 2000. Screening the precursors of strong mutagen [3-Chloro-4-(Dichloromethyl)-5-Hydroxy-2(5h)-Furanone] Mx from chlorinated water. *Water Research,* 34:225-229.

ABBREVIATIONS

AAs	amino acids
AOC	assimilable organic carbon
AOM	aquatic organic matter
APS	acylheteropolysaccharide
AWWA	American Water Works Association
AwwaRF	Awwa Research Foundation
BCAA	bromochloroacetic acid
BCAN	bromochloroacetonitrile
BDCAA	bromodichloroacetic acid
BDCM	bromodichloromethane
BMDL	the lowest confidence limit of the benchmark dose
BOD	biological oxygen demand
°C	degrees centigrade
CCL	Candidate Contaminant List
CDC	Centers for Disease Control and Prevention
CMCF	3-chloro-4-(chloromethyl)-5-hydroxy-2[5H] furanone
CP	chloropicrin
CPDB	Carcinogen Potency Database
d	day
DBAA	dibromoacetic acid
DBAN	dibromoacetonitrile
DBP/s	disinfection by-product/s
DCAA	dichloroacetic acid
DCAN	dichloroacetonitrile
DCMBQ	2,6-dichloro-3-methyl-1,4-benzoquinone
DHAA/s	dihaloacetic acid/s
DHAN	dihaloacetonitrile
dL	deciliter
DPD	diphenylene diamine
DPD-FAS	diphenylene diamine-ferrous ammonium sulfate
DNA	deoxyribonucleic acid
DOC	dissolved organic carbon
DON	dissolved organic nitrogen
EMX	(*E*)-2-chloro-3-(dichloromethyl)-4-oxobutenoic acid; also known as open MX
EPA	Environmental Protection Agency
FDA	Food and Drug Administration
g	gram

GC/MS	gas chromatography/mass spectrometry
gpm	gallons per minute
h	hour
HAA/s	haloacetic acid/s
HAA5	the five regulated HAAs (DBAA, DCAA, MBAA, MCAA, and TCAA)
HAN/s	haloacetonitrile/s
HK/s	haloketone/s
HPLC	high performance liquid chromatography
HX	hydrogen halide
IARC	International Agency for Research on Cancer
ICR	Information Collection Rule
IR	infrared
kg	kilogram
km	kilometer
L	liter
LFER	linear free energy
LOAEL	lowest-observed-adverse-effect level
MCA	mucochloric acid; 3,4(dichloro)-5-hydroxy-2[5H]-furanone
MCAA	monochloroacetic acid
MCL	maximum contaminant limit
MCLG	maximum contaminant limit goal
mg	milligram
mg/dL	milligram per deciliter
mg/L	milligram per liter
min	minute
mm	millimeter
mM	millimolar
M/M	mole/mole
mmoles	millimoles
MOE	margin of exposure
MS	mass spectrometry
MW	molecular weight
MX	3-chloro-4-(dichloromethyl)-5-hydroxy-2(5H)-furanone
μg	microgram
μM	micromolar
μmole	micromole
NA	(data) not available
NAWQA	National Water Quality Assessment
NAS	National Academy of Sciences

NCI	National Cancer Institute
N/C ratio	nitrogen to carbon ratio
ND	not detected
NDEA	N-nitroso-N-diethylamine
NDMA	N-nitroso-dimethylamine
NF	not found
NIEHS	National Institute of Environmental Health
NMR	nuclear magnetic reasonance
NOAEL	no-observable-adverse-effect level
NOM	natural organic matter
NRC	National Research Council
NTP	National Toxicology Program
NVTOC	non-volatile total organic carbon
OGWDW	Office of Ground Water and Drinking Water (of the USEPA)
OPS	Optimal prediction space
ox-EMX	E-2-chloro-3(dichloromethyl)-1,4-butenedioc acid
ox-MX	(Z)-2-chloro-3-dichloromethyl-butenedioic acid
PAO	phenylarsine oxide
ppm	parts per million
QSAR	Quantitative Structure-Activity Relationship
QTSR	Quantitative Structure-Toxicity Relationship
Red-MX	3-chloro-4-(dichloromethyl)-2-furanone; also called reduced MX
RNA	ribonucleic acid
SAR/s	Structure-Activity Relationship/s
SDWA	Safe Drinking Water Act
SPP	Soluable Plant Products
TCAA	trichloroacetic acid
TCAN	trichloroacetonitrile
THAA/s	trihaloacetic acid/s
THM/s	trihalomethane/s
THMFP	Trihalomethane formation potential
TN	total nitrogen
TOC	total organic carbon
TOX	total organic halogen
TOXFP	total organic halogen formation potential
TTHM/s	total trihalomethane/s
USEPA	United States Environmental Protection Agency
USGS	U.S. Geological Survey
UV	ultraviolet

IWA Publishing's authorised EU representative for General Product Safety Regulations is Diane D'Arras, 15 rue Duret, 75116 Paris, France, e-mail: safety@iwap.co.uk.

Printed and bound by CPI Group (UK) Ltd, Croydon, CR0 4YY
17/09/2026
02218804-0007